P9-AGK-528

SHROOM

SHROOM

A Cultural History of the *Magic Mushroom*

ANDY LETCHER

ecco

An Imprint of HarperCollins*Publishers*

 For information address HarperCollins Publishers, 10 East 53rd Street, New York, NY 10022.

HarperCollins books may be purchased for educational, business, or sales promotional use. For information please write: Special Markets Department, HarperCollins Publishers, 10 East 53rd Street, New York, NY 10022.

First published in Great Britain in 2006 by Faber and Faber Limited.

FIRST U.S. EDITION

Library of Congress Cataloging-in-Publication Data is available upon request.

ISBN: 978-0-06-082828-8
ISBN-10: 0-06-082828-5

07 08 09 10 11 ❖/RRD 10 9 8 7 6 5 4 3 2 1

For Matilda

Toadstool soup!
Toadstool soup!
Drink it singly or in a group.

Boris and his Bolshy Balalaika, 'Toadstool Soup'

Contents

Acknowledgements

Though writing a book is by necessity a solitary affair, I have been given invaluable amounts of help and support along the way. Tony Lyons of Sheffield University first awakened my interest in mycology through the brilliance of his lectures. Graham Harvey and Ronald Hutton have been, in their own individual ways, inspirational role models. Matthew Watkins patiently explained the mathematical problems with Terence McKenna's timewave theory, one starry solstice night at Avebury. David McCandless agreed to let me accompany him on what turned out to be an invaluable research trip to Holland. Edward Pope, uniquely amongst sixties survivors, is not only able to remember what happened back in the day but is also willing to talk about it – long may the spirit of '67 live on, old friend. Dave Todd alerted me to and purloined for me a hard-to-come-by copy of *Work is a Four Letter Word*, a film I would otherwise never have heard of. Alison Gill generously provided photos of her extraordinary art, and, even more generously, cycled round London delivering them for me by hand. Shawn Arthur and Stephanie Martin took me under their wing and treated me with wonderful hospitality during a research trip to Boston. And periodically my Welsh friends – you know who you are – provided me with a very necessary sense of perspective.

Many other people stepped on board to offer their assistance, whether steering my research in vital and exciting new directions, sharing their own personal experiences, providing photos, or offering tea and general encouragement. They are, in no particular order: Mike Jay, Mark Pilkington, Laura Rivel, Richard Rudgley, Toni Melechi, Michael Carmichael, Chas Clifton, Jenny Blain, Steven Harris, Peter Edge, Theo Sloot, Rumi Mohideen, Jackie Singer, Jocasta Crofts, Groovy Su, Peter Mantle, Katie Sterrett, Hafiz, Robin Williamson, George Firsoff, Anthony Henman, Hans van den Huerk at Conscious Dreams, Ananda Schouten at De Sjamaan, Runic John, Celtic Chris, Chris Spiral, Josh Ponte, Jane Griffiths, Vanessa Ryall, Amy Whitehead, the Camden Mushroom Company, Adrian Arbib, George Monbiot,

Jim and Deborah, Jason Salzman and the other organisers of Telluride Mushroom Festival, Manuel Torres, Paul Stamets, Dawn Roberts, Anthony Goodman, Richard Allen at Delerium records, The Catweazle Club, Boris and His Bolshy Balalaika, the Magic Mushroom Band, Circulus and the Ooga Bonga Tribe.

I have always found librarians to be polite, forbearing and helpful, no matter how obscure or annoying the request made, but special thanks in this regard must go to the staff at the Radcliffe Science Library and the Bodleian Library Upper Reading Room in Oxford; the Guildhall Library and the Drugscope Library in London; and the Harvard Botanical Libraries in Cambridge, Massachusetts. Lisa Decesare, in particular, deserves a medal for the tireless manner in which she handled my hectic seven day trawl through the vast Robert Gordon Wasson archive, and for providing me with photos at very short notice, just prior to Thanksgiving.

Kate Mohideen, Lydia Feodoroff and Daniel Wolverston generously translated some very dry articles for me, whilst Timea Szell, Robert Wallis, Kat Harrison and Stephen Hancock all made exacting and helpful comments on early chapter drafts. With their shrewd observations, insightful comments and polite cajoling, my editors at Faber, Julian Loose and Henry Volans, made the task of transforming my original manuscript into a publishable book a stimulating and enjoyable one. Thanks are also due to the rest of the Faber team.

It is no exaggeration to say that this book could not have been written without the assistance of Clare Brant who has acted, from the start, as my unofficial agent. She made initial introductions with Faber, read and commented upon the entire manuscript, and generally encouraged me with food and fine conversation during the last, most difficult, period of writing. I can only hope to repay the favour some day.

Henrietta Leyser has, over the years, supported me in so many ways that it is hard to know where to start or which, in particular, to mention. Perhaps most significantly, aside from commenting on early chapter drafts, she provided me with somewhere to live, rent free (again!), during the first year that I was working on the book.

But it is to her daughter – my partner – Matilda, that most thanks are due. As my biggest fan and harshest critic she embodies all the qualities of Muse, provoking me to strive for greater things even as, unswervingly, she homes in on my textual blunders and grammatical inconsistencies. It is to her, with love, that I dedicate this work.

Prologue

One evening in the late summer of 1916, an upright American surgeon from New England began to feel unwell.[1] Dr Beaman Douglass and his wife were on their way to play bridge with their neighbours when they were both overcome with preternatural waves of giddiness. Earlier they had eaten a meal of what they thought were innocuous wild mushrooms, fried in butter and served on toast, which were, by all accounts, delicious: 'We smack our lips even now over the savoury dish,' wrote Douglass a year after the event. But what neither he, nor his wife, nor the unfortunate maid who shared their meal knew was that they had in fact eaten several cupfuls of hallucinogenic fungi. They were, in other words, tripping.

By his own confession Douglass was 'not a botanist', but he was a careful amateur mycologist, competent at mushroom identification and possessed of a passion for edible wild fungi. 'For me,' he declared, 'hunting mushrooms is a pastime, eating them an adventure.' So it was with some considerable delight that he recognised a mass of mushrooms sprouting in a neighbour's cucumber bed as *Panaeolus retirugis*.[2] The definitive guidebook, Charles McIlvaine's *One Thousand American Fungi*, assured him that these were both edible and good, and Douglass had not the slightest reason to suspect otherwise.

Having cooked up the mushrooms, Douglass and his wife 'ate about one half cupful of the caps and two pieces of toast saturated with the liquor'. But a little over an hour later, when the couple had just joined their neighbours, the first peculiar symptoms came on. Mrs Douglass was the most strongly affected. Her vision was distorted, she was unable to concentrate (Douglass noted, somewhat peevishly, that she played cards badly that night), and she became increasingly dizzy. 'There was some cerebral stimulation too – a tendency to be jolly, hilarious – she laughed and talked inordinately and foolishly.' Later, however, she became depressed, could not see properly, and found it hard to breathe. Her pupils, Douglass noted, were unnaturally dilated.

As a medical man, Douglass was obviously concerned for his wife's safety, but he knew enough about mycology to be reassured that their lives were not in any immediate danger for they had not eaten any of the deadly species. Nevertheless, as he too started to be taken by the effects of the mushrooms, he began to doubt his judgement. 'I . . . thought it was time to do something for her before things might become so bad that I could not help her . . . I literally staggered back to my cottage, two hundred yards away, secured my tablets and syringe and struggled back over the road to my wife.' Eventually, with the help of his friends – who read the medicine labels, prepared the syringe, and steadied his hand as he gave the injection – Douglass was able to administer atropine, morphine and an arsenal of emetics to his now ailing wife.

All the time he fought off his own equally lurid symptoms, for he was dizzy and light-headed and his thoughts reeled around with a volition quite of their own. 'The mind was stimulated truly, but the grade and result were below normal. Thoughts flew through my brain, but they were of secondary quality. The attention was easily distracted and disturbed . . . Objects near seemed far away, sounds were diminished, muscular weakness supervened and an uncomfortable feeling of anxiety appeared.' Then he had the strongest and most improper desire 'to be noisy, to laugh and joke', and worse, his own 'trivial and foolish remarks met with *warm personal appreciation*'. It was a most unbecoming, embarrassing and disquieting experience.

In the clear light of day, when the mushrooms' effects had mercifully receded, a relieved Douglass concluded that the adventure had possessed absolutely no merits, was intellectually worthless, and was not in any sense worth repeating. He wrote it up and published it in the journal of the mycological society to which he belonged: a cautionary tale, designed to 'restrain the hazardous' and prevent others from making similar foolish mistakes. Never once did it occur to him that anyone might actually *want* to seek out the mushrooms deliberately. But just forty years later that is exactly what they started to do, and the consequences were nothing if not extraordinary. Dr Douglass would have been astonished.

SHROOM

PART ONE: AGARICUS

Know your mushrooms,
Else they'll rush you,
Stomach pump, in hospital!

Boris and his Bolshy Balalaika, 'Toadstool Soup'

1

The Mushroom People

We shall by morning
Inherit the earth.
Our foot's in the door.

Sylvia Plath, 'Mushrooms'

Magic mushrooms are becoming hard to avoid. Once they were the preserve of the psychedelic underground – of hippies, freaks and travellers – the dedicated few who may still be seen in Britain and America every autumn searching diligently for the little goblin-capped mushroom, the Liberty Cap, *Psilocybe semilanceata*. Now, however, an underground army of net-head hobbyists grows more exotic species and strains away from the public eye, in jars and terrariums secreted in basement cupboards. Young Western travellers to Indonesia, Thailand and Bali, lured by the pull of the paradisiacal full-moon beach party, buy mushroom omelettes or cola-mushroom shakes from the surreptitious locals, illicit fuel for their all-night dancing. In Holland, where liberal attitudes to such matters prevail, magic mushrooms have become big business. 'Paddos', as they are called, can be sold quite openly from market stalls, in 'head shops' and in specialist 'smart shops', and inundations of tourists flock to Amsterdam to sample these unusual wares.

For a few short years, until the legal loophole was forcibly slammed shut in July 2005, mushrooms could be bought in Britain too, provided they were fresh and unprepared. Almost overnight, it seems, they erupted onto the marketplace to become the fashionable illicit drug of choice for young and old alike. For example, at 2004's Stonehenge summer solstice gathering – that great barometer of alternative tastes, lifestyles and ideas – the principal psychoactive being peddled was not cannabis, LSD or Ecstasy, as recent trends might lead us to expect, but cultivated Mexican mushrooms. That year's Glastonbury festival – now an established mainstream cultural event, in spite of the countercultural hype – saw one wholesaler alone shifting an excessive 70 kg of fresh mushrooms, a turnover that factors out at somewhere in the region of 3,500 individual trips. You could buy DIY kits with which to grow your own or, if that was too demanding, you could find flyers advertising websites from which to order mushrooms direct, delivered to your doorstep by return of post.

The surge in mushroom consumption has not been restricted to festival-goers, hippies, clubbers, artists, musicians and the other usual bohemian suspects. I have heard of businessmen, academics, geneticists, photographers, architects, doctors, farmers, council workers and journalists who all make regular mushroom excursions. To reflect this trend, the *Oxford English Dictionary*, that great bastion of language and meaning, has been forced to add 'shroomer' to its ever-expanding lexicon.[1] From Scandinavia to Spain, from the Americas to Australia, from Ireland to Indonesia, shrooms are gathered and eaten with apparent relish, and with a total disregard for their prohibited status. Mushrooming is, well, mushrooming – and, it seems, pretty much everywhere.

From a historical point of view, interesting questions remain over how this curious state of affairs came to be, questions that this book attempts to answer: have people always consumed mushrooms, but secretly and away from the public gaze, or is this a modern phenomenon, and if so, why? Ask these questions of mushroom enthusiasts and many – at least those who are aware that mushrooms have a history at all – will tell you that psychoactive fungi have been used since ancient times.[2] With great certainty they will detail how mushrooms were used in prehistoric religious ceremonies, inspiring the building of the stone circles of Avebury and Stonehenge and the Aztec pyramids at Teotihuacán. They will tell you how Plato, amongst others, drank mushroom tea at the ancient Greek rites of Eleusis; how mushrooms were eaten by the shadowy Celts and their Druidic priests, by the Vikings to access their jingoistic rages, and then later by the medieval witches in their secretive moonlit sabbats. They will happily explain that folk memories of Siberian mushroom-shamanism gave us the figure of Father Christmas, who is, in fact, a magic mushroom in disguise. They will blame a blinkered, patriarchal and nature-hating Christianity, or perhaps the scientific machinations of the industrial revolution, for the severance of this unbroken tradition and the wilful oppression of this throwback to the stoned age. And they will claim that by reviving mushroom use they are reinstating an ancient shamanic heritage, a heritage that is their natural birthright.

This book differs from all others that have come before by breaking with this received orthodoxy, for the real and as yet untold history of the magic mushroom is at once less fanciful and far more interesting. The history of the magic mushroom is much more than a good old tripper's tale. It is intertwined with and inseparable from the social, cultural, scientific and technological changes that have occurred since the industrial revolution, the forces that have wrought the modern Western world. Because of this entanglement, the story of the magic mushroom says something rather revealing about ourselves, about the ideas, hopes, fears, aspirations and desires that shape our time: not least about our yearning for enchantment in a barren scientific world stripped of magic and meaning. That we in the West have found value in those remarkable mushroom experiences, where almost all others before us have regarded them as worthless, means that in a very real sense we could claim to be living in the Mushroom Age. We are the Mushroom People. The story of the magic mushroom therefore provides us with a window, albeit from a quite unexpected viewpoint, upon the modern condition itself.

Mushrooms may not yet have inherited the earth, as Sylvia Plath ominously predicted, but what little fossil evidence there is suggests that fungi per se have inhabited it for at least 400 million years, since the Devonian period.[3] It has been estimated that there may be as many as 1.5 million species of fungi currently in the world, of which only about 100,000 have been identified and formally described, with most of the new species being discovered in the tropics.[4] Though people often label them as plants (from a writer's point of view, it remains occasionally convenient to do so), the fungi actually constitute a distinct biological kingdom, for they contain no chlorophyll and do not reproduce with flowers. But of the four major fungal phyla, in which may be found the rusts, smuts, yeasts, moulds and mildews, only the so-called Basidiomycetes produce what we commonly think of as mushrooms and toadstools. There may be as many as 140,000 of these mushroom-producing species in the world, of which we may know only as few as 10 per cent.

Contrary to popular belief, mushrooms – or carpophores, to give them their scientific name – do not only grow at night: we just tend to notice the ones that appeared while we were sleeping. Nor is there any scientific division between mushrooms and toadstools, though following the Greeks we commonly think of the former as being the edible species, the latter the poisonous. Nor, indeed, is the mushroom the entirety of the organism, for it is merely the reproductive structure, or fruiting body, concerned with the Darwinian task of propagating genes into the next generation. The main body actually consists of a network of microscopic threads, or hyphae, which grow and branch through the species' preferred substrate, forming what is called a mycelium. Mycelia can grow to a vast size. One of the largest has been found in America, a single root fungus, *Armillaria bulbosa*. This specimen occupies an area of about fifteen hectares, weighs in the region of 10,000 kg, and is probably about fifteen hundred years old.[5]

Mushrooms and toadstools exist solely to produce spores, which they do by the million. Released and sometimes fired forcibly into the air, these microscopic 'seeds' are blown away by the slightest breeze until, if lucky, they land on a suitable substrate where, when conditions are right, they will germinate. A single hyphal thread hatches out of the spore and expands outwards, rather like the long modelling balloons used by stage conjurors, but inflated by water pressure, not air. The hypha grows, splits and branches this way and that, sensing its way to where pockets of nutriment can be found and absorbed through its semipermeable wall. At this stage the fungus is monokaryotic, that is, it contains only one nucleus and one complete set of chromosomes, and though viable for a short time, will die if it does not 'mate'. Unlike the dioecious higher organisms, fungi have many hundreds of different mating types – sexes if you will, though the analogy is not precise – to choose from. When two compatible types meet, they fuse together in such a way that each hyphal cell then contains two distinct nuclei, two complete sets of chromosomes, which coexist together harmoniously until environmental cues trigger true sexual reproduction.

In the temperate zones fungal reproduction is usually spurred by the onset of winter – frost being the greatest enemy of an organism consisting mostly of water – but elsewhere shortage of nutrients may be the trigger. When environmental cues indicate a period of stress, the ever-expanding hyphae suddenly grow together into tight balls, called pinheads. Then, from these, and occasionally rising with such force that they can dislodge stones and paving slabs, mushrooms grow up and out, an architectural triumph to rival the finest humans can offer. Indeed, the structure of the mushroom works on exactly the same principles as the fan-vaulting of our Gothic cathedrals, though unlike the fungi we lack this remarkable ability to expand our bodies to an equivalently majestic size.

It is only here, within the cells of the mushroom itself, that true sexual reproduction occurs. The two nuclei fuse, meiosis takes place, and spores are formed, which then drop away from the mushroom's gills or pores rather like confetti from a Manhattan skyscraper, caught by the breeze and blown far and wide. There the analogy ends, for confetti quickly rots (decomposed by fungi, of course), but with their tough coats spores can lie dormant for years, weathering the harshest of conditions. For such a fragile organism, spores are an eminently sensible way of ensuring that those selfish fungal genes are passed on.

The majority of fungi, whether mushroom-producing or not, are saprophytic, that is they are nature's recyclers, feeding off dead plant and animal cells. However, some are symbiotic, bonding with algae to form lichens, or together with plant roots to form complex underground networks called mycorrhizae, a 'wood-wide web'[6] without which both plant and fungus would struggle to survive. Others are parasitic (think of ringworm and athlete's foot) or pathogenic, while some are even carnivorous: remarkably, certain species twist their hyphae into spring-loaded snares, which snap shut when an unfortunate, and equally microscopic, nematode worm wriggles through. Like something out of a horror movie, the unfortunate entrapped creature is then slowly digested as the hyphae penetrate and suck out the contents of its nutrient-rich body.

Until very recently, with the advent of microscopy, all this sex and bloodshed was concealed from the human eye. Only the superficial, macroscopic mushrooms caught our attention; indeed, they are hard to miss for they come in all sizes, all shapes, all colours, all tastes and all smells. Some glow in the dark, while others are cast up in beguiling and enchanted-looking fairy rings. Some stain purple or yellow or blue when cut or bruised; others drip with a milky lactation, or dissolve away rapidly into a black slime that can double as ink. Some contain chemicals that can be used to dye cloth a vivid purple or red, while others emit hydrogen cyanide gas in quantities sufficient to be smelled.[7] One, the accurately named stinkhorn or *Phallus impudicus*, is preternaturally priapic. It springs up from the earth with

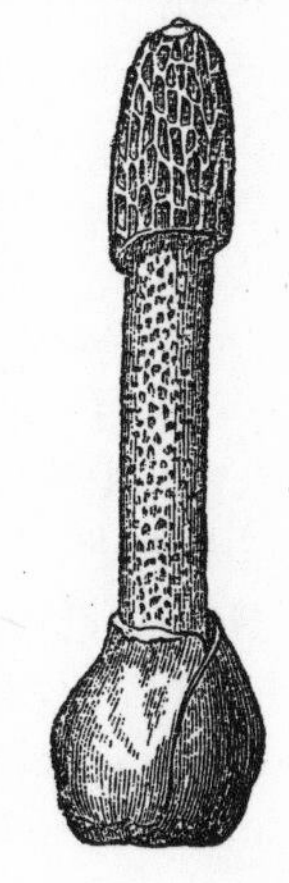

its bell-end covered in a green, fetid, spore-ridden slime that reeks of rotting meat, and is carried hungrily away by flies. (It is said that Charles Darwin's prudish granddaughter would collect and secretly burn these rude protuberances lest they corrupt passing children.[8])

Unsurprisingly, then, mushrooms of all kinds have proved an enduring source of inspiration for writers, poets, artists and musicians alike throughout Western history. Whether it is their sudden appearance, their ephemerality, their association with rot and decay, gastronomic pleasure and indigestive pain, their grotesque shapes, or just their sheer otherness, mushrooms and toadstools have always stirred and troubled our imaginations. Strange and uncertain, dark and disquieting, ruled by apparently supernatural forces, mushrooms seem to us living repositories of all that is weird, enchanted, otherworldly and uncanny. They have produced in us desire and loathing in equal measure.

For Shelley, in *The Sensitive Plant* (1820), the mushroom's stalk was like

> a murderer's stake
> Where rags of loose flesh yet tremble on high,
> Infecting the winds that wander by.

Both Charles Dickens in *Dombey and Son* (1847–8) and Edgar Allan Poe in *The Fall of the House of Usher* (1840) employed mushrooms figuratively as the obvious outward symbols of human decline and decay. It is common to find wild mushroom patches brutally trampled and kicked, as if their offensive presence will somehow drag us into ruin. Their cold, bloated fleshiness is, if not a genuine danger, then still an unwelcome reminder of mortality and death. Plath's 'foot in the door' is troubling indeed.

Different artists and writers have responded more to the mushroom's otherworldly qualities. In music, the contemporary Czech composer Vaclav Halek quite literally takes his inspiration from mushrooms, for he apparently hears eldritch orchestral music playing whenever he gazes upon one.[9] Every species produces its own unique melody, which he dutifully transcribes and incorporates into his own compositions. Though this is almost certainly caused by a bizarre medical condition called synaesthesia (more commonly a symptom of *eating* magic mushrooms), in which colours are sensed as sounds, numbers as shapes, sounds as smells and so on, mushrooms seem so strange to us that it is easy to believe Halek's claim that they really do sing out in otherworldly tones.

The Belgian sculptor Carsten Höller specialises in making accurate models of mushrooms. In one installation (*Upside Down Mushroom Room*, 2000) giant, human-sized fly-agarics appear to be growing downwards from the ceiling: a dizzying inversion guaranteed to make the hardest of heads spin. The British artist Alison Gill placed human-sized papier-mâché Liberty Caps into London's Jerwood Gallery (*Amplifier*, 1997), an installation that afforded the mushrooms a ghost-like presence. Her series of magic mushroom photos, *Fungal Kingdom Emanations* (1997) – created using a special technique known as Kirlian photography – seem to capture the very nature of these peculiar mushrooms, for they appear surrounded by a coruscating, electric-blue aura.

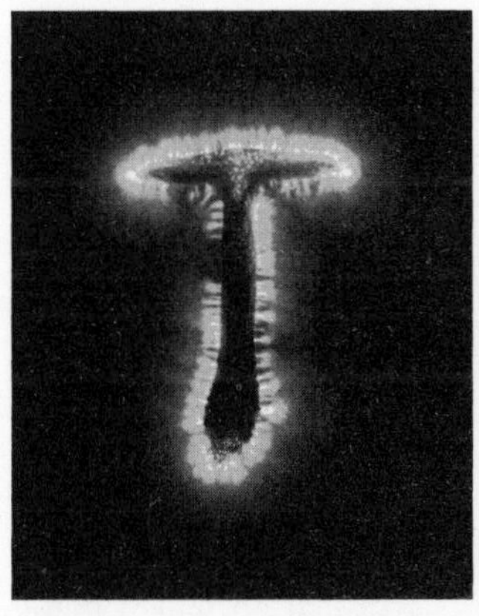

Fungi Kingdom Emanations by Alison Gill (1997)

Amplifier, Alison Gill's installation of human-sized papier-macé Liberty Caps. The Jerwood Gallery, London, 1999. (Photo by Stephen White)

Both Jules Verne and H. G. Wells imagined forests of giant mushrooms growing at the centre of the earth and on the moon respectively, as if no other plant could capture the strangeness of these imagined places. Kipling's *Puck of Pook's Hill* (1906) was accidentally summoned in a fairy ring. Shakespeare took a similarly enchanted view of mushroom-kind, placing 'demy-puppets . . . whose pastime is to make midnight mushrumps' into the magical universe of *The Tempest*. In fact, as he did so often, he was breaking with tradition here, for it was more usual in the sixteenth and seventeenth centuries to hurl the word 'mushroom' as an insult.[10] The actual mushroom's impudent and unexpected arrival, its equally rapid decay, its lack of a substantial root and its inherent love of ordure made it the perfect derogatory metaphor for lambasting the emerging mercantile middle classes. Thus, John Taylor (1580–1653) wrote the following in his *Most Horrible Satyre* of 1639:

> Consider this thou new made Mushroom man,
> Thy Life's a Blast, a Bubble, and a Span;
> And thou with all thy Gorgeous trappings gay,
> Art but a Mouldring lumpe of guilded Clay.

More recently, with all things scatological forming the cornerstone of British humour, the children's author Raymond Briggs saw fit to name his perennially favourite flatulent bogeyman Fungus. Rather like his namesakes, this Fungus lives in a shadowy, dank underworld, thriving on the things that fill us with disgust – mould, slugs, slime and pus – and only appearing in our world, somewhat grudgingly, to give substance to our nightmares. But for Briggs's comical antihero, being a bogeyman is a profession, his job to tap on windows, wake babies with a prod and leap out from behind gravestones to terrorise human passers-by. Quite why this is his lot remains for him a source of existential angst. That he turns out to be a surprisingly sensitive creature is just one of the many clever inversions that have ensured his continued popularity amongst children and adults alike.[11]

Mushrooms, then, seem to us imbued either with an intrinsic earthy, grotesque and gnomic humour, or with all the beauty, glamour, danger and charm of a faerie enchantment. How apposite it is, therefore, to find that certain species are hallucinogenic, and that when eaten they produce in us both hilarity and the most profound and otherworldly alterations of consciousness. The fact of the magic mushroom simply accords with all our expectations.

2

Science and Magic

Every drug has its own character, its unique claims to fame.

Sadie Plant[1]

There are, according to current estimates, 209 species of hallucinogenic mushroom, of which most fall into two broad groups.[2] The first contains the fly-agaric, *Amanita muscaria*, and its close relative the panther cap, *Amanita pantherina*. The former is the red-and-white-spotted mushroom so familiar to us from childhood, where it appears ubiquitously in storybook illustrations as the preferred seating of gnomes; the latter is its less well-known brown-and-white-spotted cousin. Both fungi are closely related to the deadliest of species, the Destroying Angel, *Amanita virosa*, and the Death Cap, *Amanita phalloides* (which, as their names suggest, are responsible for the majority of fatalities worldwide).

The fly-agaric, *Amanita muscaria*, from James Sowerby's *Coloured Plates of English Fungi* (1797–1815). Courtesy R. Gordon Wasson Archive, Harvard University, Cambridge, Massachusetts.

The fly-agaric is not deadly, however, but contains a fuzzy cocktail of different chemical alkaloids, including ibotenic acid (α-amino-3-hydroxy-5-isoxazoleacetic acid) and muscimol (5-(aminomethyl)-3-hydroxyisoxazole), which produce an unsteady set of symptoms: nausea, dizziness, a flushed countenance, twitchiness, increased stamina, euphoria, deep coma-like sleep, hallucinatory dreams and, occasionally, nothing but a headache the next day.

Not surprisingly given this litany, both species have been largely shunned apart from, that is, in two areas of Siberia where there is a long tradition of using fly-agaric as an intoxicant. Nevertheless, this mushroom has generated a spectacular array of myths and legends about its supposed role in the origins of shamanism and religion – so many myths, in fact, that three chapters in the middle section of this book are devoted to exploring and unpicking them. Curiously, the fly-agaric seems to be the one mushroom that most people assiduously avoid, and yet it is the one concerning which people will happily countenance all manner of moonshine. Something about its colourful and memorable form makes the mere thought of it a potent catalyst for the human imagination, but just like its genuine, capricious, psychoactive effects, these tall tales need to be approached with a good deal of caution.

The second group comprises those fungi, most commonly within the genus *Psilocybe*, that contain the active alkaloid ingredients psilocybin (4-phosphoryloxy-N, N-dimethyltryptamine) and psilocin (4-hydroxy-N, N-dimethyltryptamine): these are the ones that we typically mean by the term 'magic mushrooms' and that constitute the main subject of this book.[3] There are currently 186 known psilocybin species – the figure is rising all the time – of which 76 occur in Mexico alone.[4] To pick a mushroom at random in Mexico is to stand a very good chance of picking a hallucinogenic one, which is probably why it is the one part of the world where there is a genuinely old tradition of psilocybin mushroom usage. There are far too many psilocybin mushrooms to describe them all here, but of the plethora growing worldwide two deserve special mention.

The first is the Liberty Cap, or *Psilocybe semilanceata* (naked head, half-lance shaped). In Britain this delicate little mushroom appears in the autumn months, growing in great abundance in 'troops'. It is found in pastures across the British Isles, but especially in acid upland pastures, the wet, chilly sheep fields of Wales, the Pennines, Devon and Cornwall, and Scotland. Somewhat parochially, we think of it as 'our' magic mushroom, but in fact it grows in many temperate regions across the world. It is found across western Europe, from Scandinavia in the north to the Spanish Picos mountains in the south; from Ireland in the west to the Czech Republic and Russia in the east. Moreover, it grows across great swathes of the American Pacific Northwest, and also in New Zealand and Tasmania. Contrary to popular wisdom, it is not coprophilic, that is it is not a dung-lover, but actually grows saprophytically upon the dead root cells of certain grasses.[5] Despite its small

The Liberty Cap (*Psilocybe semilanceata*).

size – the cap is only about a centimetre across – it is home not to gnomes but to species of mycophagous sciarid flies, the grubs of which are familiar to anyone who has ever picked and dried the mushrooms.

Though originally lumped together with other small, nondescript fungi as *Agaricus glutinosus* (in the eighteenth century *Agaricus* was a catch-all generic name for mushrooms), the species was finally identified, and acquired its Latin epithet, when it was described by the great pioneer of fungal taxonomy, the Swedish mycologist Elias Magnus Fries (1794–1878). Since then it has acquired many common names, especially since it revealed itself to be a hallucinogenic species during the 1970s: mushrooms, shrooms, mushies, psillys, pixie caps, Welsh friends, Welsh tea. It is most commonly known as the Liberty Cap because of the distinctive shape of its cap, which resembles the Phrygian bonnets worn by the French revolutionaries when they stormed the Bastille. It seems to have acquired this name in Britain during the late eighteenth or early nineteenth century, perhaps in response to palpable anxieties surrounding the very real possibility of a Napoleonic invasion: the sudden overnight appearance of troops of Liberty Caps might very well have been taken as an unwelcome portent of a similar incursion by the French. These days, and somewhat tellingly, it is the mushroom's resemblance to the archetypal goblin's cap that particularly arrests attention. This iconic shape made it a potent countercultural badge during the 1980s and 1990s, when it appeared on T-shirts, postcards and album covers.

As luck would have it, this one species that grows so abundantly in the world happens also to contain a high and predictable concentration of psilocybin: about 1 per cent. It contains trivial amounts of psilocin, but significant amounts (0.36 per cent) of another psychoactive alkaloid, baeocystin (4-phosphoryloxy-N-methyltryptamine).[6] Were these concentrations not so stable, dosage would be impossible to gauge (as is the case with certain other psychoactive species) and the mushroom would probably not have been adopted as a psychoactive drug. As it is, any twenty mushrooms picked in different parts of the world will have, on average, the same concentration of active ingredients, and therefore the same pharmacological effect.

The second notable psilocybin mushroom is *Psilocybe cubensis*. It was first collected by the American mycologist Franklin Sumner Earle (1856–1929) in 1904 in Cuba, hence its species epithet (although he originally placed it in the genus *Stropharia*). It is much larger than its diminutive cousin, its distinctive golden-brown flying-saucer-shaped cap reaching sizes of up to eight centimetres across. It most definitely is coprophilic, and sprouts from the dung of bovines, or from well-manured ground, throughout the semi-tropical regions of the world. It is found in the south-east United States especially Florida, Louisiana and Mississippi; in Mexico, Cuba and the northernmost countries of South America; in Australia, notably Queensland; and in India and South East Asia, especially Vietnam and Thailand. The introduction of cattle-farming (usually through Western imperialist expansion) has undoubtedly done much to increase the frequency of occurrence of *cubensis* around the world. Whether cattle-farming actually spread the mushroom to new countries, or simply provided the already occurring species with a brand-new ecological habitat to colonise, is unknown. But what makes this species so important is that it has proved the easiest to

Cultivated specimens of *Psilocybe cubensis*. © Cordelia Molloy/SPL

cultivate: most of the magic mushrooms bought and sold, or grown in home terrariums, are *cubensis*.

Though different strains of *cubensis* exist, and cultivated mushrooms are marketed as such – for example 'Thai', 'Colombian', 'Ecuadorian' and so on – the commercial varieties are more often than not identical. Every mycelium produces several 'flushes' of mushrooms, which can be made to look different by varying the watering regime, or the time when they are harvested: selling these as different strains is simply a marketing trick to meet consumer expectations – shaped by the cannabis trade – of variety and choice.

Psilocybe cubensis is known by a variety of names. In Mexico, where it is sometimes viewed suspiciously because of its coprophilic habit, it is known as San Isidro Labrador after the patron saint of ploughing. In Mazatec it is 'di-xi-tjo-le-rra-ja', the 'divine mushroom of manure'. In Thailand it is imaginatively titled 'hed keequai', 'the mushroom that appears after the water buffalo defecates'. In Holland it is the gigglehead; in America and Australia it is the Golden Top or Golden Cap, or simply *cubensis*.[7]

On a per weight basis, *cubensis* is less potent than the Liberty Cap, with an average psilocybin content of around 0.63 per cent, psilocin 0.6 per cent and baeocystin 0.025 per cent.[8] These percentages can vary quite substantially, however, and in cultivated specimens psilocybin content has been found to be highest in mushrooms picked during the fourth flush.[9] A street dose of *cubensis* is typically in the region of fifteen to thirty grams of fresh mushrooms, though experienced users may double this. With Liberty Caps, psychoactive effects begin to be felt with ten to fifteen mushrooms, but an average dose is in the region of twenty to forty mushrooms. One hears stories of much higher, possibly heroic, doses being taken – in the hundreds – but this happens more rarely. Mushrooms of all kinds may be eaten fresh, cooked into omelettes, made into a tea (along with warming spices such as ginger to offset the typical stomach aches and gripes that accompany consumption), or dried or preserved in honey for later use. Fresh *cubensis* mushrooms have a pleasant peppery taste, whereas Liberty Caps have a greasy, rancid aftertaste (and, if dried, the texture of boot leather), which makes stomaching them a difficult task indeed.

Though the exact biosynthetic pathways have been identified, no one knows exactly why these species produce the psychoactive alkaloids psilocybin, psilocin and baeocystin. Some of the more lurid theories suggest that the mushrooms altruistically synthesise them for human benefit, to kick-start a human–mushroom symbiosis, or perhaps to open our eyes to the planet's ecological needs;[10] others, that they are a gift from God. More prosaically, they may simply be by-products from some other essential metabolic process, or have some yet-to-be-determined ecological function, perhaps in deterring mycophagous flies.

Psilocybin, its chemical relatives and the mushrooms within which they are found are formally classed as psychoactives ('activating the psyche'), psychedelics (literally 'mind-manifesting') or hallucinogens ('producing hallucinations').[11] The psychophysical effects of these mushrooms come on anywhere from fifteen minutes to an hour after consumption, depending on how empty the stomach is, and last between four and five hours; unlike, say, LSD which lasts in the region of ten to twelve hours. Psilocybin is about a hundred times less potent than LSD, but about ten times more so than mescaline, the active ingredient of the peyote cactus.

This much we know. When mushrooms are eaten most of the psilocybin is chemically converted (dephosphorylated) into the more potent psilocin by the action of stomach enzymes (though the fate of baeocystin is less well understood).[12] From there, both enter the blood, are distributed very rapidly through the body and are able to cross the blood–brain barrier, the obstacle that ordinarily prevents toxic substances from entering and damaging the precarious biochemistry of the brain. Psilocybin and psilocin belong to a class of chemicals called indole alkaloids, or tryptamines, which are structurally similar to the endogenous neurotransmitter serotonin. Alterations in the normal supply and uptake of serotonin are implicated in a range of physiological and psychological disorders, such as depression and migraine, and in the action of a variety of psychoactive drugs. For example, Ecstasy, or MDMA, works by flooding the brain with serotonin, producing intense feelings of pleasure, whereas the anti-depressant Prozac causes an even production of serotonin and a more controlled elevation of mood. The close similarity of psilocybin and psilocin to serotonin means that they latch onto the brain's so-called 5-HT_{2A} serotonin receptor sites (but, unlike LSD, psilocybin does not directly affect dopamine, the neurotransmitter implicated in schizophrenia). It is rather as if a new, alien but curiously compatible piece of software is thrown into the brain's computer, disrupting its normal operations in novel and unexpected ways.

One measurable symptom of all this chemical jiggery-pokery is that alpha-wave activity in the neocortex is completely replaced by beta-wave activity, but quite why this should produce such profound alterations of consciousness, science has yet to explain: the best it can offer is that various feedback loops between different parts of the brain, perhaps the cortex and the thalamus, are disrupted by the replacement of serotonin with psilocybin, causing an opening of the 'thalamic sensory filter'.[13] Even so, it seems quite possible that the perturbations caused by psilocybin will turn out to be extremely useful in revealing what is ordinarily going on in the complex biochemistry of the brain: they are the neurophysiological equivalent of turning over stones.

As for the subjective effects of eating magic mushrooms, these depend on a number of variables: upon the species consumed; whether the mushrooms are eaten fresh or dried, or with food; the person's body weight, psychological make-up and expectations; and the environment in which the mushrooms are taken. In addition to desirable transformations of mood and perception, mushrooms can cause ominous disturbances to the digestive system – stomach aches and nausea – and make physical movement difficult: walking can take the most intense act of will. (One British mushroom seller cautioned users not to operate any equipment more technically demanding than a spoon.) The old 1960s adage of correct set (mental preparation) and setting (a safe and supportive environment) still holds true.

At low to moderate doses, the most immediate effect is that colours seem brighter, more saturated and better defined. By contrast, at very high doses all sense of reality and one's connection to it may be severed, as the universe and one's identity dissolve into a maelstrom of colour and form. Needless to say, this can be a terrifying experience; the effects of the moderate doses that people typically take, however, are usually more pleasant. Normal vision is disrupted by juddering fractal textures and patterns that emerge out of, and are superimposed upon, the apparent world. Everything is suddenly tattooed with light, while unbidden faces may peer out from the woodwork. The bemushroomed are famously prone to hysterical fits of the giggles, for ordinary and mundane aspects of life are seen in a new, childlike and very often comic light. One habitué told me how she was once overcome with gales of laughter when the moles and freckles on her arm got up and danced away.

The epithet 'magic' appears apposite and well earned, for mushrooms create an overall ambience of earthy, Tolkienesque enchantment. The world, and especially the natural world, appears in a new light, as if some ordinarily obscured and secret aspect of it has been suddenly revealed. The smallest details – leaves, bark, cobwebs, grains of sand – appear exquisitely beautiful and heavy with meaning. Consciousness appears less bounded than it is ordinarily, for trees, plants and even rocks seem to be, in some peculiar sense, aware. However strange and unsettling this transformation, the bemushroomed may report a feeling of familiarity, of déjà vu, of having always known about this particular nook in the architecture of experience. Some say it is as if they have stepped into an archetypal space where they are actors in some ancient drama played out by fools, lovers, kings and queens. This is more than an enchanted escapade: enthusiasts report that great noetic insights and philosophical connections flash unexpectedly into the mind, so that the experience is intellectually and ontologically rewarding.

It is with eyes closed, however, that the most dramatic alterations of consciousness become apparent. Here the phantasmagoria of psychedelic patterns and visuals are most powerfully encountered, arching unbidden across the inner screen of the mind's eye; the phantasmagoria that, whether triggered by opium, LSD, hashish or mescaline, writers and poets have struggled to describe from the time of the Romantics onwards.

Aficionados say it is as if one is gazing upon the surface of an ever-changing sea, brightly coloured and turbulent, that throws up ever more dazzling combinations of pattern and colour, texture and form, with its constant boiling and seething; or perhaps upon an infinitely elastic rubber sheet, that stretches and oozes like some play-putty rolled between the fingers, luminously and languidly bulging and rippling in three dimensions, folding itself into endlessly delightful new arrangements. Here, a tessellating pattern of mushrooms glides across the inner screen like a fan of cards. There, ladders of colour and light, encrusted with millions of bejewelled eyes, spring up and away. Some users report hearing a soundtrack to complete the inner son et lumière, a mysterious timbre of undulating squelches and whoops like those created by the Radiophonic Workshop, or the early Acid House pioneers, on their creaking analogue synthesisers. While the action of the mushrooms lasts there is no end to it, and users are adamant that the lightshow *feels* as if it is originating from somewhere else, as if somehow this is the mushroom revealing itself.

Nevertheless, serious mushroom users – 'myconauts', if you will, travellers in the realm of the mushroom – distinguish between these sorts of 'visuals', which they regard as pretty but trivial, and a secondary class of 'visions' obtained at much higher doses. With these heroic quantities, the tempest of inner visuals condenses into coherent scenarios: visions of ancient civilisations or future space colonies; of elves or angels, spirits or demons. One gets the sense of being pulled deeper into trance by the mushrooms (though with great effort one can open one's eyes and momentarily escape), to a place where things of great importance will be revealed. These are the shamanic realms, myconauts claim, where autonomous, discarnate beings, the spirits of shamanism, impart information and reveal gnostic truths. Some report hearing the 'mushroom' talking to them, as clear as day. Clearly, from a scientific point of view there is much more to be discovered than is currently understood about the action of mushrooms and about how they induce such peculiar sensations: reducing all this to an increase in beta-wave activity or to a broken cortical feedback loop does not quite cut the ontological mustard.

Interestingly, users report that different psilocybin species produce quite distinctive subjective effects. For example, some say that with Liberty Caps visuals are more organic and earthy, whereas with *cubensis* they are more alien and architectural. The parsimonious explanation is that every hallucinogenic species contains its own unique composition of alkaloids which, though they may be only weakly active, synergise with the heavyweight ingredients, psilocybin and psilocin, and influence the way in which they work. Monocular science, having settled upon these two as the responsible agents, has yet to set about investigating the unique pharmacological dynamics of each and every species.

From a physiological point of view, the health risks from magic mushrooms are minimal (though no drug can be said to be entirely risk free). In mice the LD_{50}, that is the dose at which 50 per cent of the experimental subjects die, is 280 mg/kg of body weight,[14] but a high dose in humans is only 0.5 mg/kg.[15] With such a low toxicity it has been estimated that you would have to eat your own body weight in mushrooms to take a lethal dose,[16] and indeed there are no reported cases of fatalities from psilocybin mushrooms, though children may be more at risk of physical harm.[17] One foolhardy individual made himself dangerously ill by taking his mushrooms intravenously.[18]

From clinical trials we know that taking psilocybin is a literally hair-raising experience (it causes piloerection), which leaves subjects wide-eyed (it famously triggers mydriasis, a marked dilation of the pupils). Psilocybin's tendency to increase heart rate and lower blood pressure at high doses suggests that people with heart or blood pressure problems might be at risk.[19] Mushrooms are not addictive in any way, and indeed a tolerance to the active ingredients quickly builds up so that they have diminishing effects the more often they are taken. As a sign of its low toxicity, psilocybin is very quickly excreted from the body, with two-thirds of any dose removed after three hours.[20]

It is in the psychological domain that magic mushrooms become more problematic. Even experienced users know that mushroom trips can turn nasty: the visions can become hellish, the gnostic insights can be too much to take in, the fear of dying or going mad or of permanently losing one's identity can become overbearing. Quiet reassurance is usually enough to turn the trip around, for its peak is over relatively quickly, but in extreme cases sedation may be necessary. It has fortunately been appreciated by the medical community that the old-fashioned rush for the stomach pump (gastric lavage) merely makes a bad situation worse.

In only a small minority of cases do any of these symptoms persist beyond the immediate course of the trip, and there may be other aggravating circumstances: in one published British example, the man in question had taken mushrooms every day for a week while fasting and avoiding sleep, so had clearly been overdoing things.[21] There is no evidence that mushrooms can make a healthy subject psychotic,[22] but any *latent* psychosis could be triggered, for even alcohol can bring latent mental illness to the fore. PET scans of the brain have shown that most psilocin activity occurs in the areas of the brain known as the prefrontal cortex, the anterior cingulate and the temporomedial cortex.[23] Interestingly, these same areas are implicated in certain types of schizophrenia, especially the type of psychosis known as hallucinatory ego disintegration, though an exact equivalence between tripping on mushrooms and schizophrenia is far from proven. Nevertheless, it stands to reason that anyone with a personal or family history of mental illness, or who is depressed or feeling psychologically unstable in any way, should assiduously avoid mushrooms, or any other psychoactive drugs for that matter.

Although there has been an absolute moratorium on research into psychedelics since the moral panic about LSD in the 1960s, a recent softening of attitudes has opened the door to clinical trials with psilocybin. The days are long gone, however, when white-coated scientists would, say, spike individual monkeys with psilocybin to see how the other, straight members of the experimental colony would react to having a tripper in their midst (the answer was, usually, badly).[24] The political and social overtones of such experiments should be obvious: scientists wanted to 'prove' that psychoactive drugs derailed society by turning good middle-class kids into zombie-like dropouts. The fact that 'dropping out', as certain alternative lifestyle choices were labelled in the 1960s, might have been determined by cultural or sociological factors and not by deterministic psychological ones seems to have passed these scientists by.

These days most research is psychological, in the broadest sense of being to do with the mind. For example, one study showed how subjects under the influence of psilocybin were more likely to make indirect semantic associations; that is in tests they were more likely to make 'poetic' or 'creative' connections between unrelated words.[25] Anecdotal reports suggest that psilocybin may be effective in treating cluster headaches,[26] while a preliminary study has found hints that, when taken in conjunction with a course of therapy, it might be successful in curing certain obsessive–compulsive disorders.[27] If magic mushrooms do indeed make you temporarily mad, then it may be that a bit of madness is good for you.[28]

Perhaps the biggest danger surrounding magic mushrooms is the danger of picking the wrong sort. A death in Australia from renal failure was almost certainly caused by the poor unfortunate picking the wrong mushroom.[29] This danger is most acute with that other class of psychoactive fungi, the amanitas, which are closely related to the two aforementioned deadly species, *Amanita virosa* and *Amanita phalloides*. In Britain, once one is familiar with the Liberty Cap, it is difficult to confuse it with anything else (though inexperienced pickers may do so). In America, however, one psychoactive species, *Psilocybe stuntzii*, is easily confused with the deadly poisonous *Galerina autumnalis*, with which it can happily grow side by side. Wherever they are picked, accuracy of identification is *essential* and the slightest doubt should lead any would-be experimenter to err on the side of caution.[30] If harm reduction is the principal concern, then there is surely a persuasive argument for the legal sale of cultivated mushrooms to prevent people from making potentially lethal mistakes.

In the general scheme of things, then, the greatest risk to health, liberty and well-being comes from magic mushrooms' proscribed status and the fact that using them carries the stiffest legal penalties: in law, psilocybin is typically listed alongside heroin as the most dangerous of illicit drugs. In 1971, the United Nations, under considerable American pressure, introduced its Convention on Psychotropic Substances as an adjunct to its earlier Single Convention on Narcotic Drugs (1961). Though at the time illicit use of magic mushrooms in Europe and America was minimal, had not gone overground and was probably little known to the policy makers, psilocybin and psilocin were included in the Convention because of their structural similarity to the great drugs menace of the time, LSD. Signatories agreed to prohibit these synthetics but, following appeals by the Mexican government, not the plants that contained them: the Mexicans were understandably worried by the prospect of having to prevent indigenous mushroom use. As we shall see, it was this distinction that opened the legal loophole that exists in Holland, and that existed in Britain until 2005, allowing the sale of fresh mushrooms.

This legal situation is the predominant cultural force about which contemporary mushroom enthusiasts must orientate themselves, rather like iron filings in a magnetic field. One of the reasons that invented histories, supposed lineages of mushroom use stretching back to the dawn of time, are so readily believed and so trenchantly defended by aficionados is that they serve to legitimate this illicit hobby. For if great cultures, religions and philosophies were founded on bemushroomed gnosticism, then contemporary use is very far from the self-abusive, criminal, infantile or escapist act that mainstream society still deems it to be. This climate means that taking mushrooms remains a distinctly countercultural act which, for many, forms the kernel of an identity founded on a sense of alterity, or opposition to the mainstream. Clearly the contention of this book that Western mushroom use dates only to the 1950s will meet with considerable resistance from some quarters.

Aficionados claim that magic mushrooms are not just another drug but are psycho-spiritual tools that bring a greater understanding of the self, of our place in the world and of some essential 'truth'. Mainstream society, they maintain, is blinkered by the 'war on drugs' and is unfairly prejudiced against this most benign and illuminating of naturally occurring hallucinogens. In my scrutiny and revision of the story of the magic mushroom, it is fair to say that I have a certain amount of sympathy with these claims. Nevertheless, I would suggest that enthusiasts' time would be better spent arguing the case for mushrooms in terms of the culturally sanctioned criteria of our time, that is on health and medical grounds, than on the grounds of some fantastical history, dreamed up on a basis of wishful thinking and overworked evidence. In other words, enthusiasts must convincingly persuade the law makers that any risks to mental and physical health are ones that society can reasonably tolerate (as has indeed happened in Holland).

If mushrooms really are safe, if they really do deliver meaningful insights, then the fact that the practice is a modern one is neither here nor there. Indeed, I would argue that placing the history of the mushroom on a sound footing, and excising the more excessive claims made on its behalf, can only strengthen the enthusiasts' case, for it will afford them a little more credibility. As it is, the recent popularity of the magic mushroom rather suggests that popular culture has already ruled in its favour, a situation that official policy now lags breathlessly behind.

I shall return to the thorny issue of prohibition in Chapter Fourteen, but for now, our introductions over, it is time to begin telling the story of the magic mushroom and how it came to have its foot wedged firmly in the door of Western culture. We need to go back to the beginning, to look at the archaeological remains from the earliest of human cultures, to see whether civilisation really began, as so many say, in the psychedelic swirl of a bemushroomed age.

3

The Archaeology of Ecstasy

And did those feet in ancient times
Dance upon England's mountains green?
And were the holy Spores of God
On England's pleasant pastures seen?

Stephen Hancock, 'Jerusalem (Glade remix)'[1]

Imagine the scene. *A Palaeolithic hunter-gatherer bends down and picks a variety of mushroom she has never encountered before. She is struck at once by its delicate pointy cap, its distinctive and memorable shape; and yet, there is something else about it that draws her attention, for it seems almost as if the mushroom is calling to her. Nevertheless, mindful of the risks, she nibbles it gingerly; but then, finding its taste not altogether unpleasant, she swallows it and eats more. Thirty minutes later, when the colours begin and the world starts to ripple around her, she finds herself propelled into the numinous world of the ancestors and spirits who tumble towards her out of the sky and whisper songs of love and longing in her ear. 'We have been waiting for you,' they sing. Overwhelmed, she falls to the ground in ecstasy. Hours later, when it is all over, the colours and spirits long gone, she returns to her camp with a glint in her eye, and a bundle of the sacred mushrooms cupped in her hand. 'Look what I've found,' she says . . .*

Speak to any serious-minded mushroom aficionado, and this is the kind of picture of the distant past that he or she will conjure up, telling you that it would have been 'natural' or 'obvious' for prehistoric cultures to have used magic mushrooms. Our ancestors, living in much closer contact with the natural world, would have had their senses more keenly honed to the powers of plants than we do. Shamanistically inclined, they would have unquestionably welcomed and celebrated the discovery of this key to the door of the otherworld. This belief resurfaces time and again in different guises, and we will meet it many times.

Of course, this kind of argument should immediately alert the critical enquirer, for whenever anything is deemed 'natural' and 'obvious' it almost always turns out, on closer inspection, to be culturally specific, localised and historically contingent – in other words, not at all natural or obvious. The belief actually rests upon an implicit philosophical assumption: that there is some universal or essential psychedelic experience that transcends history and culture, so that anyone eating a magic mushroom will have a similar, usual spiritually inflected, experience irrespective of who, when and where they are. It will be ecstatic, boundless, oceanic, noetic, cosmic, 'far out', or any of the other superlatives that are usually applied in such circumstances. It will be wholly benign, pleasurable and desirable. Thus, goes the argument, any culture that stumbles upon a new hallucinogenic plant will embrace it with open arms.

Historians and archaeologists are now all too aware of how our views of the past are tempered by the attitudes and dispositions of the present, and how these views may say more about us than ever they can about the people who came before us. The belief in the ancientness of psychedelia is no exception for it turns out, somewhat paradoxically, to have rather recent origins: to be a product of the utopian sentiments that accompanied the psychedelic revolution of the 1950s and 1960s. In the early days of the cultural movement that would eventually bequeath us the magic mushroom, academics and intellectuals happily and legitimately took drugs such as LSD and mescaline, and then marvelled that they had discovered an ancient and secret path to the kingdom of heaven.

One of the first to do so was Aldous Huxley, who famously took mescaline in the spring of 1953, and wrote up his experiences in *The Doors of Perception*. There, alongside poetic descriptions of his spiritual ecstasies, he suggested that all 'the vegetable sedatives and narcotics, all the euphorics that grow on trees, the hallucinogens that ripen in berries or can be squeezed from roots – all, without exception, have been known and systematically used by human beings since time immemorial'. The reason, he said, was simple: the longing to transcend the drudgeries and privations of life 'is and always has been one of the principal appetites of the soul'.[2] Other intellectuals and academics, caught up in the excitement of those early days, concurred but placed that ancient appetite not in the soul but in some essentialised 'human nature', or in the genes, and argued that it is a drive that, like its better known sexual counterpart, we repress at our peril.[3]

When psychedelia went mainstream during the mid 1960s, and the drugs were hurriedly made illegal, these arguments were eagerly reiterated by hippies determined to make the case against prohibition. Even today the popular writer Paul Devereux argues that human history is one 'long trip' from which modern Western culture with its 'war on drugs' represents an anomalous aberration.[4] In the case of magic mushrooms, therefore, enthusiasts imagine an unbroken tradition of use stretching back to the Palaeolithic, which includes the hunter-gatherer shamans, the Neolithic builders of Stonehenge and Avebury, the ancient Greeks at Eleusis, the Iron Age Druids and medieval witches, and which was only severed by Christianity and the machinations of the industrial revolution.[5]

While it is true that archaeologists are coming to terms with the fact that psychoactive drugs have played a part in human culture perhaps for millennia, one of the major themes running through this book is that this essentialist view of the psychedelic experience is problematic. I argue instead that drug experiences, whatever essential elements they may or may not contain, are always culturally bound, culturally mediated.[6] Different people, let alone different cultures, have quite clearly approached the same drug differently, and the common thread linking disparate cultures in their relationship with drugs is more correctly an attitude of ambivalence. A rigorous history of the magic mushroom must therefore be built upon the evidence, and not upon a philosophical assumption, however ardently we may wish it to be true. In later chapters I shall look at the evidence for mushrooming traditions in Mexico and Siberia, but for now let us start by sifting through the material remains left by earlier, prehistoric European cultures to see whether any of them really had acquired the taste for magic mushrooms.

For any particular culture to have centralised the use of a psychoactive plant, three conditions must obtain: the plant must be available in sufficient quantities to meet demand, either growing in a plentiful enough supply or obtainable by trade; the culture must know that the plant in question, when ingested, is responsible for causing the subsequent alterations of consciousness; and finally, most importantly of all, there must exist a cultural context in which those alterations can be meaningfully apprehended, and psychologically and socially integrated.

The experience, in other words, has to be one worth repeating, its desirable effects outweighing any negative consequences. Not one of these prerequisites is ever guaranteed.

To illustrate the first point, let us consider the ecology of the Liberty Cap, *Psilocybe semilanceata*. You will recall that this mushroom favours acid upland pastures and grows in the kinds of conditions where the only viable form of agriculture is sheep- or cattle-farming. However, for much of its prehistory Britain was covered not in pasture but in dense primary forest,[7] so the mushroom here would have been uncommon or rare. It could only have been with the introduction of agriculture and the gradual clearance of the forests from the Neolithic onwards, from around 5000 BCE, that pastures of sufficient size would have been established to make psilocybin mushroom use in Britain plausible. It would be quite wrong of us to assume that just because a magic mushroom is abundant now it has been so throughout all of human history and prehistory.

The second point, that cultures must spot the causal connection between eating a plant and the unusual effects this produces, seems so obvious as hardly to need stating. In our scientific age it is easy to forget that pre-modern cultures have attributed all manner of phenomena to various agencies, not all of which proved to have substance. Take the example of ergot poisoning. At various points during the Middle Ages, and even as late as 1953 in France and Belgium, European society was periodically afflicted by a terrible plague that struck down whole regions at a time. The affected would be beset by trembling of the limbs, formication – the feeling that they were covered in crawling ants – sweats and fevers, together with terrifying visions and hallucinations. In extreme cases, victims would go permanently insane, or would develop a fatal gangrene in their extremities. This scourge was seen as a God-given punishment and named St Anthony's fire, after the saint who was said to be able to appease the Almighty and cure this terrible affliction.[8]

We now know that it was caused, if not directly by the hand of God, then by the ergot fungus, *Claviceps purpurea*, which grows parasitically upon grasses, especially species of rye. This fungus forms a hard, lumpy, rind-covered structure called a sclerotium (another fungal strategy for riding out adverse conditions), which sticks out from the grass head like a blackened ear of corn. Ergot sclerotia contain a potent array of psychoactive and toxic alkaloids, from which the infamous drug of the 1960s, LSD, may be derived and which, when accidentally milled into flour, cause gangrene and hallucinations. Remarkably, though ergot's vasoconstricting effects have long been known and employed in folk medicine as an abortifacient or to bring pregnancies to term, its role in causing St Anthony's fire went undiscovered until the modern period. It seems that medieval herbalists were quite unable to spot the causal connection.

The opposite is in fact true, in that pre-modern cultures often wrongly attribute a range of virtues to plants and potions that we now know to be inert – most aphrodisiacs providing the obvious disappointing example. But during the 1960s it was reported that the Kuma people from the Wahgi Valley of the West Highland region of New Guinea used a hallucinogenic mushroom to go into collective states of trance. The news caused a substantial amount of excitation in the West, for it was hoped that this would prove to be the first example of indigenous mushroom use outside Mexico and Siberia.[9] Under the influence of this mushroom, called *nonda* by the tribespeople, the men would dress up in ritual gear, grow tense and excitable, and run amok through their village. The women, meanwhile, would become relaxed and giggly, boast about their sexual exploits, dance provocatively and flirt indecorously with men. Both sexes acquired double vision, became shivery, and occasionally lost the power to speak. Some of the afflicted claimed to be able to see spirits and bush demons.

However, in spite of several mycological expeditions to the region in which *nonda* specimens were successfully collected and identified (it turns out that *nonda* actually comprises several different species belonging to the genus *Boletus*), no trace of any hallucinogenic compound has ever been found within them.[10] This curious omission rather explains why the mushrooms only became 'active' at certain times of the year and could otherwise be eaten with impunity, and why lying in running water apparently 'cured' the afflicted. Though it has been suggested that tobacco, which is hallucinogenic in strong doses, might have been responsible,[11] it seems more likely that the *nonda* trance was an example of an event known in anthropology as a 'rite of rebellion'.[12] Like the carnivalesque antics of Mardi Gras, or the medieval 'feast of fools', these effervescent social dramas – which invert, parody and ridicule the norms and rules of the status quo – act as safety valves and release pent-up societal tensions. None of them require the intercession of a drug, but merely an excuse – though of course intoxicants may be consumed. In this instance the symptoms of intoxication were quite spuriously attributed to a mushroom by the Kuma, illustrating that knowledge of the causality between a drug and its action can never be assumed, however obvious that connection may appear to us.

Finally, and most importantly, for a psychoactive plant to become legitimated or even institutionalised there must also exist a culturally agreed context into which the strange experiences it elicits can easily be slotted, and thus made meaningful and comprehended. But in many cultures, both current and historical, that context has been wholly lacking, and locally occurring psychoactive drugs have been known but shunned as either worthless or poisonous.

For example, it is commonly recognised that indigenous cultures in the Americas, most notably Central and South America, have long employed a whole range of psychoactive plants for ritual and recreation. The *proportion* of available psychoactive plants actually used in the Old World is substantially lower.[13] The reason for this discrepancy seems not to be a paucity of botanical knowledge but a cultural aversion towards the plants and their pharmacological effects.

In the Amazon rainforest there is one psychoactive brew, known as ayahuasca or yagé, that is used almost ubiquitously by tribes across the region and forms a potent cornerstone of indigenous spirituality. Nevertheless, and in spite of peer pressure, some tribes in ayahuasca regions regard it as so tainted with negative associations, with sorcery, warfare and strife, that they will have nothing to do with it.[14]

Similarly, it has been known since the 1920s that throughout the southern Pacific Ocean there are, rather bizarrely, several species of hallucinogenic fish, two of which are the Hawaiian manini fish, *Acanthurus triostegus sandvicensis*, and the South African brass bream, *Kyphosus vaigiensis*.[15] The unknown active ingredients seem most concentrated in the fishes' heads and brains. In Honolulu in 1927 about thirty Japanese workers from the mayor's office were struck down with strange hallucinations and a feeling of pressure upon the chest after eating a meal of one such species. But we have yet to see the emergence of a psychedelic fish craze.

Closer to home, there are several naturally occurring plant psychoactives in Britain's flora, such as henbane (*Hyoscyamus niger*) and mugwort (*Artemesia vulgaris*), but they have been largely ignored, or actively shunned, by modern psychedelic enthusiasts. Admittedly, the threshold between an active dose and a toxic dose of henbane (which contains the deliriant drug scopolamine and was used by Dr Crippen to murder his wife) is worryingly narrow. Nor are its effects supposed to be altogether pleasant: I have heard it described by more than one brave experimenter as 'the Hieronymus Bosch trip'. On the other hand, mugwort, which contains a drug called thujone, might be regarded as a little tame in comparison to modern synthetics, but this simply reinforces the point. Some drugs are seen as simply not worth the bother.

With regard to magic mushrooms, the fly-agaric is found growing across great swathes of the temperate forest regions of the world but, barring the occasional outbreak, has only been consistently used as an intoxicant in two relatively small regions of Siberia. The fly-agaric *was* consumed in Japan, but only as a food, with an elaborate set of cooking techniques employed to rid the mushroom of its psychoactive ingredients.[16] Magic, presumably psilocybin, mushrooms have been known about in China since at least the Chin Dynasty (265–420 CE), when they were written about by Chang Hua in his *Po-wu chih*, or *Record of the Investigation of Things*.[17] But for Chang Hua, and the later compilers of herbals that came after him, the mushrooms that 'made you laugh unceasingly' were poisons, to be avoided or, if accidentally eaten, to be treated with herbal remedies.

In Europe, as we shall see, psilocybin mushrooms have been known about since at least as long ago as the fifteenth century, with some writers even likening their effects to those of opium, but no one seems to have eaten them intentionally until the twentieth. And, at the risk of getting a little ahead of ourselves, contemporary statistics from Holland show that although the numbers of people who have ever taken magic mushrooms is steadily increasing, most only take them once or twice in their lifetime. If the desire to alter consciousness through drugs is a primal urge, as insistent as the drive to have sex, then we seem to have it very much under control.

With these points in mind, we can now turn to the archaeological record to see what evidence there is for prehistoric mushroom use. Immediately, we stumble upon two problems. The first is that mushrooms differ from almost all other naturally occurring psychoactives in that they do not have to be prepared in any way before they are consumed. They do not have to be roasted, fermented, pounded, boiled or infused (though they may be cooked or made into tea for palatability's sake). They do not have to be chewed in quids for hours at a time, injected, smoked, snuffed, rubbed on the skin in salves, taken by enema, or consumed by any of the other pleasant, and not so pleasant, methods by which we have learnt to introduce drugs into the body. They can simply be picked and eaten, and consequently there is nothing in the way of paraphernalia that might have been left in the archaeological record to indicate their use.

The second is that mushrooms themselves, perhaps even more so than plants, do not preserve well. Archaeological ages are defined on the basis of the skeletal remains of prehistoric cultures, upon their hard-wearing and long-lasting artefacts of bone, ceramics, metal and stone. We talk of the Stone Age, the Bronze Age, the Iron Age, but quite what wonders might have belonged to any 'Wood Age' we shall never know, for they will have long rotted away. Even more than wood – which does occasionally get preserved if dropped in acid bogs – mushrooms are delicate, evanescent and highly putrescible, often rotting away within a few hours of appearance. Nor is drying a guarantee of protection from the ravages of their hungry, saprophytic, fungal brethren. It should come as no surprise, therefore, to learn that fungi of all kinds appear extremely rarely in the archaeological record.

There are a handful of examples, however. The 'Iceman' – the Neolithic man whose wounded body was preserved where it fell five thousand years ago in the snow-capped Tyrolean Alps along the present day Italian–Austrian border – was found to have been carrying pieces of fungus strung upon a leather strap.[18] These were of *Fomes fomentarius*, a type of bracket fungus that grows out of trees in hard scallop-like shelves and has been used as tinder since prehistoric times (pieces of this fungus have been found by the hearths of Mesolithic settlements at Starr Carr in Yorkshire and Maglemose in Denmark);[19] and *Piptoporus betulinus*, the Birch polypore or razor-strop fungus, so called because it is excellent for honing the edge of a razor. For what purpose the Iceman had them is unclear (metal razors having yet to be invented in the Stone Age): they could have been used for tinder, medicine or magic, or have had some other symbolic significance that is unknown to us.

A variety of puffball, *Bovista nigrescens*, has been discovered at various British sites: from an Iron Age midden at Skara Brae in Orkney, from a Roman fortification at Stanwick in Yorkshire, and from a

Roman well in Scole in Norfolk. Like many puffballs, this species is edible when young but not by the time it has produced spores, as all these examples had. The presence of the fungus at these sites is therefore mysterious: it may have been used for magic, for medicine or even for loft insulation![20] Further afield in North America, nineteenth-century grave 'guardians', that is carved figures of faces and animals placed in indigenous graves, appear to be made from the dried remains of the mushroom *Fomitopsis officinalis*, and not from wood as was originally thought.[21]

And that, along with just one or two more examples, is pretty much the extent of it. Needless to say, there is not a single instance of a magic mushroom being preserved in the archaeological record anywhere. Supporters of the ancient mushrooming thesis have therefore invoked several other related lines of evidence to support their case but, as we shall see, none of these inferences is unequivocal, and each is open to a range of alternative explanations.

The first comes from the fact that our European prehistoric ancestors were certainly knowledgeable about a range of other psychoactive plants. Preserved poppy heads uncovered in Britain, Switzerland, and Spain – some of which were sterile, domesticated varieties – suggest that the dreamy but addictive analgesic drug opium (*Papaver somniferum*) was being deliberately produced and consumed from the Neolithic onwards.[22] Cannabis (*Cannabis sativa*) seems also to have been cultivated in Britain and Eastern Europe from perhaps the late Neolithic, and more commonly in the Iron Age.[23] The more menacing henbane (*Hyoscyamus niger*) may have been imported into Britain during the Neolithic, for seeds and pollen have been found in residues adhering to so-called Grooved Ware pottery sherds at Balfarg in Scotland.[24] Henbane seeds were found buried in a leather pouch, together with the body of a woman, in a Dark Age grave near Fyrkat in Denmark.[25] And even a quantity of ergot sclerotia were found in the stomach of one of the exquisitely preserved Iron Age 'bog-bodies', dredged up from the peat in another part of Denmark near the town of Grauballe.[26] Surely, the argument goes, psychedelic know-how was transferable, and thus if a culture employed these psychoactive plants, it would certainly have known about, and used, hallucinogenic fungi.

Unfortunately, there are two objections to this line of reasoning. The first is that the presence of these plants in the archaeological record is no guarantee that they were used for their *psychoactive* properties. Poppy seeds are a nutritious food source, while opium is a powerful painkiller and an effective treatment for respiratory and digestive complaints. Cannabis likewise has a range of medical uses, and has long been grown for its fibres, from which strong and durable cloth, canvas and rope can all be made.[27] Henbane seeds might have been purely for show, a proclamation that the owner had mastery over a poisonous plant (shamanism, particularly in Siberia, has often been accompanied by conjuring tricks – plunging hands into boiling water, swords through the body, and so on – to demonstrate superhuman powers[28]). As for Grauballe Man, his last meal might have been eaten in full knowledge of ergot's disturbingly mind-altering effects. But then again, this poor unfortunate seems to have been rather unpleasantly executed (whether as a criminal or as a willing religious sacrifice we do not know), for he had his leg broken and suffered a traumatic blow to the head, and he might have died before any effects of the ergot became apparent. The administering of a plant that was known, say, to terminate pregnancies might simply have held a terrifying symbolic force for him.

The second problem is the one of cultural specificity that I have already outlined. Cultures that use one psychoactive plant may have a socially constructed aversion towards, or be wholly ignorant of, another. Thus, when the late advocate for the magic mushroom, Terence McKenna, travelled to a region of the Amazon to investigate the use of an indigenous psychedelic plant preparation, *oo-koo-hé*, he found that the locals were quite ignorant about the magic *cubensis* mushrooms sprouting abundantly from their cattle dung.[29] Psychoactive know-how is evidently not always horizontally transferable. Mushrooms do have a distinctive form, however, and so a second, more promising line of inference has come from studying ancient and prehistoric art for obvious images of mushrooms.

Perhaps more than any other archaeological discipline, the study of ancient art is as tantalising as it is rewarding. Tracing the various ways in which ancient peoples used marks and lines, pigments and paints – often daubed straight onto smooth rock faces or cave walls – to express themselves necessarily narrows the gap between our world and theirs; yet all too often the exact meaning of their artwork eludes us. Commonly, archaeologists have to live with the uncertainty of entertaining a range of possible interpretations.[30] Of course, it is reasonable to assume that if ancient cultures *did* centralise the use of magic mushrooms they would have depicted this fact in their artwork. The trouble is that it is not always easy to tell whether something that looks to us like a mushroom really was intended to be one, and a magic one at that. The context in which the art appears is often the only thing that can help us decide, as the following example, taken not from prehistory but from the Middle Ages, demonstrates.

In Hildesheim in Germany there is a magnificent Gothic cathedral, which is renowned for its pair of cast bronze doors. Each contains eight panels depicting biblical images, but one in particular has caught the attention of mushroom enthusiasts, for there, on the right-hand door, is a panel that seems to show human figures dwarfed by what looks extremely like a giant Liberty Cap. Various writers have suggested that this is evidence for a medieval magic mushroom cult persisting secretly in spite of Christian oppression. Some mushroom-loving craftsman must have slipped the mushroom into the image as a hidden but demonstrative gesture of defiance for the benefit of other members of his secret cult.[31]

Medieval bronze door panel from the cathedral at Hildesheim, Germany. Does this reveal the mushroom's influence at the root of Christianity? In fact, in spite of its resemblance to a Liberty Cap, the plant dwarfing the human figures is a stylised fig-tree. Photo by Paul Stamets, reprinted with permission.

Sadly, careful consideration of the context in which the doors were made shows only a slight chance that this interpretation is true. The doors were commissioned by Bishop Bernward in 1015 as the finishing touch to his already magnificent cathedral. The images were extremely carefully chosen, for the eight descending panels on the left show the fall of Man, and the eight ascending panels on the right his redemption through Christ. Moreover, every image on one door appears next to, and is paired with, its matched antithetical opposite on the other. Apart from being a brilliant piece of metalworking, the doors are a coherent structural masterpiece.[32] Given that every detail of every image held significance, and that the message on the doors was so carefully constructed, it seems improbable that a magic mushroom could have been surreptitiously slipped in without anyone noticing. Nor could a secret mushroom cult have persisted and left such an emphatic mark upon such a high-profile expression of religious power and piety without there being some other evidence for its existence. But of course there is none to be found anywhere, for there was no cult. The image on the door is simply a stylised representation of that most biblical of plants, the fig-tree.[33]

Clearly it is not enough to identify an image as a representation of a magic mushroom on the basis of homology alone. The problem with prehistoric art is that very often we have only the image, and none of the contextual evidence that would tip the interpretation one way or another. For example, amateur archaeologist Reid Kaplan published an oft-cited paper in 1975 in which he argued that a recurring mushroom-shaped motif found on various Scandinavian Bronze Age razors and petroglyphs depicts the fly-agaric, which he imagined formed a part of a solar-based religion. Others have suggested, however, that the motif is the sail of a ship, a tree, or a hatchet.[34] There is no easy way of telling which, if any, is the correct interpretation.

More recently, laser scans have revealed some hitherto undiscovered Bronze Age carvings on the sarsen stones of Britain's most famous ancient monument, Stonehenge.[35] Some of them look remarkably like mushrooms in cross section, a fact that has led some archaeologists to speculate that not only do these depict *magic* mushrooms, notched up like marks on the bedpost, but that the entire monument was built to resemble a magical mushroom fairy ring.[36] The more widely accepted and prosaic theory is that the carvings are of Bronze Age axes, though their significance is unknown. Perhaps the most eye-catching artefact from the later Iron Age is the Gundestrup cauldron, a magnificent silver cauldron intricately decorated with pictures of animals, plants and even a horned figure, perhaps a deity. The archaeologist Robert Wallis has tentatively suggested that the vegetation might be a representation of a psychoactive plant, or even, because of the pointy leaves, the Liberty Cap.[37] Then again, he accepts that it could be entirely decorative.

But perhaps the most famous example comes from the abundance of rock art found upon the Tassili plateau of southern Algeria, which dates, incredibly, from the Neolithic to the start of the Common Era. Though not discovered by him, one particular Neolithic image was popularised by the aforementioned mushroom advocate Terence McKenna, who reproduced it in two of his widely read books.[38] Subsequently, it has become an icon for the psychedelic mushroom community, appearing on posters, T-shirts, postcards and, of course, liberally across the Internet. The image depicts a squat, male, human figure, standing braced against the earth as if shouldering a heavy load. He wears feathers or plumes upon his head, and a bee-shaped mask. His form is covered in a psychedelic pattern of lozenges and dots, which swirl out between his legs. But most strikingly of all, he appears to be clutching handfuls of mushrooms, which also sprout alarmingly from his legs, arms and torso. Surely here is incontrovertible evidence: a bemushroomed shaman depicted in the full force of a psilocybin-induced ecstasy?

Well, possibly. The popular image that has been so widely circulated is not a photo of the original, but a copy, a drawing made during the 1990s by McKenna's then wife and mushroom enthusiast in her own right, Kat Harrison. Neither went to see the original rock painting in Algeria, but derived their interpretation from photos in a book, *The Rock Paintings of Tassili* by Jean-Dominique Lajoux.[39] While Harrison strove accurately to copy the original, and to enhance only those aspects of it that had been damaged and obscured,[40] her representation, informed by her own mushroom experiences, subtly reinforced this shamanistic interpretation. So in her drawing the protuberances definitely look like mushrooms, whereas in the original the semblance is not quite so emphatic: they could be mushrooms or they could be, say, arrows. Similarly, in the original the plumes from the figure's headdress are not distinguishable as feathers; neither are the swirling psychedelic patterns quite so obviously swirling and psychedelic. Without any shadow of a doubt, Kat Harrison's striking

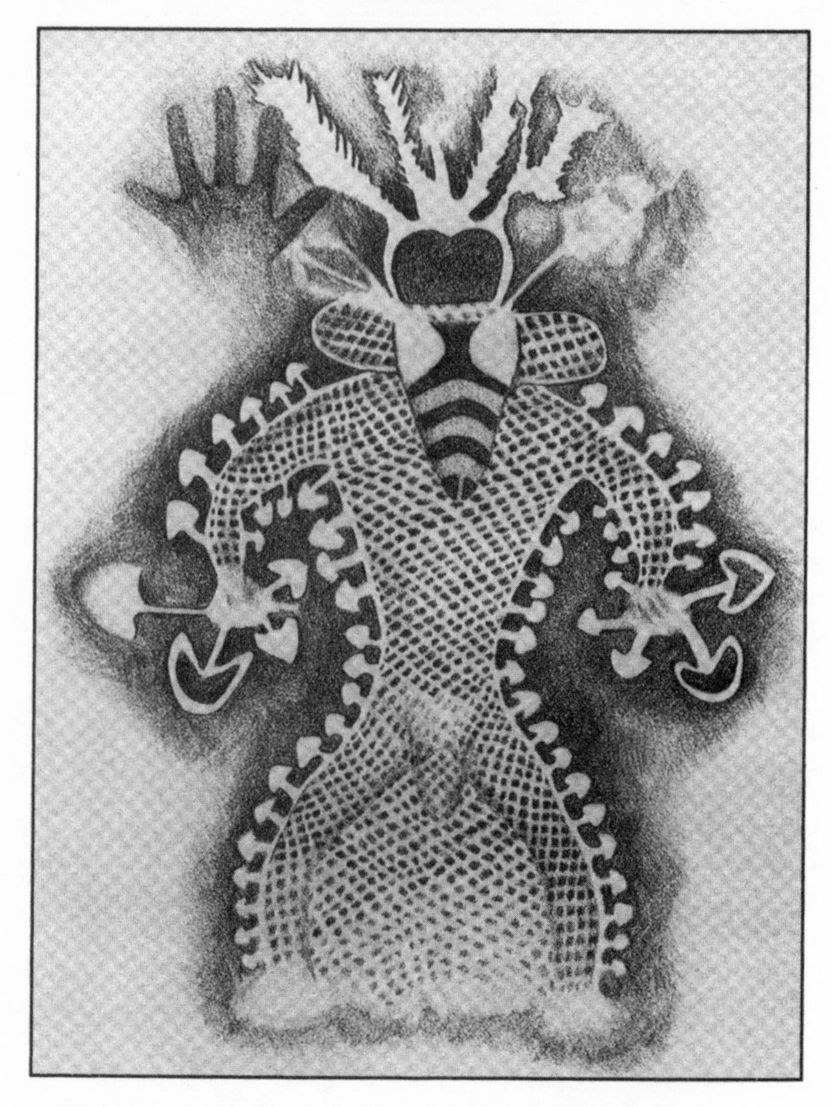

Kat Harrison's famous rendition of a piece of prehistoric rock art from Tassili, Algeria. Harrison's iconic drawing clearly depicts a bemushroomed shaman, but whether that was the intention of the original Neolithic artist(s) remains moot. Image © Kat Harrison, reprinted with permission.

facsimile depicts a bemushroomed shaman, and has rightly been adopted as such, but whether that was the intention of the original artist(s) is far from settled.

For instance, there is the contextual presence of other closely related pieces of art from the same region that depict the same 'bee-masked' figure but without any 'mushrooms'. If the mushrooms were the most important aspect of the figure, why are they absent from these other pictures? Curiously, mushroom enthusiasts have little to say about a further image of a woman bent over with 'mushrooms' apparently sprouting from her derrière.[41] Could it be that for these ancient artists mushrooms had a more scatological or sexual significance? Or that the mushroom's more typical association with decay meant that the bee-masked 'shaman' was actually some apocalyptic figure reminding the viewer of mortality and death? Then again, if the mask was really meant to look like a bee, as McKenna thought, could it not be that those 'mushrooms' were actually arrows or stings, symbols of some superhuman power? Lajoux thought the 'mushroom' protuberances incomprehensible and saw the figure as an anthropomorphised moufflon, or sheep.

McKenna also drew attention to another image from Tassili, which seems to show some anthropomorphised flat-capped mushroom figures, with faces, arms and legs, running while clutching 'mushrooms' in their hands. Admittedly, these figures are more convincing as mushrooms, but then again one is reminded of the cloth-capped jesters that appear in medieval art and have nothing to do with either mushrooms or psychedelia.[42] The point remains that all the so-called incontrovertible visual evidence produced by enthusiasts to argue for prehistoric mushroom use turns out to be open to many and varied alternative interpretations. In each case, the mushroom interpretation may be correct, but it may equally very well not be.

This is clearly an unsatisfactory state of affairs for both academics and partisan enthusiasts alike, but recently the discipline of rock-art studies was shaken up by two academics, David Lewis-Williams and Thomas Dowson, determined to set it on a more scientific and unequivocal basis. In the process they argued that some Palaeolithic rock art was an attempt to represent hallucinations obtained in trances and other altered states of consciousness, a move that has split the archaeological community ever since. Their 'three stages of trance' model is one of the most hotly contested issues in current archaeology, and has triggered more than one good old-fashioned academic spat between the various parties.[43]

They began their study by examining data from earlier neurophysiological investigations into the kinds of hallucination produced by the psychoactive drug mescaline, and by certain types of migraine.[44] From these they developed their three-stage model of trance. The first stage, a light trance, is marked by the appearance of so-called phosphenes or entoptic hallucinations: that is, hallucinations that are generated within the physical apparatus of the vision system (these can be seen, for example, when we rub our eyes a little too vigorously[45]). As they are produced by the vagaries of our shared human physiology, they are uniformly experienced across cultures. But as the trance deepens into the second level, these entoptic hallucinations become consciously or unconsciously interpreted as more culturally specific 'iconic' forms. Thus, a person in this second stage seeing a zigzag might regard it as, say, a snake or a lightning strike depending on the relative significance of each of these to his or her culture. At the third and deepest level of trance, subjects are unable to make any distinction between entoptic and iconic forms, for they are plunged into an overwhelmingly iconic experience. Thus, they might feel themselves in the presence of the deities and mythological beings specific to their world view.

According to Lewis-Williams and Dowson, each stage of trance produces its own characteristic hallucinations, but from a cross-cultural point of view, universally experienced entoptic hallucinations are easiest to recognise. Thus, they abstracted what they regarded as the six typical entoptics: grids, parallel lines, dots, zigzags, filigrees and nested curves. Their aim then was to see whether any of these 'trance signatures' appeared in examples of rock art.

Originally they tested their model on the petroglyphs of the San people of southern Africa, which were known to have been created immediately following participation in communal trance rituals. The archaeologists were excited to find both entoptic and iconic images that corresponded well with the shapes predicted by their three-stage model. Gathering momentum, they then applied the model to prehistoric art from Europe, from the Upper Palaeolithic, and again thought they found the presence of entoptic, entoptic/iconic and iconic forms. In short, it looked as if this European rock art had been produced by people who were wholly familiar with altered states of consciousness.

Prior to the three stages of trance model, a few anthropologists had wondered whether indigenous art might not have had its origins in hallucinogenically inspired visions.[46] During the 1960s, a few intrepid

Western investigators found that under the influence of Mexican magic mushrooms they saw Aztec-style patterns woven throughout their hallucinations.[47] The problem was that it might have been their cultural expectations, rather than some essential property of the mushrooms, that determined the form of these mutable inner tapestries. Subjectivity could never be entirely ruled out. Lewis-Williams and Dowson's model appeared to put the matter on a much more scientific, objective basis and it was eagerly followed by a range of academic studies purporting to have found additional evidence for trance experiences expressed through American and European prehistoric art.[48] Some of these studies argued that psilocybin mushrooms lay behind the visions.

Eye-catching headlines in the press about 'Stone Age Psychedelia' and 'Raves in the Caves'[49] did much to propel this appealing idea into popular consciousness,[50] but the equally compelling counter-arguments, and the often impassioned and vitriolic manner in which the debate has been played out, are pretty much unknown beyond academia.[51] It would take far too much space to summarise them all here but, once again, most rest on the problem of cultural specificity and the method's failure to escape subjectivity.

For example, writers and artists have struggled since the time of the Romantics to represent their natural or drug-induced hallucinations, and so the idea that anyone could reduce these to six abstracted forms seems arbitrary, subjective and unduly optimistic. And while altered states of consciousness may generate entoptic forms, not every entoptic form unambiguously indicates an altered state. There are all manner of reasons why people could have been moved to leave patterns and marks upon rocks and stones. Inspired by nature, they might have wanted to draw ripples or waves. Primitive tools might have limited artistic expression to curves and zigzags. Like doodlers everywhere, prehistoric artists might simply have found the patterns pleasing. Or they might have been bored. And even if our ancestors did make use of mushrooms or other psychoactives, there is no reason why they should have wanted to record their *entoptic* hallucinations. After all, the more intrepid mushroom enthusiasts today regard their 'entoptic' visuals as trivial compared to the deeper 'iconic' visions obtained on high doses. Academic opinion is at best divided over the model, at worst against it, so once again this particular line of inference reaches a dead end.

There is one final approach that we need to discuss, for if magic mushrooms were used in ancient times, they would have to have been used religiously (they are inherently spiritual, remember). It stands to reason, therefore, that the place to look for traces is in the relics of any ancient shamanic-inflected European religions. And the three that spring most obviously to mind, to mushroom enthusiasts at least, are Druidry, witchcraft and the ancient Greek rites of Eleusis.

For nearly two thousand years spanning ancient Greek and Roman rule, mystery rites in honour of the goddesses Demeter and Persephone were performed annually at Eleusis, near Athens, at the temple marking the place where Hades was supposed to have abducted Persephone into the underworld.[52] The September celebrations were open to all Greek speakers, excepting murderers, for the price of about a month's wages, and were attended by many thousands of people, from the lowly to the great and the good. Plato, the thinker with whom Western philosophy begins (and some say ends), is thought to have been an initiate. Celebrants walked a processional route from Athens, performing ritual ablutions along the way. They spent a night dancing in preparation and then they imbibed a potion from a sacred cup, the *kykeon*, before entering the secret *telesterion* initiation hall where a great mystery was revealed. Contemporary writers speak about the experience in hushed tones of reverential awe but, because disclosure of exactly what happened inside the temple at Eleusis was punishable by death, the climax of the ceremony has remained largely mysterious to this day. As classical writers seem, without exception, to have had a profoundly moving, ecstatic or mystical experience at Eleusis, modern enthusiasts have argued that the potion must have been psychoactive. And that, of course, means it could have contained magic mushrooms.

The poet and author Robert Graves who, as we shall see, was no stranger to mushrooms himself, was the first to suggest that plant hallucinogens might have induced the reported mystical raptures.[53] He proposed, on the basis of his idiosyncratic interpretation of the Homeric Hymn to Demeter and other Greek myths, that the potion given to all celebrants had originally contained the fly-agaric, replaced at a later stage by a more efficacious psilocybin mushroom. His American friend Robert Gordon Wasson – who played the lead role in the modern discovery of the magic mushroom – favoured ergot as the likely psychoactive agent, for its psychoactive alkaloids could have been separated from the toxic ones by soaking the sclerotia in water.[54] Others have suggested a mixture of ergot and mushrooms, or just plain old *cubensis,* as the responsible agents.[55]

Sadly, in the absence of any material residues or documentary evidence that would tell us exactly what was in the cup, all these theories must remain speculative. And while Plato could certainly have entered the realm of eternal forms after drinking mushroom tea from the *kykeon*, the history of religion is also replete with examples of mass epiphanies induced by nothing more exotic than enthusiastic fervour (contemporary charismatic Christianity being a case in point, vehemently opposed as it is to drugs of all kinds). The Eleusinian potion may have been alcoholic or, acting as placebo, may have contained nothing inebriating or psychoactive at all: the lengthy preparations and the great sense of anticipation and expectation could have been sufficient to have generated the feelings of religious awe that were so widely reported. The tendency to imagine drugs at the centre of a variety of ancient religions says rather more about *us*, and our inability to countenance religious ecstasy without the use of psychoactives, than it does about the religions concerned.[56] With so little to go on, the belief that mushrooms were consumed at Eleusis is highly tendentious, and even if it were truly the case, the Greek participants, ignorant of what was in the potion, would surely have had no idea that it was mushrooms they were taking.

The same is not supposed to be true of the shadowy Druids who, it is widely believed, knowingly consumed mushrooms as part of their pagan rites at Stonehenge. Again, it was Robert Graves who started the ball rolling. He argued, in one of his lengthy ruminations about the Greek myths, that the Druids formed part of a pan-European mushroom cult, and that they had remained on good terms with the ancient Greeks by supplying them with bundles of fly-agarics wrapped in straw, presumably for consumption at Eleusis.[57] Graves's intellectual status ensured that the idea gained a certain currency. Others followed by examining not the Greek myths but the early Irish and Welsh vernacular literature (including the Welsh *Mabinogion* and the Irish *Tain*), which are replete with references to Druids and their magical doings. Academics and enthusiasts alike have wondered whether these tales contain thinly veiled references to psychoactive potions and preparations and their consciousness-altering effects.[58] After all, isn't it natural and obvious that Celtic Europe's very own shaman-priests would have supped on the local hallucinogenic fungi?

Well, once again, alarm bells should ring, for what we actually know about the original Iron Age Druids could almost be written on the back of a postcard. They, being illiterate, left us no records of their own beliefs and practices, while the archaeological record presents us with a tantalisingly incomplete picture of Iron Age religious life.[59] We can be certain that the link with Stonehenge is spurious, an antiquarian conceit from the eighteenth and nineteenth centuries, for Stonehenge was already two thousand years old by the time of the Druids. That we know of the existence of Druids at all is because Greek and Roman classical authors chose to write about them, and it is from these second-hand accounts that the stereotypical picture of the Druids comes: the white-robed bearded philosophers, meeting in groves, harvesting mistletoe with golden sickles, and performing both animal and human sacrifices. But the extent to which the descriptive and fantastical or rhetorical elements of these accounts can be distinguished remains debatable: most of these authors had never been to Britain or Gaul, let alone encountered a Druid.[60] In any case, none mention the use of psychoactive mushrooms. Nor do any of the stories from the Celtic vernacular literature, which were written so long after the formal end of paganism, and by people so far removed from it, that they can tell us nothing meaningful about the Druids they purport to describe.[61]

If the figure of the Druid has been made to work hard by Western commentators, then that of the witch has been doubly made to do so. In particular, the episode of the medieval and early modern witch-hunts has particularly exercised, if it has not been exorcised from, the Western scholarly and popular imaginations. As the remonstrations of surly Goth teenagers everywhere will surely attest, the witches were not malevolent devil worshippers but pagans, herbalists, healers and midwives who almost certainly knew about the properties of magic mushrooms that they slipped into their 'flying ointments'.[62] A threat to patriarchy, nine million of them, mainly women, were persecuted by a vindictive Christianity and burnt at the stake.

This is the myth as it is popularly believed, but the real story, as ever, is a little more complicated. For one thing, the figure of nine million deaths is vastly overstated, and was more realistically in the region of forty thousand (though still a far from trivial number).[63] For another,

The archetypal Druid as imagined in the eighteenth century. Title page from *Antiquities of England and Wales*, 1773–87, Volume IV, by Francis Grove.

those accused as witches were most emphatically Christian, not pagan, and were, in the main, outsiders, loners and scapegoats who upset the communities in which they lived, or who took the blame for the petty hardships, travails and mishaps of ordinary life. The potent idea of witches as pagans seems to have been invented in the years between the two world wars, but was brought to widespread attention in 1921 by the British anthropologist Margaret Murray (1862–1963) in her influential book *The Witch Cult in Western Europe*. Here, and in subsequent volumes, she argued that the witches had not been devil-worshippers, as accused, but were actually members of an ancient pagan religion, existing secretly in rural backwaters since prehistory, that worshipped a a horned god and practised magic.

Although her thesis has now been demolished (the way she interpreted witch-confessions and her habit of gathering data partially and selectively are rightly regarded as unsound[64]), her ideas gripped the popular imagination. In particular, she inspired a retired colonial civil servant, Gerald Gardner (1884–1964), to set about 'reviving' (that is, creating) her witchcraft religion: currently, Wicca is the largest and most popular of the modern Pagan religions in Britain and America, followed only by revived Druidry. Wiccans are gradually accepting the scholarly revisions to their historical identity – that theirs is actually a recent, invented spirituality and none the worse for that – but one belief about early modern witchcraft remains firmly in place, particularly amongst psychedelic enthusiasts: that of the witches' flying ointment.[65]

According to contemporary accounts and recorded confessions, those accused of witchcraft would often admit to using devilish ointments with which they would smear themselves before attending their sabbats. These greasy pastes enabled them to fly or even to turn themselves into animals. The idea that these ointments were not magical but mind-altering was mooted at the time of the trials. Rationalist sceptics, refusing to countenance the idea of magic and witchcraft, advanced the hypothesis that the ointments were deliriants that merely induced the *impression* of flying, of turning into animals, of consorting with the devil and so on.[66] The idea made little impact but remained in circulation, so that ointments were depicted in eighteenth-century engravings[67] and were discussed in this light by nineteenth-century anthropologists.[68] Margaret Murray gave the idea serious consideration, and in an appendix to *The Witch Cult* her colleague A. J. Clark concluded that some of the reported ingredients, hemlock, aconite and belladonna, might have induced a feeling of vertigo or flying when introduced via the skin.[69] The idea was picked up by twentieth-century psychiatrists keen to disprove the ascendant Freudian theory that the persecutions were triggered by mass hysteria caused by repressed material surfacing from the collective psyche.[70]

It was during the 1970s, however, that the idea received its most fertile reception, when it was restated by the anthropologist Michael Harner in his widely read book *Hallucinogens and Shamanism*.[71] The idea of witches as adept users of hallucinogens appealed not only to the burgeoning sixties and seventies psychedelic counterculture, impatient to find historical antecedents that would legitimate their own practices, but also to feminists looking for empowering historical role models.[72] Harner's unfounded suggestion (apparently borrowed from the writer Michael Harrison[73]) that the witches had absorbed the poisonous ingredients of the ointment through the sensitive lining of the vagina, applied using broom-handles as dildos, allowed the witches to be reinvented as ecstatic proto-feminist separatists, sexually empowered yet independent of men. Later suggestions that the witch-hunts were an ergotism-induced mania[74] did little to dent the now popular belief that witches were adept users of psychoactive unguents, employing the poisonous tropane-containing plants belladonna (*Atropa belladonna*), henbane (*Hyoscyamus niger*) and mandrake (*Mandragora officinarum*) as their three principal agents. From there, it was not so great a leap to add magic mushrooms into this potent mixture of sex and drugs.[75]

It must be remembered that Michael Harner was an anthropologist, not a historian, but as such he argued passionately that we could only truly understand indigenous cultures and their world views if we actually partook of their psychoactive plants and brews, rather than watching from the sidelines. He took his own advice to heart, famously 'going native' while studying the Shuar Indians of Ecuador, and abandoning anthropology to undergo a gruelling shamanic training involving the repeated use of the psychoactive *ayahuasca* brew and of juice extracted from the datura plant.[76] Now datura belongs to the same family as belladonna, henbane and mandrake, the *Solanaceae* or potato family, and contains the same active ingredients as its three European cousins, namely the tropane alkaloids hyoscyamine, atropine and scopolamine. Its use in the New World is typically associated with witchcraft and sorcery, and it produces bizarre and often troubling hallucinations.[77] Harner, then, reached his conclusion about the European witches less upon the historical evidence than upon the basis of his own experiences with datura. He knew, first hand, that the tropanes were more than powerful enough to convince users that they had flown or, say, had intercourse with the devil, for his own experiences had been sufficiently terrifying and outlandish.

The fact that there was no pagan witch-cult would rather seem to pour cold water on the whole notion of the ointments, no matter how powerfully psychoactive datura and its European cousins are. Even so, scholars have accepted that occasional and localised use of tropane-containing preparations amongst the accused cannot be entirely ruled out.[78] On the other hand, the fact that the 'ointments' conveniently explained how it was that tens, sometimes hundreds, of accused witches were able to gather for their sabbats without anyone noticing, makes the story just a little too convenient. It seems most likely that it was a 'plot device' inserted to make the accusations seem more credible.[79] The idea of the ointments was already in circulation at the time of the trials, for in Apuleius' second-century ribald fable and proto-novel *The Golden Ass*, the sorceress Pamphilë smears herself with a paste before turning into an owl. The hero of the story, Lucius, attempts to copy her but finds himself turned into an ass, with all the comical results you might expect. The book was read by Boccaccio, Shakespeare and Milton amongst others, and the fact that even today Pamphilë's transformation is often taken as a literal description of a hallucinogenic experience[80] suggests that it could easily have been read as such by the witch-hunters. The origins of the ointments seem almost certainly to have been literary.

It seems that everywhere we look in the dim and distant past, those slippery and evanescent magic mushrooms simply evade our grasp. There are none to be found in the archaeological record, nor are inferences from other psychoactive plant remains easy to make. Supposed images of mushrooms are, in the absence of contextual information, quite possibly something else entirely, while purported hallucinogenically inspired rock art may be nothing of the sort. Not one single line of evidence unequivocally points to their having been used by Druids, herbalists, Stone Age architects or prehistoric shamans. Of course, absence of evidence cannot be taken as evidence of absence, and it does seem inconceivable, given the huge amounts of time in question, that someone, somewhere, in European prehistory did not intentionally reach out to pluck a psilocybin mushroom from the soil. The problem is that, if they did, they left not a single piece of evidence of having done so. The best we can do is to say that we really do not know, one way or the other, whether the ancients worshipped the holy spores of God. Though anyone is free to make that assertion, they should remember that they are doing so on the basis of wishful thinking and not established fact. But the time has come to leave behind the thorny question of archaeology and prehistoric ecstasy to ask what traces of magic mushrooms there are in the historical record. The answer is nothing if not surprising.

4

Much Disordered

The Mushroom is much eaten by the Gentry . . . but as it may delight one, so it may be Poison to another.

Randle Holme, *The Academy of Armory* (1688)

Few mushrooms are good to be eaten and most of them do suffocate and strangle the eater. Therefore, I give my advice unto those that love such strange and new fangled meates to beware of licking honey amongst thorns lest the sweetness of the one do countervaile the sharpness and pricking of the other.

Gerard, *The Herball* (1597)[1]

It is a little-known fact that the common Mexican magic mushroom species have all appeared on postage stamps in the African Republic of Benin; also, a type of 'felt' pill-box hat favoured in Romania is actually made from the flesh of a bracket fungus, *Ganoderma applanatum*; and in Zambia tortoises are believed to have a passion for mushrooms, hence the expression 'the guest of the tortoise has mushrooms for supper' (which is roughly equivalent to our 'when in Rome . . .').[2] That we know this sort of trivia – and that it is available in sufficient quantities for compilers of modern miscellanies to earn a healthy living – is due to the fact that all of it was, at some point, written down. The result is that, with the move into history, we have a much clearer picture of magic mushroom use in Europe than ever we can hope to obtain from the archaeological record. And the picture that emerges is quite surprising, for while people appear to have been eating hallucinogenic mushrooms for as long as there have been records, until the twentieth century they always did so accidentally and unintentionally.

We must, of course, exercise a little caution when interpreting early records, for writers and archivists prior to the late twentieth century did not use unambiguous terms such as 'psychedelic', 'hallucination' or 'tripping', but looser terms like 'choaking', 'vertigo' and 'giddiness'. What appear to us to be the animated antics of the bemushroomed could have struck earlier writers as 'convulsions'. Nor were they able to identify categorically the responsible species until the scientific advances of the Victorian age had revolutionised fungal taxonomy. Nevertheless, the language they used and the symptoms they describe are often more than suggestive of magic mushrooms, even if we can never be 100 per cent certain what species they were referring to.

The earliest hint of an accidental mushroom consumption, then, comes from the herbalist and botanist Albertus Magnus (1193?–1280), in his thirteenth-century treatise *De Vegetabilibus*. There he cautioned against mushrooms 'of a moist humour' that 'stop up in the head the mental passages of the creatures [that eat them] and bring on insanity'.[3] He seems not to have been referring here to the fly-agaric, for he describes it later on in the treatise. Carolus Clusius (Charles Lécluse, 1526–1609), in his *Rarorium Plantarum Historia* of 1601, described a fungus known in Germany as the *narrenschwammem*, or 'foolish fungus', which made people 'mentally upset' (unfortunately, the mushroom that most accurately fits his description is *Amanita vaginata*, but it is not psychoactive).[4] And, more strikingly, from the Netherlands, the sixteenth-century physician known as Forestus (Pieter van Foreest, 1522–1597) recorded two possible cases of accidental intoxication: in one, a countrywoman fell after eating mushrooms into a state of 'grievous disorder' from which she never recovered; in the other, a young woman was 'flung into violent convulsions and the *Risus sardonicus* by eating mushrooms'.[5] *Risus sardonicus* is a medical term that refers to the fixed rictus grin symptomatic of tetanus, but in earlier medical parlance it described a condition of pathological or uncontrollable laughter, which, as we know, is one of the well-attested characteristics of the bemushroomed.

Apart from that of Clusius, we have no records at all from the seventeenth century, though the eighteenth is more forthcoming; indeed, record-keeping and fungal taxonomy become sufficiently advanced for us to hazard attempts at identifying the species concerned. While it is generally accepted that the earliest instance of someone eating Liberty Caps occurred in Britain in 1799, there are other tantalising pieces of evidence from earlier in the century. In 1755 the Reverend Roger Pickering (*c.*1720–1755), amateur botanist and Fellow of the Royal Society, submitted 'A brief dissertation upon Fungi in general, and concerning the poisonous faculty of some species in particular . . .' to the *Gentleman's Magazine*, a journal to which Dr Johnson was a regular contributor.[6] There, apart from referencing the examples from Forestus, he distinguished two distinct types of mushroom poisoning: the one 'bringing on violent vomitings, diarrhoeas, dysentries, &c' but the other, more tellingly, affecting the nervous system with 'a sense of choaking, convulsions, *Risus sardonicus* &c'.[7]

Two years later, in 1757, the same magazine reported that a tailor, a Mr Kirk from Salisbury, together with his wife, daughter and journeyman, 'were in great danger of losing their lives in a stupor and convulsions, with which they were seized, in consequence of their eating some stewed mushrooms for supper, which they gathered the same day on the downs near Amesbury'.[8] Mr Kirk 'had [the] presence of mind, soon after he found himself affected, to call an apothecary, or 'tis imagined they must all have perished by the morning, as 'twas with the greatest difficulty the daughter and journeyman were recovered, after lying some hours quite insensible'. Clearly Kirk did not, like his more famous namesake, imagine himself boldly going where no man had gone before, but thought instead that he and his family were poisoned and very probably dying. Nevertheless, the fact that the mushrooms were picked on pastures, and that the family's alarming symptoms wore off after some hours, are consistent at least with the Liberty Cap.

Some fifteen years later, the physician W. Heberden also wrote to the *Gentleman's Magazine* to warn readers of 'the noxious effects of some Funguses'.[9] He described how a man and his family picked and ate some 'champignons' but were, some five minutes later, 'all much disordered'. The man 'was unable to shut his eyes and was so giddy he could hardly stand; the woman felt the same symptoms in a more violent degree; and the child, who had but just tasted them, had convulsive agitations in its arms'. The family responded well to the emetics that were typically prescribed in such cases of poisoning; the mother, who had been unable to drink anything, made a complete recovery some days later without this treatment.

Heberden was not content to leave the matter there as he wanted to broadcast the identity of the culpable mushrooms pro bono publico.

He suspected two varieties, which he labelled as 'the fungus parvus pediculo oblongo, pileolo hemisphaerico, ex albide subluteus' and 'the fungus minimus e cinereo albicans, tenui et praelongo pediculo, paucis subtus striis'. He obtained these snappy, pre-Linnaean Latin descriptions from John Ray's *Synopsis*, the authoritative flora of the time,[10] but clearly, by modern standards, the experts' unwieldy attempts to identify the mushrooms were scarcely better than those of the family who had mistakenly eaten them: we cannot identify the species in question.[11]

One further incident, however, from 1799, has proved to be more conclusive. Early one October morning Dr Everard Brande, a physician, was summoned urgently to a house in London's Piccadilly where a family had been taken ill. The father, known only as J.S., had gone out to Green Park to gather wild mushrooms for his family's breakfast, as he was accustomed to do every morning. He cooked up a broth in an iron pot, and served the mushroom soup with tea. But one hour later strange symptoms began to manifest themselves in the children. The doctor later described how the youngest son, Edward, aged eight, 'was attacked with fits of immoderate laughter, nor could the threats of his father or mother restrain him'. Shocking as this was, it was succeeded by attacks of vertigo, then stupor, and when roused Edward answered yes or no willy-nilly to a series of questions put to determine how he was feeling. His pupils, Brande observed dispassionately, were 'dilated to nearly the circumference of the cornea'.[12]

Shortly afterwards his father, now equally wide-eyed, also started experiencing vertigo, then complained that everything had 'gone black'. The world was restored after ten minutes or so, but this was not sufficient to elevate his mood for he became convinced that he was dying. His other children – Martha, eighteen, Harriet, twelve, and Charlotte, ten – were all affected to a greater or lesser extent. Charlotte, in particular, was delirious, her sight impaired. All of them had variable pulse rates, a feeling of coldness in the body, and loss of voluntary motion. Dr Brande, joined by his colleague Dr Burges, immediately suspected poisoning, and proceeded then to apply emetics and cathartics to his patients; he remained with them all day until the distressing symptoms had passed.

Gathering the remains of the mushrooms from the broth, and instructing the father (who claimed that he had picked and eaten the same sort of mushrooms for years, with no deleterious effects) to collect some more, Brande arranged to have the samples sent to a Dr

Williams, Botany Professor at Oxford, for identification. Williams consulted with another expert, James Sowerby (1757–1822), and together they concluded that the mushroom was a variety of the recently described species *Agaricus glutinosus*. The botanists noted that the other experts in the field, Dr William Withering (1741–1799) and Dr William Curtis (1746–1799) (the latter author of the magnificent *Flora Londinensis*), had listed this particular species as harmless. Brande, a belt and braces man, obtained a second opinion from a Mr Wheeler, Demonstrator of Botany for the Apothecary's Company, who confirmed the identification. Brande confidently wrote up the incident for publication in the *London Medical and Physical Journal*, concluding with a description of the offending variety of *Agaricus glutinosus* so that further incidents might be avoided.

It is thanks to Brande's diligence and perspicacity that later mycologists have been able, in the light of the nineteenth century's taxonomical revolution, to reclassify this species as *Psilocybe semilanceata*. For on hearing about the case, James Sowerby hurriedly postponed publication of his *Coloured Figures of English Fungi or Mushrooms*, which had been due for release in January 1800, to insert an account of the incident and an additional plate illustrating the offending species.[13] It is from his paintings, and from Brande's description, that the fresh identification was made in the late 1960s. The American mycologist Dr Rolf Singer stumbled across the episode in Sowerby's book and immediately identified the mushroom as the Liberty Cap.[14]

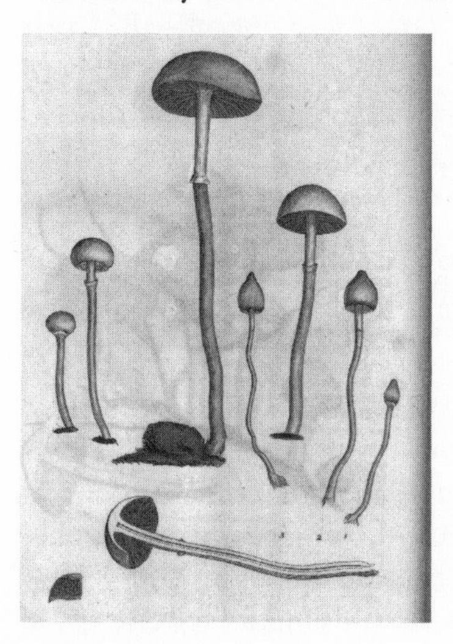

Agaricus glutinosus, from James Sowerby's *Coloured Plates of English Fungi* (1797–1815). The larger, dome-headed, mushrooms are probably *Stropharia semiglobatus*, the Dung Roundhead (which is not psychoactive), whilst the smaller, goblin-capped, mushrooms are almost certainly Liberty Caps, *Psilocybe semilanceata*. Sowerby's was the earliest illustrated book to warn of their intoxicating properties. Courtesy R. Gordon Wasson Archive, Harvard University, Cambridge, Massachusetts.

Back in the early nineteenth century, however, fungal taxonomy was still far from an exact science, and very often several species were lumped together as different varieties, as was the case with *Agaricus glutinosus*. Binomial classification had only just been introduced, following Linnaeus, and even then it was assumed that all mushrooms belonged to the same homogenous genus, *Agaricus* (named from an ancient Greek word for mushroom). Nevertheless, the improvements in classification, such as they were, meant that further cases of this kind of poisoning were swiftly recognised. Sowerby himself reported two more, from Mitcham in Surrey and Christ Church in Hampshire.[15] Some thirty years later his son, James Junior (1787–1871), stepping into his father's shoes, remarked that *Agaricus glutinosus* was well known as an injurious species.[16] He noted that a 'poor family at Lambeth, in 1830, having eaten but few for supper, were taken very ill and continued so all night; but by means of the stomach-pump and medicines, were in a day or two recovered'.[17]

In 1816 a Dr G. Glen was called to Knightsbridge to assist a man who was suffering very badly indeed.[18] The man had gone out in the morning to Hyde Park to pick mushrooms from his favourite spot, the trees behind Horse-Guard Barracks. He cooked up his pickings in a delicious stew, but ten minutes after eating them he was 'suddenly seized with a dimness or mist before his eyes, lightness and giddiness of his head, with a general trembling and sudden loss of power'. The poor man forgot who and where he was, and was so stunned he nearly fell off his chair. Recovering slightly, he staggered out onto the street to find assistance (as luck would have it, the doctor lived only five hundred yards away), but then forgot who he was and where he was going. A neighbour came to his rescue, and by the time he reached Glen his 'countenance betrayed great anxiety; he could scarcely stand, but reeled about somewhat like a drunken man; he spoke with hesitation and reluctance; he complained of no pain . . . [but] he suffered much from giddiness'.[19] After a day's treatment he appeared to recover completely.

Glen took samples of the mushrooms to his friend and mentor, William Salisbury of Sloane Street, who identified them as *Agaricus campanulatus* (now called *Panaeolus campanulatus*) because he had previously written to the *Gentleman's Magazine* to record two further cases of poisoning with this species.[20] His identification could have been correct, for *Panaeolus* species have been found to contain psilocybin, albeit only occasionally and with great variability of concentration.[21] It is possible, however, that Salisbury was mistaken, for his diagnosis was disputed shortly afterwards.[22] It may well have been that what he thought was *Panaeolus campanulatus* was actually *Psilocybe semilanceata.*

Eager mushroom pickers were not the only ones affected, for children left playing unattended were particularly at risk. In September 1879, a three-year-old boy, 'B.J.', was brought to see Mr J. Ernest Bullock, the resident medical officer of the Western General Dispensary on London's Marylebone Road. The poor boy was in a fit, and could neither speak nor stand. According to his older sister, he had gone out to play in Hyde Park, then picked and eaten a few mushrooms, after which he had collapsed and been carried home by a passer-by. His teeth were so firmly clenched that no emetic would pass his lips, but by the next day he had recovered completely, with no memory at all of what had happened. This, and other similar cases, led the respected naturalist Mordecai Cubitt Cooke (1825–1914) to issue a stern admonition to 'parents and guardians, that children at play in the fields may be warned against putting in their mouths any of the little "toadstools" which grow amongst the grass'.[23]

Nor were these incidents restricted to England. In 1883 a Dr E. Downes wrote to *The Lancet* from India to describe how several of his 'coolies' had been taken ill after eating mushrooms. None of them seemed to know what they were saying or doing, and they charged around as if drunk. Neither salt water nor strong tea made the slightest difference to their condition, so the doctor had them tied down lest they roll down the precarious Indian hills during the night, never to be seen again. Their symptoms – falling into a deep slumber, before rising up and rushing around madly – strongly suggest that this was a rare case of accidental fly-agaric intoxication.

But of all the nineteenth-century accounts, the following will strike the modern reader as the most recognisable and, because of the contrast between the doctor's punctilious manner and the patients' outrageous symptoms, the most delightfully comic. In late August 1830, D. O. Edwards, surgeon at Westminster Hospital, was called to assist a young family – Frederick Bickerton, aged twenty-five; his wife, Anne, twenty-three; and their child George Bickerton, aged four – who were restlessly pacing around the waiting room. At first Edwards was rather annoyed at the inconvenience because he thought they were drunk and wasting his time. However, closer inspection convinced him 'that the inebriation proceeded from no ordinary cause'.[24] The Bickertons were taken with the 'highest hilarity' and, giggling hysterically, they strode around 'in continual motion, either dancing or throwing themselves into grotesque attitudes'. Remarkably, 'their consciousness was quite unclouded' and on 'being charged with drunkenness, the adults exhibited the most lively indignation'. Frederick 'was most vividly affected by the poison; his eyes glistened; the pupils expanded; the pulse was full and frequent . . . He conversed without embarrassment, and said that he was intelligent of everything that passed around him.' It was, for Edwards, a most perplexing case; that is, until he discovered the provenance of the symptoms.

The Bickertons, resident at No. 2, Providence Court, Peter Street, Westminster, had fallen on hard times. Frederick was a labourer and a chancer, but out of work for some time; the family had neither food nor money. Ever enterprising, they determined to pick mushrooms in Hyde Park and Kensington Gardens, and then sell them for people to make ketchup, or 'catsup', which was a popular condiment at the time. Perhaps the mushroom market was saturated, or perhaps the general public had become alert to the dangers of buying wild mushrooms from dubious-looking street sellers, but whatever the reason, the family could not rid themselves of a single punnet. Of course, if they had, they might have been responsible for the biggest outbreak of magic mushroom poisoning on record, but as it was, tired, hungry, and dejected, they went home and cooked up a broth with the morning's pickings. Between them they ate the lot, washed down with water. Edwards records that they had picked 'rather more than a quart of them', which, even if only a small proportion had been Liberty Caps, would have made for a sizeable dose, especially when taken on an empty stomach.

Their symptoms came on rapidly. First Frederick was 'affected with giddiness; this gradually increased until a dimness of sight supervened'. Then he saw himself surrounded by 'flames'; his hearing became painfully acute and everyday objects 'became confused to the eye. He occasionally felt a sentiment of uncontrollable gladness, which prompted him to the muscular movements. Yet he remained fully conscious that he was in a state of preternatural excitement.' Anne's symptoms were identical, but 'the condition of the child could only be gathered from the obviously excited irritability'. Once they had found their way to the hospital, Edwards's emetics immediately had a beneficial effect upon mother and child, but made little impression on Frederick, who was by this stage overcome with it all and virtually unconscious on the floor. Edwards immediately employed the stomach pump, after which 'the patient vomited a quantity of water' and 'half-digested mushrooms'. 'By this process,' concluded Edwards, 'the man was quickly relieved,' and with a few judicious 'leeches applied to the forehead, he was, as well as his wife and child, sufficiently recovered to be discharged next day.' And with that, the Bickertons disappear from history.

These, then, are the records which, while not conclusive, strongly suggest that magic mushrooms had been eaten. Amusing as they are to the knowing modern reader, they show that until the end of the nineteenth century both amateurs and experts were woefully ignorant about how to identify mushrooms, and that victims and medics alike took any strange symptoms as evidence of poisoning. Not even the Bickertons, who arguably had the most pleasant time of it, disagreed with this diagnosis; though the medic's eagerness to reach for the leeches and the stomach pump potentially contributed to an unpleasant ending to their trip. This widespread confusion about mushrooms and their unexpected effects is surprising, particularly when compared with the more advanced understanding of plants at the time. Remarkably, to understand the cause of this bewilderment we need to go all the way back to the ancient classical world where natural philosophers, baffled by these strange autumnal fungal growths, created a Gordian knot of confusion and disorder that would not be sliced apart until the nineteenth century.

For country dwellers and metropolitan philosophers alike, mushrooms have always seemed mysterious and problematic. Their sudden appearance and equally rapid decay, their often grotesque shapes, smells and textures, their association with rot and decomposition, have all contributed to their becoming objects of fascination and occasional abhorrence. Perhaps not surprisingly, these slippery forms managed to elude all classical attempts to categorise them. They were clearly not animals, yet neither were they plants. They grew in great profusion but appeared to have no flowers and to shed no seeds. They were sometimes exquisite to eat, sometimes deadly poisonous. Such a puzzle, thought classical writers, demanded philosophical, not empirical, inquiry and thus they attempted to *reason* the nature of mushrooms and thereby to address the two fundamental problems: first, exactly what mushrooms were and how they reproduced; and second, how poisonous and edible varieties could be distinguished.

A common answer to the first was that mushrooms were spontaneously produced by thunder, where lightning struck the ground. The great Roman natural historian Pliny the Elder (23–79 CE) thought, rather, that they were produced from mud and the acrid juices of moist earth, especially that found beneath acorn-bearing trees.[25] Then again, the writer Nicander (fl. 197–130 BCE) thought mushrooms 'the evil ferment of the earth', warmed by the heat rising from its core.[26]

Regarding the more pressing second question, classical writers were well aware that while some fungi were gastronomic delights, others were disagreeable to the digestion, or worse, poisonous. Euripides and Hippocrates both recorded instances of mushroom fatalities, while the Roman Emperor Claudius was famously murdered by his wife, Agrippina, to ensure the succession of her son Nero. She prepared for him a dish of his favourite mushrooms laced with the Death Cap, *Amanita phalloides*: he died slowly and horribly. But writers were wide of the mark when attempting to distinguish the edible from the noxious.

Pliny rather alarmingly thought that poisonous mushrooms were livid in colour and edible ones red (several poisonous species are in fact red). Nicander thought that fungi growing on oak and olive trees were poisonous, those on fig-trees safe.[27] Dioscorides (40–90 CE) – a Greek-born physician, botanist, pharmacologist, and surgeon to Nero's army, who described about five hundred plants and their healing properties in his *De Materia Medica* (On Medical Matters) – thought mushrooms edible unless they grew over rusty iron or near serpents' dens.[28]

The great herbals of the medieval and early modern periods, the tomes and compendiums of herbal remedies and lore produced by Gerard, Clusius, Culpeper and others, took much of their knowledge directly from these classical authorities, especially Dioscorides, and so retained their confusion about fungi. Albertus Magnus in the thirteenth century and Gerard in the seventeenth thought fungi 'earthie excrescences',[29] while others thought mushrooms a product of decomposition, not its cause.[30] Later, in the eighteenth century, thinkers gave their imaginations free rein and variously attributed mushrooms to the fermentation of oil and nitrous salt, the coagulation of snails' slime trails, or vegetable crystallisation – whatever that may have been.[31] Even the great Swedish botanist Linnaeus (1707–1778) – who overturned the kind of unwieldy plant classification used by John Ray with his revolutionary binomial system – was persuaded by another naturalist, Otto von Munchausen, that fungi were created as dwellings by insects. He quite overlooked the observation of Pier Antonio Micheli (1679–1737), published in 1729, that fungi did in fact reproduce by a kind of 'seed'.[32]

It is hard for us to imagine anyone taking the classical beliefs seriously, but because of the lasting Western obsession with all things to do with the ancient world, their influence endured way beyond their time. In particular, Dioscorides' binary division of *Fungi esculenti* (edible) from *Fungi pernicosi* (poisonous) became the template by which all fungi were categorised and led to the misguided search for generic rules rather than specific identifying features to distinguish the two. Thus, as late as the nineteenth century, and even in the early years of the twentieth, rules such as poisonous mushrooms 'will blacken silver or turn onions brown', or 'can easily be peeled', or 'do not grow in meadows, open fields and roadsides', were published in august journals like *The Times* and *The Lancet*.[33] These rules, absolutely spurious but adhering impeccably to classical reasoning, made mushroom-eating a game of Russian roulette and caused more accidents than they prevented.

Indeed, so entrenched was the classical paradigm that it was not until the nineteenth century that botanists began to see what had been staring them in the face since the invention of the microscope some two hundred years previously: that fungi were indeed living species, reproducing sexually by means of spores, and that variations in microscopic features enabled species, poisonous or otherwise, to be distinguished. Nevertheless, Dioscorides' binary division remained stubbornly instilled in Western consciousness as the single, overriding framework for understanding all unusual mushroom-induced symptoms. It ensured that when magic mushrooms were accidentally consumed, as the various accounts suggest, their psychoactive effects were regarded as at best meaningless and at worst life-threatening or injurious to sanity. 'It is certain that some [mushroom] species have an intoxicating quality,' wrote the eighteenth-century Swiss botanist Albrecht von Haller (1708–1777), 'followed often by deliriums, tremblings, watchings, faintings, apoplexies, cold sweats, and death itself.'[34] It would take the psychedelic revolution of the twentieth century finally to overturn this pervasive framework.

Its tenacity is perplexing, however. As the Age of Enlightenment gave way to the Age of Empire, Western society became exposed to, and made use of, a series of ever more exotic psychoactives, used primarily in medicine but also, to a certain extent, recreationally. It is no exaggeration to say that the cogs of the industrial revolution were oiled with opium, which was available over the counter at pharmacies in a variety of potions, powders and quack remedies, and which was prescribed for almost every minor ailment of body, mind and spirit. Its dreamy, visionary pleasures and nightmarish pains, were well known by the end of the eighteenth century and could easily have provided an alternative framework for understanding mushrooms. In fact the connection was spotted as early as 1744 by the naturalist and Fellow of the Royal Society William Watson (1715–1787). Watson was engaged in an intellectual spat with the aforementioned Roger Pickering about the origins of mushroom poisons, for he disagreed vehemently with Pickering's belief that burrowing insects were what turned esculent mushrooms noxious. But in an aside about the early European treatises on fungi, he noted that in 'most of these authors we find instances of mischievous effects from the pernicious kinds; which property some of them have equal to Opium, Aconite, or Henbane . . .'.[35] He could not have made the link more explicitly.

Samuel Taylor Coleridge (1772–1834) wrote his supposedly opium-inspired visionary masterpiece 'Kubla Khan' in or around 1799, the year that Dr Brande was called to attend the young Edward who was giggling hysterically in the kitchen.[36] Coleridge's struggles with opium addiction were well known, those of Thomas de Quincey (1785–1859), the English opium eater, even more so, and other famous writers were afflicted too: Louisa May Alcott (1832–1888), author of the children's classic *Little Women* (1868–9), was one such. Meanwhile, the chemist Humphry Davy found spiritual enlightenment in an oiled silk bag of nitrous oxide, the French psychiatrist Jacques-Joseph Moreau de Tours dined on candies of hashish, the American psychiatrist Weir Mitchell chewed buttons of the peyote cactus, and later, Sigmund Freud considered the pathways of the unconscious while injecting himself with the occasional syringe of cocaine.[37] In other words, there were all manner of narcotic perspectives in which the mischievous effects of mushrooms could have been reframed, but not one made the slightest impact: mushrooms were poisonous, and that was that. The question is, why?

Some twentieth-century scholars have, as we shall see, argued that this amounts to evidence of a lingering primal taboo against the eating of hallucinogenic mushrooms, but we need not invoke such a tenuous solution. For although mushroom-induced fatalities were rare, they were always reported sensationally and in gruesome detail. Thus, the alarming story of two soldiers and their cat who went mad and then died in agony after a mushroom repast was related in the *Gentleman's Magazine* in 1757.[38] Nearly a hundred years later, stories about a soldier in Bruges whose convulsions were so appalling that he broke his back before expiring,[39] and of an old man whose body was so black and swollen that his face was no longer recognisable, appeared in *The Times*.[40] In other words, the mental connection between unusual mushroom-induced symptoms and the possibility of a horrible death was constantly reinforced, and would have been uppermost in people's minds if and when any unexpected post-prandial symptoms materialised. The mycologist Charles Badham (1806–57) complained about a general climate of hysteria in this regard:

> Persons have fancied themselves poisoned when they were not; indigestion produced by mushrooms is looked upon with fear and suspicion, and if a medical man be called in, the stomach-pump used, and relief obtained, nothing will persuade either patient or practitioner that this has not been a case of poisoning. 'You have saved my life,' says one. 'I think you will not be persuaded to eat any more mushrooms for some time,' says the other: and so they part, each under the impression that he knows more about mushrooms than anyone else can tell him.[41]

Added to this was the problem that mushrooms were often all regarded as being essentially the same. The man J.S., for example, believed that he had been picking *mushrooms*, not a *particular* mushroom.[42] Bemushroomed 'patients' typically had little idea of what exactly they had been eating, while the medical and botanical experts called in to assist them were rarely any the wiser. As the exasperated naturalist William Hay complained, the general fear of poisoning arose therefore from

> the singular popular incapacity for individualizing fungus species. They are all confounded together in the mind, and are not regarded separately, each kind by and for itself, as are other plants. An ordinary Englishman's only idea of gathering wild Fungi is to make a heterogeneous collection of everything fungoid that comes in his way. Put into practice this will obviously result in mistakes . . . Then the doctors will blame, not the stupidity or carelessness of the gatherer, but the mushrooms indiscriminately.[43]

It was only when mycologists recognised that fungi were of distinct species, and learnt to demarcate the deadly ones, that this alarmingly inaccurate framework was abandoned.

Charles Badham was one who lamented this woeful state of affairs, and he complained about the 'apparently clashing testimonies of authors respecting the same species, who not infrequently describe, under a *common* name, a fungus which some of them assert to be esculent, some doubtful, and others altogether poisonous in its qualities'.[44] Badham's concerns were shared, and it was during the nineteenth century that fungal taxonomy was finally systematised and placed on an equal footing with that of plants, firstly by the Swede Elias Magnus Fries, who was the first to separate species on the basis of spore colour, and later by the German Paul Kummer (1834–1912), who revised Fries's work on the gilled fungi. Between them, they finally extracted *Psilocybe semilanceata* from the heterogeneous tribe *Agaricus glutinosus*, as part of a continual process of ever closer reticulation by which the number of recognised species doubled between 1860 and 1890.[45]

It inevitably took time for this new way of looking at and thinking about mushrooms to disseminate beyond the borders of academe. However, the Victorian vogue for natural history ensured its eventual transmission, for this was the era of the amateur naturalist club, a trend that began in the 1830s and peaked some thirty years later. The most influential was surely the Woolhope Naturalists' Field Club in Hereford, of which most of the pioneering British mycologists were members. It began organising 'fungus forays' in 1868, excursions on which members would stroll genteelly through the autumn countryside, basket in one hand, mushrooming knife in the other, collecting and identifying whatever fungi could be found along the way. Then, for the benefit of interested locals, the day's trophies would be displayed in the nearest village hall and discussed over a well-earned cup of tea.[46]

THE GREAT FUNGUS MEETING AT HEREFORD.

By the turn of the century therefore, as a result of the work of this and similar clubs in publicising the new advances in taxonomy, the modern understanding of mushrooms as distinct species began to filter through to the popular consciousness. The old folk wisdom about how to distinguish 'mushrooms' from 'toadstools' gradually gave way to positive identification of species, or more commonly, to an avoidance of wild mushrooms altogether. It is not hard to see why these upright Victorian gentleman mycologists would have found the idea of eating magic mushrooms abhorrent. One of these early mycologists and a member of the Woolhope Club deserves special mention, however, for of all of them he came closest to starting a Victorian magic mushroom craze.

Mordecai Cubitt Cooke is generally remembered, if at all, as an early science writer, author of tomes such as *Rust, Smut, Mildew and Mould* (1865) and the more popular *Plain and Easy Account of the British Fungi* (1862), and editor of the improbably named magazine *Science Gossip*. However, he began his writing career with an exuberant and incautious treatise on drugs, *The Seven Sisters of Sleep* (1860), written while he was in his mid thirties.[47] Not only was this a defence of

Founding members of the British Mycological Society *c.*1896. Mordecai Cubitt Cooke is seated in the middle of the front row.

Cooke's tobacco consumption (he was a lifelong smoker), but it also argued for the moral equivalence, and therefore legitimacy, of the world's regional drug preferences.

He built the book upon the conceit that the Queen of Sleep, in order to fend off the political ambitions of her seven jealous sisters, gave to each a narcotic drug capable of producing exotic visions and thereby having power over man's waking hours. Each sister then took control of one part of the world – 'Morphina' taking Tartary and Mongolia, for example, and 'Amanita' (fly-agaric) Siberia. In other words, Cooke was arguing, in the racial discourse of the time, that each of the human groups naturally utilised the local narcotic to which it was drawn as a result of geography and innate physical constitution: an idea he almost certainly plagiarised from James Johnston's earlier best-seller *The Chemistry of Common Life* (1855). If Johnston's tone was disapproving, Cooke's was, as one reviewer put it, 'joyous and thoroughly irresponsible',[48] and he was clearly targeting what he saw as the hypocritical mores of the emerging bourgeois classes – of people like Johnston, in fact. 'Philanthropists at crowded assemblies,' complained Cooke, 'denounce, in no measured terms, "the iniquities of the opium trade", and then go home to their pipe or cigar, thinking them perfectly legitimate.'[49] No wonder the book has come to be seen as an early 'drug classic'.

Of the seven plants described (tobacco, opium, hashish, betel, coca, belladonna and fly-agaric), Cooke devoted the most space to his own beloved tobacco. Even though he almost certainly encountered the effects of opium addiction first hand while working as a teenage apprentice apothecary, he wrote approvingly of the drug's effects and remained a lifelong fan of De Quincey.[50] Of most interest here is Cooke's description of the fly-agaric: not only did he popularise the myths surrounding the mushroom's use in Siberia, but he also retold the story of Mr Glen's encounter with the bemushroomed man staggering through the streets of Knightsbridge. No fool, he realised that this must have been caused by a mushroom similar in its intoxicating properties to fly-agaric, for he wrote that 'if future generations do not deem it desirable to indulge in a narcotic of this kind for the purposes of producing pleasurable sensations, or to smother the carking cares of life, yet they may learn more than we do at present know of the peculiar characteristics which distinguish this [mushroom] from all the others of the "Seven Sisters of Sleep"'.[51] The implication was plain: future generations would indeed find local intoxicating mushrooms desirable, and Cooke knew exactly why.

Even though Cooke, a brilliant naturalist, knew all about the Liberty Cap and how to identify it, and seemed, here, to be interpreting its effects favourably, it is extremely unlikely that he consumed the mushroom himself. For one thing, the logic of his argument was that he, an Englishman, was not racially constituted to use mushrooms: that ability belonged only to the Siberians. For another, his career was, apart from this one book, distinguished by diligence, probity, and sobriety. Born of humble origins, Cooke had to work hard all his life as a botanist and science writer, a precarious occupation that necessitated the continual search for paid work.[52] In spite of his great achievements in the field, his origins prevented him from entering the ranks of the gentlemanly classes who were busy establishing and institutionalising the discipline of mycology. Unlike, say, his contemporary the Reverend M. J. Berkeley (1803–1889), Cooke always remained an outsider. Perhaps not surprisingly, therefore, he distanced himself from this earlier irresponsible work, born of a youthful naivety, as his career developed.

By the time Cooke published one of his last works, *Edible and Poisonous Mushrooms,* in 1894, he had overturned his earlier opinions completely. In this classic work of Victorian philanthropy – a good deed performed for the greater public good, published by the Society for Promoting Christian Knowledge (SPCK) – he warned people to avoid Liberty Caps, and rather stuffily cautioned against overindulgence in all its manifestations: 'punishment will follow inordinate indulgence in any of the good things of this life, and those who disregard reason, and are intemperate in eating fungi, must expect to suffer from repletion and indigestion'.[53] The one person, then, who might have triggered a Victorian magic mushroom craze by instilling a 'narcotic' discourse into the public consciousness, turned instead to

'The mental effects of mushroom poisoning'. Detail from a larger cartoon, 'A week with the Hereford fungus eaters,' by the mycologist Worthington Smith, published in *The Graphic*, 15 November 1873. This is the first indisputable pictorial representation of the effects of magic mushrooms in the West. It was drawn by someone who had never personally taken them and who, in keeping with the times, believed them to be toxic. Courtesy The Guildhall Library, London.

his career, and to temperance and moderation, and so left the popular fear of mushroom-poisoning firmly in place.

Clearly, then, localised and episodic incidents of accidental magic mushroom consumption have been happening since records began. Without any kind of framework within which to interpret the experiences favourably, however, the victims, who believed themselves to have been poisoned, would have been most unlikely to repeat them. (As every enthusiast knows, the experience of taking magic mushrooms can be harrowing enough, even without the belief that one is dying.) The classical approach of lumping all mushrooms together meant that they would have had only a very confused idea of what type of mushroom to pick, even if they had wanted to. Intentional consumption therefore seems extremely unlikely.

The increasing incidence of 'poisoning' episodes throughout the nineteenth century did not represent a Victorian mushroom craze, but was simply a product of the better record-keeping of the emerging institutions of medicine and science. Of course, we have no first-hand accounts from the victims, who were mostly illiterate, but only from the physicians who treated them, so it is impossible to prove that mushrooms were *never* knowingly consumed. Inadvertent consumption, however, seems more likely, with mushrooms eaten either by children who did not know any better or by families in such abject poverty that they were reduced to gathering whatever wild food they could find in order to put breakfast on the table.

This, then, was the situation that prevailed at the end of the nineteenth century. The Liberty Cap was known to be poisonous, and was advertised as such in popular guides to fungi, but was nonetheless eaten occasionally. But then, perplexingly, its listing as a poisonous species began to disappear from the guidebooks until by the 1940s it was wholly absent. For example, the respected mycologist John Ramsbottom (1885–1974) listed it as 'suspicious' in 1923, but omitted it from both his *Poisonous Fungi* of 1945 and his widely read *Mushrooms and Toadstools* of 1953.[54] Worthington Smith regarded it as poisonous in 1891 but similarly omitted it from his guide to poisonous fungi of 1910,[55] as did Mordecai Cooke from his *Plain and Easy Account of British Fungi* of 1904.[56] It is as if a collective amnesia fell upon the founding fathers of British mycology, with the result that the mischievous Liberty Cap was quite forgotten.

Most probably this was because of the dramatic decrease in incidents of mushroom poisoning in the early twentieth century: if no one was being admitted to hospital having eaten Liberty Caps, there was a less pressing need to alert the public to their dangers. It seems that a gradual erosion of confidence in old wives' tales and folk wisdom, together with an ever-present fear of poisoning, led to a consensus that the only sensible thing to do was to avoid eating wild mushrooms altogether: people no longer trusted fungi, nor anyone selling them at market. *The Lancet*, reporting the introduction of official mushroom inspectors in France in 1934, suggested that in Britain 'consumption of mushrooms would go up by leaps and bounds if the public felt sure that toadstools were never offered in their place'.[57] Perhaps this amnesia is also attributable to the hiatus of two world wars, in which some of the brightest minds of each generation lost their lives. Nevertheless, and even though advances in taxonomy and biochemistry had shown that the Liberty Cap need not be listed amongst the deadly species,[58] it stands out as a rather inexplicable episode of sloppy scholarship in the history of modern mycology.

Without this lacuna, the story of the eventual rediscovery of the Liberty Cap might well have taken an entirely different course, or might not have happened at all. For when British hippies started looking for indigenous magic mushrooms in the early 1970s, they found that none of the authoritative field guides – such as *The Collins Field Guide to Mushrooms and Toadstools* – listed any psilocybe species as poisonous.[59] The way was clear for them to start picking.

Here, then, for the first time in its history, the Liberty Cap's effects were seen in an entirely different frame from the one that had dominated for at least seven hundred years: it was no longer 'poisonous' but 'hallucinogenic', which meant that its effects were, for the first time, seen as desirable. This radical shift in cognition and understanding was triggered by a series of events that, unbeknown to the British medics and mycologists, had been unfolding for some five hundred years far away across the Atlantic.

5

Feasts and Revelations

He who eats many, many things sees.

Bernardino de Sahagún, *Historia General de las Cosas de Nueva España*

Nothing that the mushrooms show should be feared.

María Sabina, *The Life*[1]

In July 1914, as Europe was sliding ineluctably into war, an upheaval of a very different kind was taking place in Oxford County, Maine. A man known to us only as Mr W. went mushroom-picking in the morning and 'gathered a good mess of the mushrooms (*Panaeolus papilionaceus*)' which he prepared for his lunch.[2] He shared the harvest with his niece by marriage, a Mrs Y., and between them they ate about a pound of the mushrooms fried in butter. His nephew, Mr Y., declined to partake and was able to witness the dramatic events that followed unaffected. Writing up the incident a week later, Mr W. described how he was rapidly overcome with peculiar symptoms after finishing the meal: he became drowsy and felt his thought processes to be rather more impaired than was usual following a large lunch. Then, getting up to cross the room, he had the distinct impression that time had slowed down and that it was taking him far longer than was reasonable to reach the door. Having finally crossed the room, he discovered that Mrs Y. was similarly affected, and from then on their symptoms grew rapidly more intense.

They stepped outside and both were immediately struck by the brightness of colours. For Mr W., a 'field of redtop grass seemed to be in horizontal stripes of bright red and green, and a peculiar green haze spread itself over all the landscape'. Mrs Y. yelped with surprise on finding that her fingers had become delicately undulating snakes. The pair became clumsy, and even though it was hard for them to stand, let alone walk, they were overcome with the childlike urge to run and leap about. 'We became very hilarious,' wrote Mr W., 'with an irresistible impulse to laugh and joke immoderately, and almost hysterically at times. The laughing could be controlled only with great difficulty; at the same time we were indulging extravagantly in joking and what seemed to us funny or witty remarks.' The poor sober and bemused Mr Y. noted dryly that some of the jokes were more successful than others.

Things took a slightly darker turn when the party returned indoors. Mr W. noticed that 'the irregular figures on the wall-paper seemed to have creepy and crawling motions, contracting and expanding continually, though not changing their forms; finally they began to project from the wall and grew out toward me from it with uncanny motions.' He was struck by a bouquet of roses arranged on the table: 'the room seemed to become filled with roses of various colours and of all sizes, in great bunches, wreaths and chains, and with regular banks of them, all around me, but mixed with some green foliage, as in the real bouquets'. This rather beautiful vision lasted only a short time before Mr W. was overwhelmed with 'a decided rush of blood to [the] head' and a dizzying barrage of hideous human faces that leered at him from all sides: 'The faces appeared in all sorts of bright and even intense colours – so intense that I could only liken them to flames of fire, in red, purple, green and yellow colours, like fireworks . . . They were all grimacing rapidly and horribly and undergoing contortions, all the time growing more and more hideous.' Looking down at his hands, he saw that they had become 'small, emaciated, shrunken and bony, like those of a mummy'. It was at this point that Mr W. decided to call for a doctor.

However, by the time the doctor arrived the worst of the symptoms had passed, and so, unlike his British counterparts, this physician felt unable to administer any palliative or useful treatment. Before the day was out the bemushroomed pair had further fits of the giggles, and at one point the peculiar sensation that their bodies were elongating: 'I grew far up, like Jack's bean-stalk,' wrote Mr W., and then 'collapsed back to my natural height'. Mrs Y. said that her 'hands and arms seemed to grow unnaturally long'. Mr W. also had the distinct impression that he had become clairvoyant and could read the thoughts of those around him, but shortly afterwards the effects of the intoxication began to wear off until by six o'clock in the evening they both felt quite normal again.

There remains a degree of uncertainty over exactly which species Mr W. and Mrs Y. had been eating, for modern studies have found *Panaeolus papilionaceus* to be only capriciously psychoactive.[3]

Nevertheless, this litany of symptoms seems wholly in keeping with those of psilocybin. The fact that the episode was written up by Yale biologist Addison Emery Verrill (1839–1926), and published in the major journal *Science*, suggests that mushroom 'poisoning' of this kind was as uncommon in the US, and therefore as noteworthy, as equivalent incidents had been in Britain. In fact, the first half of the twentieth century saw a few sporadic cases of accidental poisonings recorded in the US (and also in Japan and Australia).[4] It was the account of Mr Glen's bemushroomed stagger through the streets of Knightsbridge in the *London Medical and Physical Journal* of 1816, and William Salisbury's subsequent identification of the offending species, that first alerted American botanists to the unusual qualities of certain mushrooms, especially those of the genus *Panaeolus*.[5] The earliest American guidebook, *One Thousand American Fungi*, published earlier in 1900, corroborated that *Panaeolus papilionaceus* had indeed been observed to cause intoxication, if only capriciously.

Full acknowledgement must be given to the author of this pioneering guidebook, the courageous Captain Charles McIlvaine (1840–1909), who on finding that very little was known about the edibility of American fungi determined to resolve the matter by eating each and every one himself. He clearly must have avoided the deadly varieties, for he lived long enough to catalogue the edibility of over a thousand species. His persistence was surely stoical at times, for he wrote (with, one detects, a certain degree of understatement) that 'while often wishing I had not undertaken the work because of the unpleasant results from personally testing fungi which proved to be poisonous, my reward has been generous in the discovery of many delicacies among more than seven hundred edible varieties I have found'.[6] A thorough man, he tested *Panaeolus papilionaceus* on himself, but found that thirty mushrooms produced no unusual effects whatsoever. He nevertheless concluded that the species was probably best avoided.

As for Mr W., though he knew his mushrooms, he obviously had no reason to suspect *Panaeolus papilionaceus* to be anything other than edible and delicious. But what is so striking about his account is that, even though he was forced to call a doctor when the visions became oppressive, at no time did he consider himself *poisoned*. Indeed, as the effects of the mushrooms came on he immediately recognised that they were similar to those attributed to other drugs, for he was familiar with Thomas de Quincey's writings on opium, Fitz Hugh Ludlow's on hashish and Weir Mitchell's on peyote.[7] His account is therefore somewhat groundbreaking. It stands as the first intimation that the old framework of dividing edible 'mushrooms' from poisonous 'toadstools', which had been in place for nigh on two millennia, was about to be overturned. Within the space of forty years a third category would be added – 'hallucinogenic mushrooms' – and, within this new, revolutionary, psychedelic discourse, the effects of these mushrooms, like those of other visionary plants, would be seen as wholly desirable.

Notwithstanding the importance of this break with the past, there is nothing to indicate that Mr W. ate the mushrooms purposefully for their visionary effects: like all the unlucky characters in the previous chapter, he entered into the trip quite unwittingly. Neither did he attempt to repeat the experience for, though he was a vigorous middle-aged botanist with a specialist interest in fungi, he was 'strictly temperate in his habits'.[8] However, just forty years later another middle-aged, temperate American gentleman with a keen amateur interest in fungi *did* go actively seeking psychoactive mushrooms, with the express intention of experiencing their effects first hand. His search took him many thousands of miles to the south, to the remote mountain regions of Mexico, where the situation regarding the knowledge and understanding of psilocybin mushrooms could not have been more different from that which had prevailed for so long in the West. How and why he did so, and what happened subsequently, form the central subject of this and the following chapter, for his arrival in Mexico had dramatic consequences that led directly to the unstoppable rise of the magic mushroom.

Perhaps uniquely in the world, there exists in the region we now call Mexico a genuine history of intentional psilocybin mushroom consumption that extends back at least five hundred years to the time of the Spanish conquest, but may go back much further. The West's encounter with this wholly other way of being began when Hernán Cortés (1485–1547) landed on the Mexican coast, near present-day Vera Cruz, on Good Friday 1519. Enflamed by rumours of riches and gold whispered to him by his translator and lover, Malinche (1505?–1529?) – the slave woman presented to him as a tribute by the coastal indigenes – the low-born Spaniard advanced inland with the intention of wresting whatever territory and gold he could from this 'new world'. With just 550 men, some horses and dogs, and a single cannon, Cortés led a charmed offensive against the ruling Aztec civilisation, which, remarkably, fell to him a mere two years later.

The Spanish invaders and settlers were shocked and fascinated in equal measure by the world they encountered, marvelling at its architecture, writing and laws and at the same time recoiling from its religious and cultural practices. While it was the Aztecs' predilection for human sacrifice – the relentless offering of still-beating human hearts, thousands at a time, to voracious gods – that was chiefly castigated, their application of a range of psychoactive plants and fungi in a variety of religious, secular, and prophylactic contexts[9] caused the horrified Spanish no less offence. In this particular encounter the West quite literally demonised the other, attributing the alleged curative and divinatory powers of these plants to the work of demons or the Devil. They duly began the process of imposing Christian, 'civilising' values upon the pagans. Nevertheless, it is thanks to the endeavours of the sixteenth-century Spanish chroniclers, fascinated enough by what they encountered to make written records, that we have a picture of indigenous mushroom usage, however distorted it may be by the religious, imperialist and primitivist ideology through which the Spanish viewed this alien world.

A demon hovers menacingly above a cluster of mushrooms in Bernardino de Sahagún's sixteenth-century account of Aztec life, the *Florentine Codex*. The Catholic Conquistadors attempted to eradicate indigenous mushroom use, believing it to be a diabolical practice. Courtesy R. Gordon Wasson Archive, Harvard University, Cambridge, Massachusetts.

Several sixteenth-century accounts describe the use of mushrooms at the highest social levels.[10] The Dominican friar Diego Durán (d. 1588) wrote a detailed treatise, the *Historia de las Indias de la Nueva España*, based in large part upon an indigenous written source, now lost. From this source, Durán relates the story of how at the coronation of the Aztec sovereign Tizoc in 1481 inebriating mushrooms were provided for the guests. 'And all the lords and grandees of the provinces rose and, to solemnise further the festivities, they all ate of some woodland mushrooms, which they say make you lose your senses, and thus they sallied forth all primed for the dance.'[11] It is hard to imagine today's royals sharing out magic mushrooms at their nuptials but, according to Durán, Aztec royalty rarely consumed anything else, and especially not alcoholic drinks. They preferred 'the woodland mushrooms which they ate raw, with which . . . they would rejoice and grow merry and become somewhat tipsy'.[12] The evident delight with which the Aztecs took mushrooms to accentuate the pleasure of dancing seems thoroughly modern.

The Aztec Lord of the Underworld, Mictlantecuhtli, appears behind a man eating what are presumably magic mushrooms. *Magliabechiano Codex*, sixteenth century. Courtesy R. Gordon Wasson Archive, Harvard University, Cambridge, Massachusetts.

Moctezuma II (1466–1520), the last and ill-fated Aztec ruler defeated by Cortés, employed a coterie of old priests whose job it was to consume mushrooms for divinatory purposes, especially to prognosticate the outcome of battles. Anyone predicting defeat was, however, hastily executed – perhaps a clue to why the Aztec ruler was so swiftly conquered. Moctezuma also appeased his traditional enemies with an annual mushroom feast, the 'Feast of Revelations', a custom that may have originated at his own coronation in 1502. Spies from the rival Tlascalan tribe were captured, pardoned, and magnanimously regaled with mushrooms. This gift may have been double-edged for, if Durán is to be believed, 'they all lost their senses and ended up in a state worse than if they had drunk much wine; so drunk and senseless were they that many of them took their own lives, and by dint of those mushrooms, they saw visions and the future was revealed unto them, the Devil speaking to them in that drunken state'.[13]

Another more hard-line Franciscan friar, Toribio de Benavente (d. 1569), otherwise known as Motolinía, wrote a treatise on indigenous religious customs – his *Ritos Antiguos, Sacrificios e Idolatrías de los Indios de la Nueva España* – in which he described the Aztecs eating bitter-tasting mushrooms sweetened with honey. Less sympathetic than Durán, he claimed that the mushrooms merely sharpened the Aztecs' already savage cruelty.[14] After eating the mushrooms the locals 'would see a thousand visions and especially snakes; and as they completely lost their senses, it would seem to them that their legs and body were full of worms eating them alive, and thus half raving they would go forth from their houses, wanting someone to kill them.' Motolinía's disapproving tone is perhaps attributable to his religious sensibilities, which were affronted by the unfortunate resonances of these practices with the Christian communion. The mushrooms were known in the local Nahuatl language as *teunamacatlth*, which meant 'God's flesh', and which to Motolinía would have seemed blasphemous.

But of all the accounts, the most comprehensive, and relatively speaking the most sympathetic, was that compiled by the Franciscan friar Bernardino de Sahagún (1499–1590), who spent some sixty years studying the indigenous populations. In his voluminous work the *Historia General de las Cosas de Nueva España* (the so-called *Florentine Codex*), he presented a summation of his many conversations with locals, written in the predominant native language, Nahuatl, with a parallel Spanish translation. He recorded the use not only of mushrooms but of other psychoactive plants such as peyote, tobacco and Morning Glory seeds. His informants called the mushrooms *teonanacatl* – the name by which mushrooms subsequently became famous in North America – a variant of Motolinía's *teunamacatlth,* translated likewise as 'God's flesh'. The mushrooms, he said, were to be found growing in grassland, in fields, moors and waste places, and had a round cap and a thin stem. They were so bitter as to hurt the throat, and only two or three were eaten at one sitting. They caused palpitations of the heart, excited lust, and induced laughter and terror in equal measure. At 'the hour of blowing conches and flutes', chocolate would be drunk and mushrooms eaten, with honey taken to ease the bitterness. After dancing and weeping, bemushroomed participants would have visions in which they might see their destiny, or even the manner of their death, before falling into a stupor. Later, they would sit around discussing the meaning of what they had seen (again, a thoroughly familiar scene to the modern mushroom enthusiast).

Clearly, then, at the time of the Spanish invasion, psychoactive mushrooms were being consumed in a variety of religious, secular, recreational and even diplomatic contexts within the dominant Mesoamerican Aztec civilisation. There exists material evidence, however, to suggest that the practice may be very much older. Several Mesoamerican codices, the indigenous texts written in symbolic picture language, portray mushrooms.[15] For example, the *Codex Vindobonensis* (the *Vienna Codex*), a Mixtec work depicting the mythological origins of the world, shows several gods and goddesses, including the feathered serpent god Quetzalcóatl, clutching or eating mushrooms. Another Mixtec codex, the *Lienzo de Zacatepec No. 1*, shows a man with mushrooms in his hair, while the *Magliabechiano Codex* shows a man who seems to be eating mushrooms while a 'supernatural' figure stands behind him, possibly Mictlantecuhtli, the Lord of the Underworld.[16] All are, in the light of the chronicles, at least suggestive of mushroom use that predates the conquest.

Furthermore, throughout Central America, about three hundred stone and pottery 'mushroom' effigies have been uncovered.[17] Although originating mainly in the highlands of Guatemala, a few have been found in the southernmost Mexican states. These sculptures are about a foot high and free-standing, are mushroom-shaped with a thin stem and a domed cap, and often depict a carved human, animal or supernatural figure squatting at the base. The pottery figures are much simpler, so that for a long time they were considered to be simply vessels. The earliest figures date from the pre-classic period, that is the second millennium BCE; the latest, and simplest, from the late classic 600–900

CE. Quite what they were for is uncertain: they have been variously and plausibly interpreted as phallic symbols, boundary markers, seats even, or as being connected in some way to the Mesoamerican 'ball-game' (a sort of gladiatorial football) beloved of the Aztecs and the civilisations that preceded them.[18] Nevertheless, given the historical evidence, and the discovery of other metal and pottery figurines from the region depicting elated-looking humans with mushrooms, it is extremely likely that these figures *were* connected with mushroom consumption, albeit in some unspecified and unknowable way. This would suggest that the practice extended back not just hundreds of years, to the time of the Spanish invasion, but thousands of years to the pre-classic period. The contexts in which mushrooms were consumed prior to the Aztecs remain mysterious, however.

Moving forward in time to the period of Spanish rule, the historical and material evidence all but disappears. The Spanish chronicles cluster together in the sixteenth century, with only one further account from the seventeenth. Here, the bullish cleric Jacinto de la Serna contrived to eradicate the persistent 'idolatrous' practices of the indigenes by writing a guidebook on how to minister to them correctly, his *Manual de Ministros de Indios para el Conocimiento de sus Idolatrías y Extirpación de Ellas*. In a section devoted to local 'witchcraft', that is medical practices and midwifery, he describes how an Indian by the name of Juan Chichitón ('little dog'), a 'great maestro of superstitions', performed a mushroom ceremony in Serna's own village. On investigating the matter, Serna discovered that the mushrooms were customarily picked by the locals after a night of 'prayer and superstitious entreaties', were gathered once the dawn breezes had subsided and, whether eaten or drunk, intoxicated the locals, deprived them of their senses and made them 'believe a thousand foolish things'.[19]

Chichitón's crime, it seems, had been to celebrate a saint's day with this solemn mushroom vigil 'after the manner of Communion', so that all present 'went out of their heads'. Having incurred the priest's wrath, Chichitón went on the run and evaded all Serna's attempts to have him captured and censured. But as the 'little dog' melted away into the Mexican foothills, so too did the knowledge of the mushrooms which, whether concealed from or simply unnoticed by the Western gaze, was to remain unrecorded for another four hundred years.

It therefore remains unclear how – or indeed whether – mushrooming practices continued in an unbroken fashion to modern times. Nevertheless, the discovery in the early and mid twentieth century that indigenous Mexicans were still using mushrooms in ways not so different from Chichitón's all-night syncretic religious observances was to have a much more dramatic impact on Western culture. This time the West was primed to accept a radically different way of thinking about psychoactive mushrooms and their strange effects.

The Western rediscovery of Mexican mushrooming practices began, ironically, with a vigorous scholarly denial that they had ever existed.[20] A year after Mr W.'s unexpected mushroom visions, the respected American botanist William Safford (1859–1926), oblivious of such shenanigans so close to home, published a paper on the identity of the supposed *teonanacatl* of the Aztecs in which he stated emphatically that Sahagún and his native informants had been wrong.[21] They had mistakenly confused dried plant fragments for a fungus, and *teonanacatl*, revealed Safford, had been none other than the infamous peyote cactus (*Lophophora williamsii*). After reviewing the records of the early Spanish chroniclers, and the fruitless results of his own field enquiries, Safford reported that 'three centuries of investigation [had] failed to reveal an endemic fungus used as an intoxicant in Mexico'.[22] He bolstered his argument by claiming that peyote 'resembles a dried mushroom so remarkably that at first glance it will even deceive a trained mycologist'.[23] He was wrong on both counts.

Regarding the latter, the ethnobotanist Richard Evans Schultes (1915–2001) drew attention to how extremely unlikely it would be for anyone to confuse peyote with mushrooms, so different are they in appearance.[24] As for the former, Safford's failure to find any psychoactive fungi in Mexico was because, having misread Sahagún, he had been looking in the wrong place. A cursory consideration of the hydrophilic nature of fungi might have alerted Safford to the implausibility of finding mushrooms growing in the arid north of Mexico, but he remained implacable. Given that the rest of his work on Mexican botany was first class, his findings became widely accepted in scientific circles.[25]

Nevertheless, it is the finality with which he delivered his conclusions that prompted others, better placed in the field, to challenge them. An Austrian physician and amateur botanist who had settled in

Oaxaca, Blas Pablo Reko (d. 1953), was the first to challenge Safford, in an article published in the journal *El México Antiguo* in 1919.[26] He repeated his assertions in a letter to the United States National Herbarium in 1923, writing that *teonanacatl* 'is actually as Sahagún states, a fungus which grows on dung-heaps and which is still used under the same old name by the Indians of the Sierra Juárez in Oaxaca in their religious feasts'.[27] His cousin, Victor A. Reko, took up the cause, but the cousins' objections did little to dent scholarly opinion, persuaded as it was by the eminent Safford's assertion.

In 1936, however, Robert J. Weitlaner (1883–1968), a linguist living in Mexico City and an expert on Indian culture, set out to conduct field research amongst the Mazatec Indians in a small, remote and otherwise insignificant town in Oaxaca. The town's name was Huautla de Jiménez, unheard of beyond its immediate hinterland but later to become famous as the epicentre of a global magic mushroom craze. There, Weitlaner discovered through a good friend and local merchant that mushrooms were indeed used in curative divinatory rites.[28] He managed to procure a sample of the mushrooms, which he had arranged to be sent to Reko, who forwarded them to the Botanical Museum at Harvard, but by this stage the fragments were too badly decomposed for formal identification.

Reko's determination to prove Safford wrong eventually paid off, however, for in 1938 he was joined by the brilliant young Harvard ethnobotanist Richard Evans Schultes. Together, and in spite of their acute political differences (Reko was a Nazi sympathiser), they collected samples from which Schultes was able to make the first formal identification: *Panaeolus campanulatus* L. var. *sphinctrinus* (Fries) Bresodola.[29] Nearly twenty years later it transpired that although Schultes had identified the samples correctly, *Pan. campanulatus* was neither psychoactive nor was it used in indigenous ceremonies, a mistake that arose perhaps either from misdirection by his informants in the field, or as a result of the labels becoming muddled at the Harvard herbarium.[30]

However, in the same year that Reko and Schultes were collecting in Huautla, a party of linguists and anthropologists, led by Jean Bassett Johnson (1916–1944)[31] and inspired by Weitlaner, became the first Westerners to witness an indigenous mushroom curing ceremony.[32] 'The *curandero* (healer),' wrote Johnson, 'while under the influence of the hallucinogenic mushroom, divined the patient's illness; and it was the mushroom that gave instructions on how the sick person should be cured.'[33] The mounting evidence was incontrovertible and Safford's peyote hypothesis was finally abandoned. And there the matter might have ended – an obscure academic debate, of little concern to the wider world, that had finally been resolved – had it not come to the attention of one man, the temperate, middle-aged American gentleman previously mentioned.

Robert Gordon Wasson (1898–1986) – or Gordon, as he preferred to be known – was born on 22 September 1898 in Great Falls, Montana, the son of an Episcopalian minister.[34] After seeing active service in the First World War, he graduated from Harvard with a bachelor's degree in literature and began his career as a journalist, eventually working in the financial news department of the *Herald Tribune*. He moved across into banking, at first with Guaranty Company of New York, and then with J. P. Morgan & Co., where he was vice-president from 1943 until his retirement twenty years later. On paper there could not have been a more unlikely person than this upright and hard-nosed Wall Street banker to investigate and popularise the hallucinogenic mushrooms of Mexico, but remarkably, over the years, he had developed a passionate and consuming interest in all things to do with the role of mushrooms and fungi in human cultures. Once he heard that the mythological *teonanacatl* of the Aztecs was not only a mushroom but that it was still in use, it was inevitable that he would make the journey to Mexico to investigate the matter for himself.

This absorbing interest in the cultural history of mushrooms was stimulated by a widely re-told incident from his honeymoon.[35] He married Valentina Pavlovna Guercken (1901–1958), a Russian émigré training to be a paediatrician, in 1926. The newlyweds took their honeymoon in the Catskill Mountains, famous for the beauty of their forests in autumn, and while strolling through the woodlands they spotted some wild mushrooms. His horror and revulsion were matched in intensity by his bride's evident delight, for she ran about gathering as many mushrooms as she could stuff into the makeshift carrier of her knotted skirt. Convinced that the pickings were toadstools, he would not touch the meal she prepared, and expressed alarm that he would wake up a widower. He did not, and in the clear light of day the couple wondered what could possibly lie behind their extreme opposite reactions: clearly there was a major difference between Russian and

American attitudes towards mushrooms. Their curiosity aroused, they then embarked on what was to become a lifelong quest to investigate and understand the varied cultural relationships between humans and fungi.

It must be said that while this hoary old story has become something of a foundational myth for modern mushroom enthusiasts,[36] it was repeated by Wasson *ad infinitum* and grew ever taller in the telling. Its growing resemblance to a Hollywood movie script made Wasson's own daughter, Masha, question whether it had any substance at all, but eventually she conceded that the incident had genuinely occurred, however embroidered the story had become over the years.[37] This is itself telling for, as we shall see, Wasson had a knack of overworking dry empirical facts in the interests of a good story.

Whatever the details, it is clear that the Wassons began shortly thereafter to accumulate a vast amount of evidence, from philology, etymology, ethnography, folklore and fairy tale, to explain the provenance of their different reactions to all things fungal. They eventually presented their findings in a massive, privately published, two-volume opus, *Mushrooms, Russia and History* (1957),[38] a work that had itself mushroomed from its original conception as a Russian cookbook. After Valentina died in 1958, Wasson continued their investigations alone, going on to publish a large number of papers and a further five lavish books before he too died in 1986. Amongst other ideas, he claimed to have solved the mystery of the identity of the Soma plant mentioned throughout the Rig Veda – that it was the fly-agaric mushroom, *Amanita muscaria* – and to have established that the potion at the heart of the Eleusinian mysteries contained an infusion of the ergot fungus, *Claviceps purpurea*. These ideas came later, however, built upon the discoveries made in Mexico for which he is principally remembered, and which, inserted hurriedly into *Mushrooms, Russia and History* just prior to its publication, transformed this otherwise laborious text into a classic of its genre.

In 1952 Wasson received two letters, arriving virtually in the same post, that alerted him to the existence of the Mesoamerican mushroom stones, and to the fact that *teonanacatl* had been a mushroom.[39] Immediately, he wrote to Reko expressing his interest in the matter: Reko, by then an old man, forwarded his letter to a Eunice Pike, a missionary resident in Huautla. Pike, exasperated by her inability to sway the locals from their heathen practices,[40] was able to confirm that mushrooms were indeed used in curing rituals. She added incredulously that the Mazatecs believed the mushroom to have a 'personality', which spoke through the *curanderos*, the healers, divining the cause and cure of illnesses and revealing the location of lost or stolen property. This was exactly what Wasson wanted to hear and, wasting no time, he began to organise an expedition. In the late summer of 1953 (the rainy, mushroom season), accompanied by his wife and daughter, he made what was to be the first of ten successive trips to Mexico on the trail of the psychoactive mushrooms.

Reko had died in the interim, but the Wassons were accompanied on this first trip by Weitlaner, who provided invaluable local knowledge and acted as translator. Together, they made their way – initially by car over the appalling roads and then with donkeys over narrow and vertiginous mountain passes – to the town of Huautla. Late in the day, after much fruitless searching during which the taciturn Mazatec locals proved more than a little reticent on the subject of mushrooms, they discovered to their surprise that their local guide, the one-eyed Aurelio Carreras, was himself a *curandero*. After a little persuasion he agreed to hold a *velada*, Wasson's term for the indigenous all-night mushroom vigils, to determine the health and well-being of Wasson's son, Peter. Though only Aurelio consumed mushrooms, Wasson was at last able to witness, and make detailed records of, an authentic ceremony. He politely humoured Aurelio when the results of the mushroom-assisted divination revealed Peter to be not in Boston, as believed, but in New York, and in a state of some emotional turmoil. On returning home, however, Wasson found things to be exactly as predicted: the enterprising Peter had made use of his father's absence to hold a party in their New York apartment, and was indeed in a state of adolescent turmoil, having been spurned by his girlfriend. Aurelio, Wasson noted, had scored a 'palpable hit'.

The American's second trip to Mexico, in 1954 – this time to the mountainous Mijería region of north-east Oaxaca – generated further ethnographic data on the indigenous use of mushrooms, but is not noteworthy here. His third trip, in 1955, with society photographer Allan Richardson (they were only joined later by Valentina and Masha), proved far more momentous. Returning to Huautla, he had his now infamous meeting with the *curandera* María Sabina (1894–1985): not only was she a locally respected and charismatic healer but, most importantly of all, she agreed to let both Wasson and

Gordon and Valentina Wasson during their momentous visit to Huautla, 1955. Photo by Allan Richardson. Courtesy R. Gordon Wasson Archive, Harvard University, Cambridge, Massachusetts.

Richardson eat the 'sacred' mushrooms. Thus, they became the first Westerners ever intentionally to do so.

María Sabina, a Mazatec Indian, was born in Huautla on 17 March 1894,[41] the year that Mordecai Cooke was fulminating against irresponsible parents for letting their unattended children pick and eat the 'poisonous' Liberty Caps. Very different strictures operated in Huautla, however, for in her autobiography Sabina recorded how she intentionally ate her first mushrooms while still a child of about six or seven.[42] She grew up in a culture in which, although it was not openly talked about, it was a given that the highest class of *curanderos*, the 'Wise Men' and 'Wise Women', derived their healing powers from the 'sacred' mushrooms. According to *curanderos*, mushrooms granted access to, or were literally seen as, spirits with whom the healers could forge beneficial relationships. If a *curandero* proved worthy, the *saint children*, as Sabina called them, would impart information, or speak through the healer in improvised, poetic chants that were believed to have healing power.[43] As a child, Sabina watched a healing ceremony in which her uncle was cured by the bemushroomed utterances of just such a Wise Man, so when she stumbled across the same mushrooms growing wild in the woods she knew that, were she to eat them, she too would 'sing beautifully'.[44]

She ate her mushrooms sitting on the mountainside, supposedly tending her family's goats. Very soon she felt propelled into another world full of temples and golden palaces. The mushroom spirits were waiting for her there, appearing as clownlike dwarfs and 'children with trumpets, children that sang and danced, children tender like the flesh of flowers. And the mushrooms talked, and I talked to the mushrooms . . . And the *teo-nanacatl* answered . . . saying . . . that when we needed something, we should go to them and they would give it.'[45] Discovering that the mushrooms made her feel content, 'like a new hope in life', and secondarily that they assuaged her almost permanent hunger, she ate them many times with her sister. The two of them were often found sprawled or kneeling in the woods. Their parents, knowing that upsetting children in this state could cause permanent ill effects, carried them home gently. Sabina's grandmother believed the girl was destined to become a great *curandera*.

According to an interview she gave for the Italian magazine *L'Europeo*,[46] Sabina conducted her first healing ceremony when she was only eight years old. But famously she found her vocation when her sister, Ana María, became seriously ill some years later. Moved by sisterly affection, Sabina ate more mushrooms, and went deeper into the mushroomic world than ever she had before. A *duende*, or mushroom spirit, appeared to her and presented her with a stark question: 'But what do you wish to become, you, María Sabina?' She replied 'A saint' – an answer that granted her access to a great book which, though she could not read, filled her with the knowledge of how to cure. 'When I came to myself . . . I looked for the herbs that the Book had indicated to me, and I did exactly what I had learned from the Book. And also Ana María got well.'[47]

Thus, when Wasson met her in 1955 she was, within the local indigenous community, a renowned '*curandera de primera*, of the highest quality, *una Señora sin mancha*, a woman without stain'.[48] Wasson was immediately struck by her. 'She had presence,' he wrote, and was 'a woman of rare moral and spiritual power, dedicated in her vocation, an artist in her mastery of the techniques of her office'.[49]

Arriving in Huautla, Wasson won the confidence of a town official, Cayetano García, who arranged the meeting with Sabina, and again Wasson used the well-being of his son as a pretext to partake in a *velada*. According to Wasson, García and the Americans picked a boxful of the mushrooms that were growing from a rubbish tip at the bottom of the valley and presented them to a delighted and grateful Sabina, who immediately agreed to hold a ceremony that night, 29 June.[50] According to Sabina's autobiography, however, she only condescended in deference to García, and the implication is that this was against her better judgement. Nevertheless, she held not just one ceremony but two, the second on 2 July, with Wasson consuming the mushrooms on both occasions, Richardson only the once.

Shortly after eating six pairs of the grubby, acrid mushrooms,[51] Wasson felt as if his soul had been scooped out of his body.[52] Visions of geometric patterns gave way to 'architectural structures, with colonnades and architraves, patios of regal splendour, the stone-work all in brilliant colours, gold and onyx and ebony, all most harmoniously and ingeniously contrived, in richest magnificence extending beyond the reach of sight . . .'. He felt himself a 'disembodied eye' floating above strange, new landscapes, and then, somehow, a witness to the Platonic realm of forms. For a moment he felt as though he understood the true, awful meaning of the word 'ecstasy', and then:

> it seemed as though the visions themselves were about to be transcended, and dark gates reaching upward beyond sight were about to part, and we were to find ourselves in the presence of the Ultimate. We seemed to be flying at the dark gates as a swallow at a dazzling lighthouse, and the gates were to part and admit us. But they did not open and with a thud we fell back, gasping. We felt disappointed, but also frightened and half-relieved, that we had not entered into the presence of the ineffable, whence it seemed to us at the time, we might not have returned, for we had sensed that a willing extinction in the divine radiance had been awaiting us.[53]

While the influence of earlier drug writers upon Wasson's prose is apparent, he stands independently alongside figures such as Thomas de Quincey and Aldous Huxley in being able to convey vividly the full enormity of his drug-induced ecstasies: his accounts have rightly become classic pieces of 'trip-lit'. But what makes them interesting is not so much their stylistic pedigree as the fact that they show Wasson had, and anticipated having, a religious or mystical experience. He had, in other words, already abandoned the classical framework in favour of the psychedelic discourse, and went to Mexico quite convinced that the mushroom's effects were benign and desirable – perhaps the first Westerner to believe as much. Clearly, he had come a long way since the events of his honeymoon.

Like Huxley's, Wasson's writing became extremely influential. The 512 privately published copies of *Mushrooms, Russia and History* were priced well beyond the reach of all but private collectors and university libraries, but he popularised his discoveries in a piece for the middle-brow magazine *Life*, published on 13 May 1957.[54] This famous article was eye-catchingly captioned 'Seeking the Magic Mushroom: A New York banker goes to Mexico's mountains to participate in the age-old rituals of Indians who chew strange growths that produce visions'. Here, for the first time, a positive way of understanding the psychoactive properties of fungi was broadcast to the Western world, captured in the snappy and memorable phrase 'the Magic Mushroom'. The article was read by millions.

Wasson, meanwhile, made many more return visits to Mexico to assess the extent of mushroom usage, and occasionally to consume mushrooms himself. He met Sabina several more times, and made two separate recordings of her *veladas*, one of which was released for public consumption (*Mushroom Ceremony of the Mazatec Indians of Mexico*, New York: Folkways Records); the other, for an academic audience, came with an accompanying book in which the entire evening was linguistically and musically transcribed, and meticulously cross-referenced and annotated.[55] Produced off his own bat, and taking sixteen years, this last was a labour of love, and remained the work of which Wasson was most proud.

Wasson enlisted in his endeavours the help of (amongst others) two notable scientists, Roger Heim (1900–1979) and Albert Hofmann (1906–). Roger Heim, Professor of Mycology and director of the prestigious Muséum National d'Histoire Naturelle in Paris, set about the task of identifying and describing the mushroom species employed: several species, primarily of the genus *Psilocybe* (but also *Stropharia* and *Conocybe*), were used according to seasonal availability.[56] Sufficient samples of the mushrooms were cultivated back in the laboratory for Albert Hofmann, the Swiss chemist responsible for the discovery of LSD, to isolate the psychoactive alkaloids: in 1958, while working for the Basle-based Sandoz pharmaceutical company, he successfully synthesised psilocybin and psilocin, which he named after the mushrooms' Latin epithet.[57]

The team made further discoveries regarding the psychoactive plants employed by the *curanderos* outside of the mushroom season. Wasson collected samples of what the Aztecs had called *Ololiuhqui*, and these were identified as seeds of the Morning Glory, *Turbina (Rivea) corymbosa*: Hofmann swiftly isolated their LSD-like active compounds.[58] Wasson and Hofmann discovered the Mexican *Hierba de la Pastora* or *Hierba de la Virgen* ('herb of the shepherdess' or 'herb of the Virgin') to be a variety of mint, *Salvia divinorum*[59] (like magic mushrooms the bizarrely hallucinogenic *Salvia* has found its way into contemporary Western drug markets). Heim, meanwhile, turned his attention away from Mexico to the psychoactive fungi in Europe, about which much more will be said later.

Within a very short space of time, then, the team made significant ethnobotanical advances inside and outside Mexico, while Wasson himself went on to promote his influential ideas about Soma, Eleusis and the religious nature of the mushroom experience. Consequently, for mushroom enthusiasts, Wasson has come to be seen as an intellectual giant who bestrides the history of the magic mushroom, bringing necessary scholarly gravitas to an otherwise flaky subject. He is rightly remembered for his Mexican discoveries, but twenty years after his death many of his ideas have yet to be subjected to any kind of formal scrutiny. It is time to examine them to see what truth, if any, lies within.

6

Wasson

What pre-historic religion or tabu is finding expression when the English governess, with a facial spasm and shudder, grinds a delectable mushroom under her heel and warns away her charge?

Gordon Wasson[1]

It is in the nature of an hypothesis, when once a man has conceived it, that it assimilates every thing to itself, as proper nourishment; and, from the first moment of your begetting it, it generally grows the stronger by every thing you see, hear, read, or understand.

Lawrence Sterne, *Tristram Shandy* (1759–67)[2]

In the early 1950s, when Gordon Wasson first went to Mexico, travel to and from the Mazatec hinterlands was a considerable undertaking. Roads into Oaxaca were rudimentary – or non-existent – and were frequently blocked by muddy landslides and rockfalls. Though it was possible to travel some of the way by light aircraft, the only airstrips were those that had been cut precipitously into the sides of mountains, which made landing a bumpy and perilous process. From then on the only way forward was by foot, or on the back of a mule. If the weather turned bad you could be stranded in one place for weeks, forced to subsist on a local diet of eggs, cornbread and beans. And, reliant as you were upon the goodwill of the indigenous inhabitants for guidance, mules, food and lodgings, the chances of being caught in the crossfire of internecine feuds, or even of being ambushed by bandits, were worryingly high. None of this seems remotely to have bothered Gordon Wasson.

Starting in 1953, Wasson embarked on his ten successive annual trips to Mexico with an enthusiasm that was not always shared by other expedition members. Albert Hofmann, for example, seems to have found the idea of straying off the well-beaten tourist trail a cause of some considerable anxiety, as a series of apprehensive letters to Wasson attest.[3] The monotony of trekking through Mexico led photographer Allan Richardson to bow out after only a handful of voyages.[4] But Wasson was indefatigable and positively thrived on the privations and hardships of voyaging to the remote, pre-modern parts of this extraordinary country.

Of course, Wasson was possessed of an assured self-confidence that only great wealth and high status – of the kind that comes from being vice-president of a major US bank – can bring. He feared no obstacles for, as he would admit many times, he was used to getting what he wanted. But his temerity, his love of travel and even his slightly obsessive interest in all things mushroomic do not really explain what motivated him to keep going on these gruelling journeys: something more than just the quest to find the magic mushroom was driving him on. Indeed, he repeatedly made the trip to Mexico because he was looking for evidence, evidence that would prove a revolutionary new theory he had been developing ever since his passion for ethnomycology had first been aroused. If true, he knew that it would both overturn established thought about the history and origins of religion and ensure for ever his reputation as a scholar.

Gordon Wasson was convinced that the human religious impulse itself had been awakened by a Palaeolithic magic mushroom cult. The great pillars of Western civilisation, our religious traditions, owed their origins to a time when our distant ancestors freely ate from the original tree of knowledge, the divine magic mushroom. And though long extinct, traces and legacies of the original cult could still be detected in certain attitudes and dispositions that obtained today.

But what really spurred Wasson on was the hope that the cult had not died out entirely. What if, hidden away in the remote mountains of the Sierra Mazateca, it had clung on in some half-remembered form, of which the indigenous mushroom ceremonies were, in fact, the very last vestige? As he followed the mule trails ever deeper into the mountains, and as he left the modern world further and further behind, Wasson became ever more convinced that the answer was positive, and that his hoped-for religion lived on. The hardships of travel were nothing compared with the magnitude of this thrilling possibility. And then he made the biggest and most exhilarating find of all: María Sabina, the last living priestess of the ancient mushroom cult.

To find the origins of Gordon Wasson's radical theory we need to look not for some hitherto unappreciated archaeological discovery turned up in the course of his researches, but at his horrified reaction to the wild mushrooms of the Catskills, way back on his honeymoon. The force with which he was repelled, and with which Valentina was attracted, suggested to the couple that there must be an underlying reason or original cause for their respective revulsion and delight. They began to amass references to mushrooms from folklore, mythology, art history, etymology and philology, and noticed patterns within them, such as the paucity of names for mushrooms in the English language compared to the Russian. They coined the terms mycophilic, or mushroom-loving, and mycophobic, or mushroom-hating, and separated the nations of the world accordingly, placing Russia, for example, in the former category, and Britain in the latter.[5] Soon they had divided up the countries of Europe according to this strange dialectical axis.

As the couple perused their unorthodox map, a troubling idea kept surfacing. What if the two reactions, mycophilia and mycophobia, were residual and half-forgotten cultural memories of some ancient strictures or taboos placed upon the eating of mushrooms? Surely only something as powerful as a religious prohibition would have been awe-imposing enough to have endured for so long down the ages, in which case could not the axis be the last surviving cultural trace of an ancient mushroom-worshipping religion? Surprising as the notion seemed, the more the couple looked – and the more Wasson continued his researches after Valentina's death – the more the evidence kept on accruing. Wasson's Mexican discoveries seemed to put the matter beyond doubt.

He came to imagine the following scenario. Our Palaeolithic European ancestors had consumed magic mushrooms – most probably the fly-agaric, but perhaps later psilocybin mushrooms – in an original and archaic form of shamanism. As this loose spirituality gradually became institutionalised, mushroom consumption was restricted to, and jealously guarded by, a powerful priesthood that placed a terrible taboo upon its profane usage. The cult spread globally, but eventually gave way to what would become the non-mushrooming world religions. Nevertheless, the taboo lingered in a vestigial fashion so that fear of, or fascination with, the magic mushroom lastingly determined people's attitudes to mushrooms in general. The magic mushroom, wrote Wasson, 'must have been hedged about with all the sanctions that attend sacred things in primitive societies

. . . it must have been instilled with mana, an object of awe, of terror, of adoration'.[6] No wonder he recoiled from the Catskill mushrooms so violently.

At first sight Wasson's extraordinary thesis seems incredible, more like something from the novels of H. Rider Haggard than a serious academic proposition. But what made Wasson's conclusions plausible was that he modelled his theoretical and methodological approach explicitly upon the foundational works of anthropology, and consequently a diverse range of distinguished academics were persuaded that Wasson was absolutely correct.

The theoretical paradigm that steered Wasson's hand, and that, ever the autodidact, he had soaked up through his reading, is now termed 'cultural evolution'.[7] As a systematising intellectual framework, it reached its peak in the early years of the twentieth century, and was espoused by the first anthropologists, such as Edward Burnett Tylor (1832–1917) and Sir James Frazer (1854–1941). It emerged out of a tempestuous intellectual climate in which debates raged about Darwinian (and other) ideas of evolution, about Freudian psychoanalytic theories of the unconscious, and about the beliefs and practices of the 'primitive' peoples encountered through colonial expansion.

Ideas of evolution and progress suffused the nineteenth century quite independently of Darwin's intellectual breakthrough in naming natural selection as the mechanism of biological diversity and change. Thinkers as various as the pioneering French sociologist Auguste Comte (1798–1857) and the British writer Herbert Spencer (1820–1903) proposed that it was not just plants and animals that evolved, but human societies as well. Comte, for example, thought that there was a natural progression by which societies moved from a theological, to a metaphysical, to a scientific stage; Spencer, that evolution pushed societies into ever more complex forms of organisation. The colonial encounter with 'savages' living beyond the edge of the civilised world seemed confirmation of this view, the logic of which was that Western civilisation represented the pinnacle of the evolutionary process, the goal to which all other cultures naturally aspired. Primitive cultures were just those caught in stasis, in the evolutionary backwaters as it were, away from the driving current that had swept the West onwards to civilisation and 'supremacy'. As an idea, cultural evolution gave intellectual justification for the imposition of Western 'civilising' values upon colonised cultures as a means of accelerating their 'arrested development'. But it also presupposed a singularity from which all the cultures of the world had diverged, and so justified the search for origins: the academic discipline of anthropology was founded on precisely this basis – the search for the 'origins of man' through the study of the primitive 'other'.

That search was brought closer to home by the ideas of another revolutionary nineteenth-century thinker, Sigmund Freud (1856–1939), whose radical psychological theories sent ripples through polite society. According to Freud, within the Pandora's box of the unconscious lodged terrible scatological and oedipal desires which, when repressed, led inevitably to neurosis or psychosis. The realisation that the primitive could be found not just 'out there', but also 'in here', raised troubling anxieties. What if, beneath the veneer of Western civilisation, and malingering unconsciously like repressed Freudian desires, were to be found atavistic traces of our own evolutionary lineage, our own savage past? If primitive societies were living fossils from an earlier stage of human evolution, what vestigial myths and rituals might remain caught in our language, customs, folklore and religion?

Edward Tylor named this idea the 'doctrine of survivals' in his monumental *Primitive Cultures* of 1871, and by the beginning of the twentieth century the hunt for survivals in the hinterlands of our own culture had begun in earnest. Interest was suddenly expressed in rural populations and their traditional rites and folk customs, previously dismissed as examples of rustic, uncultured backwardness (as they are so often today). So while Cecil Sharpe (1859–1924) wandered through summer meadows collecting and preserving folk songs in his bowdlerising aspic,[8] other metropolitan folklorists and anthropologists lifted the stones on morris dancing and mummers' plays, in the hope of uncovering the remains of fertility rites from our own pagan past.[9] Margaret Murray scrutinised the early modern witch-confessions for traces of her imagined 'Old Religion', while James Frazer collated twelve volumes of data – published in three ever-expanding editions as *The Golden Bough* – in support of his notion that all religions, including Christianity, were actually expressions of a universal motif, that of the dying and resurrected vegetation god. As someone vehemently opposed to religion, his hope was that he would propel the West into Comte's scientific stage of development, banishing irrationality once and for all.

Alas, these works proved to be rather more titillating than repellent to the general public, and exerted a lasting influence upon the popular imagination: the abridged 1922 version of Frazer's *The Golden Bough*, for example, has never been out of print. Frazer's legacy, in particular, can be seen in the work of various writers – including T. S. Eliot (1888–1965), W. B. Yeats (1865–1939), D. H. Lawrence (1885–1930), and especially Robert Graves (1895–1985) – and more recently in the cult films *The Wicker Man* (1973), *Apocalypse Now* (1979) and *The Lair of the White Worm* (1988), which portray the disturbing consequences of a modern return to the primitive. Far from ushering in a scientific age, *The Golden Bough* was one of the factors that prompted people to go about reviving the supposed pagan religions described within it, leading to the emergence of modern Druidry and Wicca.[10]

Seen in this light, the provenance of Gordon Wasson's obscure mushroom thesis starts to become clearer, for he too was influenced by Frazer, both directly and via a long correspondence and friendship with Robert Graves. Wasson's emphases on the importance of taboo – which he thought a lasting and 'deep-seated emotional attitude' – and of survivals – 'fossil meanings, fossil sayings, fossilised folkloric bits of our various languages'[11] – were both borrowed from Frazer. He religiously copied Frazer's comparative methodology, amassing evidence from as many different sources as he could find (as a consequence of which his books are all, quite literally, weighty). And from Graves he took the belief that myths, however fantastical they seem, have their origins in actual historical events: hence any myths or folktales involving mushrooms could be read as a vestigial memory of the supposed mushroom cult.

The only problem with all this was that by the time Gordon and Valentina were bickering about the edibility of the Catskills' mycoflora, the arguments about the merits of cultural evolution and the Frazerian comparative method were long over: both tropes had been soundly rejected within anthropology. One of the staunchest critics was pioneering American anthropologist Franz Boas (1858–1942) who, preferring fieldwork to library-based research, could find no evidence for the sorts of evolutionary cultural progressions outlined by Comte, Spencer, Tylor and others.[12] Without the rationale of cultural evolution, searching for survivals was meaningless. Certainly cultural forms and artefacts can linger way beyond their original historical moment – one only has to think of classical architecture, for example – but anthropologists like Boas rejected the idea that it is possible to work backwards from these alone to make inferences about earlier cultures. Later, by the 1970s, revisionist work undertaken by a new generation of folklorists and historians specifically demolished the idea that British folk customs were pagan survivals, and the Gravesian notion that myths inevitably had historical and not literary origins.[13] In anthropology at least, cultural evolution was cast aside as an intellectual dead end.

The principal objection to the Frazerian comparative method that went with it remains that it attempts to make data fit theory and not vice versa. Tylor, Frazer, Graves and Wasson all attempted to prove their particular, preconceived hypotheses by accumulating massive amounts of apparently supporting evidence while ignoring all that was contradictory. The selection process inevitably reflected the inherent class, gender and cultural biases of these scholars – a point enlarged upon by later Marxist, feminist and postmodernist critics – while cherry-picking evidence from sources widely separated by distance and time stripped it of the all-important context in which it occurred. Boas believed the comparative method could only ever produce a distorted picture of the subject, which is why he argued so forcibly for detailed, localised fieldwork – ethnography, in other words – to become the basis of anthropological enquiry.

Later, in the 1960s, perhaps as a reflection of social upheavals at home, anthropologists began to notice the importance of transgression, of rule-breaking, in the structuring of both traditional and modern societies, and so challenged the great foundation of Wasson's thesis: namely, the Frazerian notion of taboo as an enduring and binding stricture, silently obeyed by all those subject to it. The many 'rituals of rebellion' discovered by anthropologists, and designed either to heal the social rifts caused by transgressive acts, or to permit them to occur within tolerable parameters (as with the Kuma people's *nonda* 'mushroom trance'), showed that taboos have always been there for the breaking. In any case, declaring that a hostile reaction to, say, a mushroom is evidence that it must have been taboo at some earlier stage of cultural development is to make an unwarranted inductive leap. To take this line of reasoning to its absurd limit, one could argue that Lenten restrictions are evidence of an early Christian taboo against eggs, which must have been secretly worshiped by a jealous priesthood.

Indeed, the very premise that prompted Wasson's investigation – the notion that peoples are either mycophilic or mycophobic – now appears too problematic to be of use. The term 'phobia' implies an overwhelming and irrational panic reaction, a force that bears down upon the sufferer and has its origins deep within the unconscious, whereas anxieties about mushrooms are wholly rational. In the absence of any reliable methods for distinguishing the edible from the poisonous, and given the high price of getting the decision wrong (a slow and painful death in the case of the Death Cap, *Amanita phalloides*), blanket avoidance of all mushrooms is the most sensible and reasonable option. Fear of mushrooms is actually the fear of being *poisoned* by mushrooms, and the way this expresses itself is no more mysterious than our aversion to other life-threatening dangers, such as snakes (which may or may not be venomous). Once someone is bitten, we all remain shy.

Empirical evidence bears this out. A long-term study conducted during the 1970s in Colorado, where there is no widespread custom of gathering wild mushrooms, found that on average only six cases of accidental poisoning per year required hospital treatment. Fatalities were so rare as to be all but non-existent. By contrast, in the Polish province of Poznan, which has a similar population size to Colorado but a long-standing folk tradition of mushroom-picking, an average of fifty or so incidents of poisoning every year required hospital treatment, of which around 10 per cent proved fatal.[14] In other words, the risk of being killed by mushrooms was substantially higher in the culture that was nominally mycophilic. This runs counter to Wasson's thesis and suggests that folk wisdom regarding mushroom identification is worryingly unreliable: without a decent field guide, the safest option remains not to go mushroom-picking at all.

And, of course, patterns of mycophagy are not set in stone as Wasson supposed, but change with time. During the nineteenth century it was common for wild mushrooms to be sold at (supposedly mycophobic) British markets, but fear of poisoning – no doubt encouraged by the antics of ne'er-do-wells like the Bickerton family – meant that this had all but died out by the twentieth. One small part of the British Second World War effort, therefore, was directed at encouraging people to learn how to identify edible mushrooms safely in order to prevent this nutritious food source going to waste. The campaign was successful, and wild mushroom consumption duly increased, until after the war when commercial mushroom production made foraging unnecessary.[15] This is a far more complex situation than can ever be encompassed by Wasson's blanket designation of Britain as a mycophobic nation. In any case, it is rather meaningless to try to designate the eating preferences of an entire people as one thing or another, for taste is, by definition, a fickle thing.

So even though the terms mycophobia and mycophilia have achieved a certain currency within mycological circles, they are often applied inaccurately, are based upon misleading assumptions, and should be abandoned as scientific terms. Likewise, though the archaeological record is sufficiently vague for the possibility of prehistoric magic mushroom use to remain open, Wasson's labours cannot demonstrate that this was definitely the case. For while it is true that anxieties about mushrooms have been expressed since classical times, there is nothing to suggest that these originated in anything more exalted than a fear of poisoning. (Indeed the pressing question is not why some cultures shun mushrooms, but why others overlooked the frequency of accidental poisonings and carried on eating mushrooms regardless.) In the evolutionary race that is the history of ideas, Wasson's hypothesis would seem to have had its day.

Of course none of this would be of any great significance outside academic circles were it not for the fact that Wasson's theories were to have a rather profound impact upon the people he studied. In particular, María Sabina was to have her life turned upside down by Wasson in his pursuit of his ancient mushrooming cult. It is for this reason that any balanced assessment of Wasson must consider the man, and not just his unusual ideas.

Though Gordon Wasson had little time for hippy culture, hated being treated as a psychedelic guru, and expressed nothing but contempt for Timothy Leary (the Harvard psychologist who, in the sixties, became notorious for his profligate use and advocacy of LSD),[16] his 1957 *Life* article 'Seeking the Magic Mushroom' played a significant role in kick-starting the whole psychedelic revolution. Many of the key figures of that movement, including Leary, experimented with psychedelics as a direct result. That Wasson wanted widespread recognition for his Mexican discoveries is clear but, perhaps because of the distance he kept between himself and the hoi polloi, he underestimated the extent to which readers would want to follow in his footsteps, find Sabina and try the mushrooms themselves. He disguised her name and whereabouts in the *Life* article but not in *Mushrooms, Russia and History*, so very quickly they became public knowledge: within months of the *Life* article going to press, Westerners were knocking on Sabina's door. The trickle of visitors in the early 1960s became a flood, until Oaxaca was awash with hippies on the magic mushroom trail. The trip became fashionable, and with various rock stars – Pete Townshend, John Lennon and Bob Dylan, amongst others – rumoured to have made the pilgrimage, this was perceived, true or not, as a celebrity endorsement, making Huautla an essential destination.[17]

Of course, for anyone who made the trip to, and with, María Sabina, it was easy to believe that she really was some priestess of an ancient cult. She held her *veladas* in her basement on a compacted dirt floor and in front of a rudimentary altar upon which were placed pictures of Christian saints and a bowl of burning copal: the Mazatec resin used as incense. By candlelight she would bless the mushrooms, passing them through the smoke before commencing a lengthy series of prayers to various saints. Then, gauging the appropriate dosage for each participant, she would pass round pairs of mushrooms, reserving the largest dose for herself. The candles would be extinguished so that the room was pitch black, and as the effects of the mushrooms came on she would begin to sing the improvised chants for which she was famous. Long lines, sung in a low monotone and interspersed with rhythmic claps and shouts, marked out the contours of the *velada*, guided the trip, and almost invariably Western visitors – famous or otherwise – wrote of having had intensely spiritual experiences under her expert guidance.

The trouble was that hippy culture arrived with a very different set of ideas and outlooks from those that prevailed in traditional indigenous Oaxaca. For hippies, the mushrooms were 'psychedelic', which meant that they were bound up with notions of authenticity, freedom, individualism, bohemianism and rebellion. Used to seeing psychedelics as drugs not deities, commodities not conscious entities, many expected to be able to buy and consume mushrooms as and when they desired, irrespective of local sensibilities. They certainly did not want to be bound to the *curandero*-led *velada*, or to the archaic mores and strictures of an animist peasant culture. One American hippy visitor was recorded as saying: 'Look, man. You can go for that *curandero* shit if you like but it's not my bag. I don't need an old hag mumbling in Mazatecan to turn me on. I don't dig this Indian doctor jazz. I turn myself on. It's not my culture. You just score the mushrooms . . . we'll do the rest.'[18] Two others smashed up the furniture in their lodgings simply so they could hang up a hammock, while others complained volubly about having to pay for what they considered bad food and worse accommodation.[19] One tactless visitor tried to pay Sabina for her services with a pet dog. 'I told him that I didn't want a dog, that I didn't have the money to maintain it,' Sabina recalled. 'What was the animal going to eat? Shit?'[20]

Of course, the ill-considered behaviour of the few necessarily tarnishes the reputation of the many, but post-colonial expectations – the assumption that locals would simply roll over and offer up their most intimate practices for the tourist dollar – still suffused the attitudes of all those who went, souring whatever 'good vibes' they had intended to spread. Sabina, though she deplored the arrogance of the more thoughtless visitors, suffered the inevitable backlash against this unsought fame. Jealous rivals burnt down her house and shop so that she lost both home and livelihood. She was falsely accused of peddling cannabis, with the result that she was briefly imprisoned and had her meagre possessions confiscated.[21] When anthropologist Joan Halifax visited her during the 1970s, Sabina was in a very sorry state. She was dressed in rags and covered in bite marks, having been attacked by an envious relative.[22] Wasson, as you might expect, denounced this, branding the hippies as 'riff-raff';[23] but while he was certainly not responsible for the behaviour of those who came after him, the question of the extent to which he was to blame for them going in the first place remains.

He was certainly sensitive to the accusation that he had brought ill fortune upon Sabina and the other residents of Huautla, so much so that in 1970 he published an anguished defence in the *New York Times*.[24] There and elsewhere he sought to make a clear distinction between his actions and those of the 'thrill mongers' who followed him.[25] He 'winced' at the hurt he had inadvertently caused to Sabina and her profession,[26] saying he had gone to Mexico with the best of intentions: to track down the last living vestige of an ancient mushroom cult and record it before it was eroded away by the inevitable onslaught of modernity. His was a salvage operation conducted for that most noble of purposes: contributing to the sum of human knowledge. 'What else,' he implored, 'could we have done'?[27]

Doubtless this sophianic justification was heartfelt, but in several ways his behaviour failed to live up to this noble image of himself as dispassionate observer. It is clear that his motivations became more complex, and more muddied, once he had actually eaten the mushrooms in 1955. Thereafter, his banker's instincts seem to have taken over so that he came home with every intention of profiting from his discoveries. In a shrewd move, he had already acquired the rights to all of Allan Richardson's photos, in exchange for funding the photographer's travel and subsistence costs.[28] But within months of returning he had a meeting with top executives of the Merck Sharp & Dohme pharmaceutical company to discuss rights to the mushrooms' potential active ingredients. It was his hope that some psychiatric use could be found for them.[29] As we have seen, it was Albert Hofmann's team at Sandoz in Basle who eventually isolated psilocybin and psilocin, but nevertheless, when Sandoz put its patented brand *Indocybin* on the market, Wasson appears to have been rewarded for his part in its discovery with a directorship of one of its American subsidiaries.[30]

He made several other attempts to profit from his story by offering it to various magazines, including *National Geographic*, but after a chance meeting he opted for *Life* instead. Knowing that it would serve as a convenient and timely advertisement for *Mushrooms, Russia and History*, Wasson urged the editors to bring the article out in the May edition of 1957, and not later in the year as had been proposed. The editors acceded to his demands – and advertised the mushroom edition of *Life* extensively on television – with the effect that his book, published a few months later in a limited edition of 512 copies and retailing at $125, doubled in price before all the copies were sold. The article itself netted Wasson the then extraordinary sum of $6,000, which he used to cover the cost of sending out a hundred complimentary review copies of the book.[31]

The deal with *Life* was agreed before Wasson returned to Huautla and Sabina in 1956. Though the editors were to write the headlines and straplines (and in the process would give the world the term 'Magic Mushrooms'), Wasson retained editorial control over the all-important body of the text. Nevertheless, he took advice on how to structure the piece, and on how Richardson could capture more compelling and magazine-friendly images.[32] Wasson also brought sound-recording equipment with him to Mexico so that he could tape Sabina and sell the recordings (they were subsequently released on the

Smithsonian Folkways label). The return visit to Sabina appears to have been much less about the scientific acquisition of knowledge, and far more about securing the scoop.

Despite his protestations, then, it seems that he went south in possession of the very same post-colonial attitudes for which he would later berate the hippies. 'I have often taken the sacred mushrooms,' he declared rather pompously, 'but never for a "kick" or for "recreation". Knowing as I did from the outset the lofty regard in which they are held by those who believe in them, I would not, could not, so profane them.'[33] But this is not borne out by his actions, for whatever 'lofty regard' he felt for local practices, customs and proscriptions, he seems to have been more than happy to ignore them if they impeded his own private agenda. During his first trip to Mexico in 1953, though he had been able to pick some of the mushrooms, he could not initially find a *curandero* willing to host a *velada* for him. Throwing caution to the wind, and disrespectful of local custom, he ate his pickings anyway. 'Bitter to the taste, they were not sufficient to cause psychic symptoms.'[34] Knowing that *veladas* were only ever conducted for healing or to discover the whereabouts of lost or stolen possessions, he eventually persuaded the various *curanderos* he did find, including Sabina, to host sessions on the pretext of concern for the well-being of his son. Later he admitted that this had been a deception performed simply to get him access to the ceremonies.[35]

Then, in 1955, after his two bemushroomed epiphanies with Sabina, he was joined in Huautla by Valentina and his daughter Masha, but the party became marooned after a fortnight of torrential rain prevented their plane from landing. Bored and with little to do, the idea of eating some mushrooms seemed to them a way to brighten their spirits. 'We were damp, chilled through and miserable,' recalled Valentina. 'A few hallucinations, we decided, would be a great help.'[36] Wasson justified this particular transgression of local custom on the basis that he was conducting a 'scientific' investigation.

Returning to New York with a collection of dried mushrooms and bags of excitement, Wasson did what so many others would later do after their first psychedelic epiphany: he 'turned on' his friends. In his Manhattan pied-à-terre, he held a series of rather solemn *veladas* of his own, in which he invited guests to achieve a state of grace, before handing out pills made from dried mushrooms, supplied by Sandoz. The trippers were presented with a son et lumière show of Richardson's photos and Wasson's sound recordings of Sabina in full bemushroomed flow. Robert Graves and the proprietor of *Life*, Henry Luce, were just two of the luminaries turned on in this way. Though Wasson felt that he was recreating the special atmosphere of the *velada*, these soirées were something else entirely: cod Mazatec ceremonies, with Wasson assuming a priestly role.

Most surprisingly, when US federal prohibition made obtaining and using mushrooms a criminal offence during the 1970s, Wasson was happy to emulate the hippies he so despised by surreptitiously importing psilocybin into America for his own use. He managed to persuade Albert Hofmann to risk his job by posting at least two consignments of psilocybin pills, wrapped in foil to avoid detection, from the Sandoz factory in Switzerland to a care-of address in California.[37] Wasson may have been a pillar of society, but in this regard he seems to have considered himself above the law.

To add to this, it now appears that the picture of Mazatec religion he presented within the pages of *Life* and in his other writings was slight, superficial and in many ways inaccurate; and this misleading picture contributed significantly to the flow of people travelling to Huautla. His exploits provide us with a model of how not to go about ethnographic fieldwork: he made lightning visits of only a few weeks at a time, rather than the months and years deemed necessary to build the essential trust and 'rapport' with another culture; he never learnt the Mazatec language, was always reliant on translators, and correspondingly overlooked or missed many of the subtleties of the Mazatec world view.[38] His single-minded focus upon the mushrooms meant that he failed, for example, to tease out the complex relationships between the famously taciturn Mazatec and their healing plants, many of which have no psychoactive properties but are nonetheless considered powerful healing agents in their own right.[39] It was as if, presented with a beautiful and many-tiered cake, Wasson had eyes only for the cherry on the top. But, most problematic of all, he was so obsessed with the issue of the mushrooms and with fitting them into his ancient survival thesis that he completely misunderstood the true nature of what Sabina was doing in her mushroom *veladas*. He failed to spot that it wasn't a cherry at all.

María Sabina was born just a few years before Wasson, and although her life and circumstances could not have been more different, the two shared one thing in common: they were both, in their own ways, charismatic. Many of the pilgrims who came to her after Wasson agreed that she possessed that mysterious quality, which in less secular times would have been labelled holiness. Álvaro Estrada, the Mazatec who assisted her in writing her autobiography, said that she had a look 'charged with mystery and light'; Joan Halifax that 'a light seemed to emanate from the eyes of this woman, a quality that imparts to those who are with her a sense of the divine awakened'.[40]

That Wasson was captivated by her seems understandable, given these qualities. What is less understandable is why he dismissed the other *curanderos* he encountered as second rate, practitioners of a degenerate tradition: their standing was as high as Sabina's within their respective communities. For example, Sabina recognised the one-eyed Aurelio Carreras as a healer of distinction, and in Huautla he was considered a 'wise one', while amongst the Zapotecs Aristeo Matías was regarded so highly that he sat on a committee of *curandero* elders. But for Wasson, both were inferior to Sabina; Matías's bemushroomed utterances were, by comparison, 'feeble'.[41]

It is possible that Wasson's perspective was skewed by his having seen only Sabina through bemushroomed eyes. After taking mushrooms with her, he wrote that she was 'Religion Incarnate. She was the hierophant, the thaumaturge, the psychopompos in whom the troubles and aspirations of countless generations of the family of mankind had found, were still finding, their relief.'[42] There may have been gender issues at play here as well, for this was the 1950s after all and Wasson was, by all accounts, a man of his time. But I think it most compelling that Wasson alighted upon Sabina as 'the archetypal shaman' because she neatly fitted his preconceptions about what a priestess of his old religion should look like. On finding Sabina, his heart must have skipped a beat, for in her he saw the missing link for which he had been looking.

What is striking now is the extent to which he constructed his ancient mushroom cult in Christian terms. Though he always considered himself deeply religious, Wasson was, until his latter years, a rather non-committal Christian.[43] But he had been brought up an Episcopalian by his father, a minister, and had read the Bible twice by the age of eleven. Something of this remained, for he seems to have pictured his ancient mushroom cult as a form of proto-Christian mysticism, with mushrooms foreshadowing the sacrament of bread and wine and elevating the communicant to a direct encounter with God.[44]

Whereas the *veladas* of the other *curanderos* were cluttered with pagan symbolism and practices – such as a complicated divination system using husks of maize – Sabina employed very little paraphernalia, preferring the mushrooms to 'sing through her' in a form of improvised poetic glossolalia, a song of songs. Devout, suffering, compassionate, generous, humble, a loving and devoted mother, a mystic, a woman without stain: she was a most Mary-like figure, and as such she slotted so very easily into Wasson's High Church expectations. Through his writings, and with the help of Richardson's literally iconic photos – cropped and reprinted in grainy black and white, the better to show Sabina supplicating to heaven and illuminated from above – Wasson beatified her. Then, through the *Life* article, he presented her to the world as the last living saint of his ancient mushrooming religion.

A bemushroomed María Sabina hosting a *velada* for Gordon Wasson in Huautla, 1956. This is one of Allan Richardson's iconic pictures which inadvertently contributed to Sabina's unsought celebrity. Courtesy R. Gordon Wasson Archive, Harvard University, Cambridge, Massachusetts.

The only problem with all of this was that the *veladas* were not religious ceremonies. That is, though each was framed within a unique and adaptive blend of Catholic and pagan ritual actions – prayers to Christian saints and Mazatec spirits, for example – they were not performed as an act of worship, or to induce mystical experiences of God: they were performed for the serious and pragmatic purposes of healing. Sabina was clear on the matter: 'the vigils weren't done for the simple desire to find God, but were done with the sole purpose of curing the sicknesses that our people suffer from'.[45] To find God, Sabina – like all good Catholics – went to Mass.

The solemnity and seriousness with which *veladas* were conducted (and which Wasson misinterpreted as being purely devotional) were due to the fact that, as the mainstay of Mazatec medicine, they were literally matters of life and death. In the Mazatec world view, sickness – which, given poor diets and sanitation, occurred with debilitating regularity – was believed to be caused by the loss of one's soul through a sudden fright, or through magical interference by malevolent spirits and human sorcerers. When Sabina took mushrooms, what she called the *saint children*, or mushroom spirits, would appear and inform her whether or not the patient could be cured. Through visions, these benevolent mushroom spirits would show her the cause of the illness – where and why, for example, the person's soul had been lost – and she would then be able to heal the patient through the power of her singing. 'Language makes the dying return to life,' she said. 'The sick recover their health when they hear the words taught by the *saint children*. There is no mortal who can teach this Language.'[46] But the spirits might equally tell her that the patient was beyond help and would necessarily die, no matter how intensely she sang. A trip to the *curandera* was always a sobering experience.

The anthropological heresy that Wasson committed, therefore, was that he forced these complex indigenous healing practices to fit in with his own preconceptions rather than attempt the more difficult task of trying to understand them on their own terms. He went to Mexico with his vision of the ancient mushrooming cult fully formed, and projected the priestly role onto Sabina – Carreras and Matías, fitting less well, were simply dismissed – then returned home triumphantly with her as his trophy, the final proof of the thesis that would ensure his place in the academic firmament. And it just so happened that this image of a mushrooming religion, so much a product of its time, proved absolutely enticing to a generation of young Americans hungry for spiritual satisfaction.

What are we to conclude from this episode? Whatever his intellectual aspirations, Wasson's claims to have respected local custom, to have behaved differently from the hippies, and to have acted out of some inviolable desire for enlightenment certainly ring a little hollow. He went to Mexico more as a prospector digging for gold than a philosopher looking for knowledge and truth, and ever had his eye upon how the name of Gordon Wasson would be remembered. It was this vainglorious streak that prompted him to publish his findings in *Life*, a middle-brow magazine with a massive global readership, with barely a thought for how the article might affect anyone but himself.

Of course, had he written an academic monograph or published his findings in a peer-reviewed academic journal, the news about the strange Mexican mushrooms would almost certainly have filtered out, and people would still have made the pilgrimage. But far fewer would have gone, and their impact would have been diluted as they would not all have headed for Huautla and Sabina. It is because of this decision, motivated primarily out of self-interest, and the inaccurate way he singled out Sabina as somehow more saintly and more special than any of the other *curanderos*, that Wasson must, I think, take a substantial proportion of the blame for what subsequently happened. His pride led directly to Sabina's downfall.

That said, there is one more myth surrounding this historical episode that needs to be tackled, and that is the idea of Mexico as some primitive, prelapsarian paradise violated and desecrated by the civilised invaders. It is a powerful and coercive myth, etched into our very notion of the 'New World', and almost all writers about Wasson, and visitors to Huautla, have reiterated it: Huautla is now a dirty, commercial, corrupt town, where before it was an idyllic Eden.

Without wishing to deny the impact that mushroom tourism had upon the indigenous population, I want to challenge this myth. For one thing, it absolves each successive wave of visitors of the need to address their own impact on the indigenous population. 'The damage is done,' they say, 'the fall has already happened, so I can do as I please.' For another, while the mountains of Oaxaca are undoubtedly beautiful and picturesque, the myth misrepresents life before Wasson: it was never Edenic nor paradisiacal.

At the time of Wasson's arrival, most of the indigenous population eked out a living from subsistence farming or by selling a little coffee.

Sabina recalled being almost permanently hungry during her childhood, and Mazatec society was marked by petty blood feuds, vendettas, and alcohol-related problems (one of Sabina's sons was stabbed to death during a drunken brawl), not to mention a widespread belief in malicious sorcery.

Nor was it an unchanged, traditional way of life insulated from Mexico's struggles to become an independent nation. One of Sabina's three husbands, for example, was a guerrilla who fought in the Carracista and Zapatista insurrections.[47] This oppressed rural Indian underclass existed on the margins of a modern metropolitan industrial nation, which, during the 1960s and 1970s, developed its own urban counterculture, *La Onda*, modelled upon the American hippy movement. Many of the hippies who went to Huautla were rather Mexican *jipitecas*, and as culturally removed from Mazatec society as their American counterparts.[48] The Mexican authorities, troubled by the effects of mushroom tourism, intervened in 1967 by gaoling or deporting the mushroom seekers, and criminalising the use of mushrooms a few years later.[49] In this complex scenario, then, the impact of internal sociopolitical factors upon the indigenes was as important as the arrival of American and European outsiders.

And, of course, the myth of the desecrated paradise has the subtle but coercive effect of casting the indigenes as helpless and passive primitives, able only to respond to the intrusion of foreigners: it strips them of agency, in other words. It is rare for Western writers to question María Sabina's role in the Wasson affair, but in Mexico some commentators have portrayed her as a modern-day Malinche, the slave woman who betrayed the Aztecs for her love of Cortés.[50] Was the Wasson affair really some ghastly replay of this foundational episode in Mexican history? While there is a little merit in this reading, and while Sabina most definitely had a hand in what happened, I think it too harsh.

Certainly Sabina did what no other *curandero* had done, and allowed Western outsiders to share the mushrooms; and, in spite of her denials, she was happy to charge for her services. At any time she could have refused, but anyone in such dire circumstances would have found this sudden unexpected source of income hard to turn down. Besides, she did not know all the facts and could have had no comprehension of the likely consequences of her decision. By agreeing to service Wasson, she was unwittingly brought into contact with that great Western malaise, celebrity, and as someone insulated from modernity she lacked the wherewithal to deal with it. Had she known, she would undoubtedly have remained tight-lipped as so many of the other *curanderos* did. In later life, she lamented that the power of the mushrooms had been corrupted and was dwindling away, but I rather suspect that this was an expression of the guilt she felt at what she had done.

Unfortunately, gauging exactly what Sabina felt and thought is extremely difficult, for if there are precious few women's voices in the history of the magic mushroom, then Sabina's is the most marginalised of all. It is true that we have her autobiography, recently republished, but she was illiterate and in none of the 'writings' attributed to her are we looking at her unedited words. The well-known interview that appeared in the Italian magazine *L'Europeo*[51] during the 1970s was originally conducted in Mazatec, and translated first into Spanish and thence into Italian: a veritable chain of Chinese whispers. As for her autobiography, this was compiled by Álvaro Estrada, a Mazatec resident of Huautla, from a set of interviews conducted over the course of a year between 1975 and 1976. Young and under-confident, Estrada sent his transcripts to Wasson for comment. Wasson passed them on to his friend, the Mexican poet Octavio Paz (1914–1998) who, while expressing admiration for the work, suggested that Estrada edit it to make it sound more 'primitive' and more 'in keeping' with Sabina's personality. We can only guess at the corrections that Wasson proposed. Whatever they were, Estrada dutifully complied, and so what we possess is an edited representation of Sabina's words, tailored to Westernised primitivist expectations. Estrada could therefore provide a great service to scholarship by publishing the transcripts of his interviews verbatim, and it is regrettable that the recent academic reprint of *The Life* did not seize the opportunity to do so.

In truth, María Sabina has always been misrepresented by Western observers. For Eunice Pike, the missionary in Huautla who advised Wasson before his first trip to Mexico, her bemushroomed utterances were the work of the devil.[52] Wasson inverted this reading to make her a saint. For legions of mushroom enthusiasts across the world, she has become the quintessential psychedelic shaman – a term she never used – while in certain academic circles she is honoured as an ethno-poet of some distinction. She has inspired feminist poetry – the American Ann Waldman freely reworked Sabina's transcribed utterances into her poem 'Fast Speaking Woman'[53] – and even a modernist opera, performed at the Carnegie Hall.[54]

In Mexico the comparison with Malinche may be popular amongst intellectuals, but others see her as more heroic. She has been the subject of documentaries, had comic strips written about her, rock bands and theme pubs named after her. Her image has been used, somewhat bizarrely, in a campaign against drugs, and even more bizarrely, given her three marriages, she has been reinvented as a lesbian icon.[55] She seems set fair to make the journey from icon to brand, and so if the knowing interplay of signs and surfaces continues to define the postmodern world, we can be certain that with the passing of time the real María Sabina will slip ever further from our grasp.

Will the real María Sabina please stand up? A rare and intriguing photo of María Sabina, taken in Huautla during the 1980s, which suggests that the *curandera* was a much more complex character than is ordinarily supposed. Western commentators, preferring to see her as a primitive priestess, have failed to mention, for instance, that she played the guitar. Photo by Álvaro Estrada. Courtesy R. Gordon Wasson Archive, Harvard University, Cambridge, Massachusetts.

Depressing as this is, there is at least one cause for comfort. Hidden in the mountains, away from Huautla and the hippy tourist trail, and pretty well closed to outsiders, the indigenous healing practices, of which the mushroom ceremonies were only ever one part, still go on.[56] For example, a Mazatec taxi driver in Mexico City told a friend of mine how a visit to a mushroom *curandero* had cured him of a chronic stomach complaint that repeated trips to the hospital had done little to alleviate. How long these traditional practices will survive the relentless pressures of the modern world remains to be seen, but for the moment we can be grateful that Wasson's prediction of an immanent decline was, like so many of his pronouncements, an exaggeration.

Returning to Wasson, what are we to make of this extraordinary man, his work and ideas? If his academic reputation in later life is anything to go by, we should think of him highly, for he was honoured with several accolades and awards. In 1960 he was elected an honorary research fellow of the Harvard Botanical Museum, a position that he proudly occupied until his death in 1986. A Harvard graduate himself, he felt this particular reward was the ultimate vindication of his life's work. Then in 1983 he was awarded Yale University's prestigious Addison Emery Verrill Medal for having made an outstanding contribution to the study of natural history. Ten years earlier, and for identical reasons, he was elected a fellow of the Linnaean Society of London. By all accounts, it appeared that his gruelling field trips and years of research had paid off, and that his groundbreaking ideas had gained acceptance within academia.

Twenty years after his death, his scholarly reputation, at least within certain circles, is still high. Those who knew him eagerly report the details of their acquaintance, and stress his qualities as a scholar and a gentleman. He was, they say, a generous host and a tireless correspondent,[57] a raconteur possessed of a rich and witty sense of humour, a bibliophile, an unremitting grammarian, and a diligent and meticulous worker.[58] According to his protégé Jonathan Ott, he was 'a brilliant thinker, a superb writer, a patient teacher, a kind and decent man, a loyal friend'. Anthropologist Joan Halifax called him 'a twentieth-century Darwin', while Ott concluded that 'the world will not soon see his equal'.[59] Amongst mushroom enthusiasts, Wasson is seen as a founding father, not only of psychedelia but also of a particular, rigorous approach to the study of this most colourful of subjects.

That Wasson possessed some admirable personal qualities seems beyond question, but today, viewed from a sufficient distance, these claims and acclamations read as just a little excessive – some might say overblown – for his ideas have yet to shake the world in quite the same way as Darwin's. This is by no means to detract from his ethnobotanical discoveries in Mexico, which were genuinely groundbreaking. By taking the brave, perhaps foolhardy step of eating the mushrooms himself, he was able to bear personal witness to their powerful hallucinogenic effects, and thus to put the matter of the identity of *teonanacatl* beyond all doubt. His status as a writer of 'trip-lit' is similarly beyond question. But his extraordinary ideas about the origins of religion, and the way he went about trying to prove them, now seem very hard to countenance.

The key to understanding Wasson lies, I think, in the fact that however much he wanted to be taken seriously by scholars, however much he longed to be one of them, he always was, and always relished being, the outsider. On the one hand, he expressed what could be termed 'scholar-envy'. Always self-deprecating about his own scholarly abilities, he greatly enjoyed erudition and the company of intellectuals, and was delighted when they condescended – his view – to assist his research. 'What guardian angel had me in his keeping,' he wrote, somewhat ingratiatingly, 'when . . . I ascended the steps of Roger Heim's laboratory in Paris to meet him for the first time, a stranger, an American, an ignoramus in the intricate, the vast, the exacting discipline that was his! At once he made me feel at home . . .'[60] On the other hand, he felt that his discoveries could only have been made by someone unblinkered by the disciplinary boundaries and specialisations of academe. Though he possessed the means and the influence to have opted for an academic training late in life, and indeed persuaded others of a similar age to do so, he preferred to remain an amateur.

Free from the pressure to publish and so justify ever more stretched research budgets, he was never obliged to adhere to the usual checks and balances that operate upon scholars. That is, he only ever published a smattering of peer-reviewed articles – and never in the mainstream anthropological literature that would have been the obvious home for his ideas – preferring instead to disseminate his theories through his lavish, expensive and privately published books. This back-door approach means that it is hard to assess the extent to which his radical ideas were really accepted within the academy.

For example, his ancient mushrooming thesis received short shrift when presented to E. E. Evans-Pritchard (1902–1973) and the other members of the Anthropology Department at Oxford University,[61] but, dismissing it as the fringe theory of an amateur, none of the Oxford anthropologists saw the need to publish a rebuttal. The scholars who admired his work, and felt compelled publicly to say so, were usually those still wedded to the armchair scholarship of the comparative method, such as anthropologist Claude Lévi-Strauss (1908–) and scholar of religion Huston Smith (1919–). (To be fair to Wasson, the comparative method lingered much longer in American scholarship, especially in comparative religion and the work of Mircea Eliade (1907–1986).) A more revealing measure of Wasson's academic achievements than all the awards, medals, and accolades heaped upon him in later life is perhaps the fact that he was consistently refused research funding from grant-awarding bodies.[62] His honorary fellowship at the Harvard Botanical Museum came without stipend and was self-funded.

This anomalous approach to intellectual endeavours begins to make sense when we consider that for most of his life Wasson was one of America's leading financiers. In many ways this served him well. His considerable wealth meant that he could afford both the time and the expense of travelling the world in pursuit of his hobbies and interests. A mover and shaker used to making the hard sell, he had the charisma, the self-belief and the sheer chutzpah necessary to pursue his research and to assemble the most prestigious team available to him. If he came across an aspect of knowledge beyond his own expertise or understanding, he would have no hesitation in writing to the recognised world expert on the subject – no matter how lofty their academic reputation – where others might have settled for a more junior colleague: this is how he began his collaboration with Heim. As vice-president of J. P. Morgan & Co., he had no difficulty in obtaining the permits required for travel through the unstable hinterlands of Mexico, nor in procuring the use of the Banco Nacional de Mexico's private plane. Wealth, status and charisma allowed him to conduct and publish his research with ease.

The flip side to this businessman-like approach, however, was that he went about his research rather as if he were conducting a major financial deal. He treated knowledge as a commodity, something that could be bought, sold or treated as an investment – quite literally offering money for information – with all the concomitant effects on Sabina and others that that was to have. He invested considerable effort in networking, entertaining and pressing palms, so as to build up alliances with supporters and potential supporters – for example with Dick Schultes, Roger Heim and Albert Hofmann. (At the same time, he was a little too quick to dismiss detractors like Evans-Pritchard out of hand as inherently mycophobic.) And in this vein he rather used his books as 'sweeteners' to persuade recipients of the merits of his arguments. A substantial number of copies of all his books were given away to scholars and prospective reviewers, but these rare and extremely valuable gifts were double-edged for they made gainsaying their forceful author that much harder. Wasson's persuasiveness stemmed less from the clarity or originality of his thinking than from force of personality and his sheer belief in the product being sold. He was a brilliant salesman.

The books were, of course, a shrewd way of sidestepping the tiresome process of peer review, and to a non-academic audience it is the books, as rare and beautiful objects, that have afforded him credibility. Like medieval alchemical grimoires, they came wrapped in an aura of authenticity, and that, in popular terms, was sufficient to grant Wasson's ideas widespread acceptance.

How we judge Wasson, then, ultimately depends on whether or not we choose to regard him as a scholar. If we do, and judge him according to the standards by which he so dearly wanted to be measured, I think we must now come down on him rather harshly. He was guided by an intellectual tradition – cultural evolution – that was thirty years out of date. His subsequently formed ideas concerning the origins of religion and the division of peoples into mycophobes and mycophiles are unfeasible, and should be rejected. His ethnography, though detailed in places, was poorly conducted, sketchy overall, and misrepresented the practices he observed. More worryingly, it had a hugely detrimental effect upon the people he studied. And he surrounded himself with alliances of yes-men rather than submitting his ideas to peer review. Even by the standards of the time, this is a poor record.

If we see him for what he was, however – a Wall Street banker who developed a passionate interest in hallucinogenic fungi – then I think we can absolve him from some of these criticisms, see his genuine achievements in a more favourable light, and draw attention to his positive qualities: his charm, his manners, his skills as a writer and raconteur, and so on. Though intelligent, he was never an academic, and consequently he was insulated from the process of self-criticism and self-reflection that the discipline of anthropology underwent from the 1960s onwards, and which might have alerted him more strongly to the ethical dimensions of his research. There is no reason why a banker should be expected to have behaved any differently.

Gordon Wasson was, then, a literate, well-educated and somewhat eccentric hobbyist who, driven by a personal obsession, made remarkable discoveries and then cleverly broadcast them to the world. He wanted to be remembered as the man who, under his own steam, discovered that the true origins of religion lay in a prehistoric mushroom cult. His claim was misplaced, but what he actually succeeded in doing was all the more extraordinary. Thanks to Wasson, people began looking for and eating psilocybin mushrooms in just about every country of the world, armed for the first time with a favourable way of understanding their effects. He may have failed in his attempt to prove the existence of an ancient mushrooming cult, but in doing so he gave the world a modern one. For this amateur, this upright banker, this conservative member of the establishment, it was a truly remarkable achievement.

Gordon Wasson as he wished to be remembered: the amateur scholar who discovered that the origins of religion lay in an ancient magic mushroom cult. Photo by Jane Reed. Courtesy R. Gordon Wasson Archive, Harvard University, Cambridge, Massachusetts.

PART TWO: AMANITA

Amanita!
You no eat her!
Else you'll end up,
On the slab!

Boris and his Bolshy Balalaika, 'Toadstool Soup'

7

The Fly-Agaric

Oh you little gold-striped fly-agaric, chao-chao-chao,
Such tidings you brought me, chao-chao-chao,
Little patterned-stiped fly-agaric, chao-chao-chao,
Many messages, many words you have, chao-chao-chao.

Khanty Song from western Siberia[1]

Dear children,
Be not deceived
By the red toadstools!

Kobayashi Issa, Japanese haiku (1763–1827)[2]

It is impossible to go on with the story of the magic mushroom without pausing, rewinding somewhat, and considering the cultural history of another altogether different fungus. Psilocybin mushrooms tend to be brown and nondescript and often require an expert eye to distinguish them, but they have a distant and unruly cousin, which is by contrast lurid and eye-catching, and positively demands attention: the notorious fly-agaric, *Amanita muscaria*. With its milk-white gills and stem, and its wart-covered scarlet cap – which can grow to the size of a small plate – the fly-agaric is striking, alluring, strangely beautiful, and yet everything about it seems to scream 'danger'!

No wonder, then, that it has become the iconic, archetypal mushroom, everywhere representing all of its kind and the peculiar qualities with which we imbue them. It adorns the covers of field guides, a guarantee of sales. It appears in children's books, cartoons, and illustrated fairy tales, a shorthand signifier of otherworldliness, enchantment and the uncanny. It is found on good-luck cards, Christmas tree baubles, and kitsch 'collectable' china.[3] And because of its sheer presence, not to mention its genuinely bizarre psychoactive properties, this mushroom has exercised the Western imagination for at least three hundred years in such a way that its story is inseparably intertwined with that of its visually drab, but in other ways more interesting, psilocybin-containing relatives.

Found almost ubiquitously around the world growing in symbiotic association with birch, pine and fir, and often culturally associated with flies, the mushroom has acquired many names. In Germany, Italy, Spain and Russia it is respectively *Fliegenpilz*, *moscario*, *hongo mosquero* and *mukhomor*, the 'fly-mushroom'. In France it is *amanite tue-mouche*, the 'fly-killer'. Most delightfully, in Japan it is *beni-tengu-take*, the 'long-nosed goblin mushroom'.[4] With a few notable exceptions, it has been widely shunned as the worst kind of toadstool. But while its close relatives the Death Cap, *Amanita phalloides*, and the Destroying Angel, *Amanita virosa*, are both deadly poisonous and responsible for the majority of mushroom-induced fatalities, the belief that the fly-agaric is similarly injurious turns out to be misplaced (in its long history, only a few deaths have been attributed to it, and in most cases the victims were already in poor health[5]). Certainly it can cause vomiting, headaches and even unconsciousness when eaten, and is dangerous if taken in excess, but it is, as we shall see, a potent psychoactive in its own right, a fact that has merely added to the air of mystery surrounding it.

Teasing apart fact and fiction in the cultural history of this mushroom is, however, made especially difficult because myths cling to it in greater profusion than the white spots that adorn its crown. Indeed, the ever more implausible stories concocted about the mushroom have become almost a literary genre in their own right. One of the most tenacious in a long line of academic hypotheses and urban myths is that Jesus was an amanita-eater, breaking fly-agaric toadstools rather than the more usually accepted bread and wine at the Last Supper.[6] But it is also widely believed that the shamans of Siberia embarked on their spirit-journeys buoyed up by the mushroom; that they drank each other's urine – not to mention that of their mushroom-loving reindeer – to prolong the mushroom's effects; that the myth of Father Christmas (with his red-and-white coat and flying reindeer) is nothing less than a folk memory of fly-agaric shamanism; that Viking warriors entered their frenzied berserker battle trance with the help of the mushroom; that the mushroom, crushed in milk, provides an effective fly-killer (which explains its ubiquitous association with flies); that the mysterious Soma, praised throughout the ecstatic hymns of the Rig Veda, was the fly-agaric; and last, that secretive and jealously guarded fly-agaric cults lie at the origins of most of the world religions. Such is the body of extraordinary folklore that has built up around the fly-agaric: what truth, if any, resides within these stories? And if the answer is none or very little, what is it about the mushroom that has caused such fevered speculation?

Although its supposed insecticidal properties were recorded by Albertus Magnus in the thirteenth century and repeated three hundred years later by Clusius, the cultural history of the fly-agaric properly begins in the late seventeenth and early eighteenth centuries. It was then that travellers' tales about the peculiar habits of the so-called primitive peoples living to the east of the ever-expanding Russian empire, in Siberia, began to filter into polite Western society. These tales were shocking, and hence grippingly enticing, for they revealed that lying just over the edge of the civilised Western world were people quite other in their lifestyle, habits, demeanour and religion. As disturbing as the revelation that some Siberian tribes appeared to shun washing entirely was the discovery that others delighted in eating a fish stew so gamey that its ingredients had to be not just well matured but half-rotted (the fish were buried in pits and allowed to decompose for several weeks before being cooked).[7] But two stories in particular had a more significant impact upon Western culture: the first concerned that mysterious class of people known as shamans; the second, the Siberians' habit of intoxicating themselves with a mushroom believed throughout Europe to be deadly poisonous.

According to the early travelogues, the many animistic Siberian tribes all supported a few privileged persons, called shamans, who acted as intermediaries between the world of humans and the world of spirits. These strange, otherworldly figures were called upon to heal the sick, to prognosticate the future or divine the whereabouts of lost or stolen property, and to interpret dreams and omens. And this they did in the most dramatic fashion. Donning elaborate costumes of tatters and feathers, they would dance themselves into a trance-like stupor with the aid of the monotonous beat of an enormous drum. Then they would fall to the ground insensible, foaming at the mouth, occasionally muttering strange calls and cries, which they said were the utterances of spirits. In this condition shamans were thought to undertake a flight to the spirit-world, to demand assistance from the spirits if prognostication were required, or their appeasement should they be causing human illness or misfortune. For the audiences, spellbound by these lengthy fire-lit performances and the spirit-battles they depicted,

An eighteenth-century depiction of a Siberian shaman. Nicholaes Witsen, *Nord en Oost Tartarye*, 1705.

the climax came when the shamans, demonstrating the extent of their supernatural powers, would plunge blades through their bodies, or reach into boiling water, all without apparent harm.[8]

For the earliest Western travellers reaching Siberia in the sixteenth century, the shamans were clearly involved in diabolical or heathen practices, which were roundly condemned. But as Enlightenment thinking took hold in the eighteenth, and with it the reification of reason and rationalism, later travellers were more concerned with demonstrating that shamans were fraudulent manipulators of credulity, conjurors possessing nothing more than a knack for ventriloquism and stage magic, charlatans whose mumbo-jumbo contained no curative powers whatsoever. Nonetheless, even though shamans were frequently held up as exotic exhibits from the far side of the world, atavistic curiosities to be ridiculed along with their credulous audiences, something about them resonated with Western sensibilities.

Shamanism enthused and titillated the newly emerging and socially mobile middle classes, to the extent that the average literate person would have been unlikely not to have heard of their strange doings. Shamans were the talk of the coffee shops and salons; they inspired trends in fashion and formed the subject of operas; and they became a legitimate topic for essayists, writers and thinkers of the calibre of Voltaire and Kant. Physicians and philosophers alike wondered about the authenticity of the shaman's trance, and whether there might not be something of philosophical or physiological value contained within it. Everyone but everyone held an opinion about the shamans of Siberia.[9]

As the century progressed, a shift of emphasis occurred in which the shaman ceased to be considered as the embodiment of irrationality and came to be seen as a privileged individual whose creative trance gave him a unique vantage point from which to view the world. Contemporaneously, the emerging Romantic movement produced its own inflated idea of the 'artist', with creativity viewed as something mysterious, God-given, and beyond the reach of ordinary men. Artists were outsiders, set apart by their unique, inherent sensibility or genius, suffering for their art but only so that they could affect, transform and momentarily elevate audiences to their own level through the brilliance of their vision and the creative force that flowed through them. Writers, poets, composers and painters, in other words, had become the shamans of the civilised West.[10]

While shamans were being elevated as Rousseauesque noble-savages, the ordinary Siberian tribespeople were being denigrated as backward and barbaric primitives, in part because of the second influential theme that was emerging from the Siberian travelogues. In 1736 a book by the Swedish colonel Philip von Strahlenberg (1713–1755), his *Historico-Geographical Description of the North and Eastern Parts of Europe and Asia*, was translated into and published in English. It described his ill-fated reconnaissance expedition, in which he had been taken captive and held as a prisoner of war for twelve years. During this time he frequently observed members of the Koryak tribe in the far north-east of Siberia getting 'drunk' upon an unspecified mushroom known only as *mukhomor* in Russian.

Though this was interesting – a curiosity, in fact – it was not particularly shocking. Von Strahlenberg, however, went on to say that as the highly prized mushrooms were in short supply, the poor could not always afford to purchase them from the Russian and other traders who attempted to meet this local demand. Instead, members of the lower social strata would loiter outside the dwellings of the rich whenever a mushroom feast was taking place, waiting for the moment when the guests would come out to relieve themselves. Collecting the urine in wooden bowls, they would then drink it down 'greedily, as having still some Virtue of the mushroom in it, and by this way they [would] also get Drunk'.[11]

It was this simple act, this unusual and, to polite Western sensibilities, revolting, dissolute and transgressive act of drinking another's urine, that ensured that, as the myth of the fly-agaric trickled into the West, it very rapidly made a splash. Its scatological content ensured that it intoxicated the Western popular and literary imagination. As we shall see, just like the fly-agaric's active ingredients, the myth of the mushroom has been endlessly recycled and filtered from one source to another through a chain of retellings. But whereas the Siberians found that mushroom-tainted waters eventually diminished in strength, diluted with every micturition, the reservoir of mushroom myths that their habits generated seems only to have gained in potency with the passage of time.

Von Strahlenberg's account was immediately seized upon and brought to popular attention by the struggling writer, satirist and friend of Dr Johnson, Oliver Goldsmith (1730–1774), best known for his 1766 novel *The Vicar of Wakefield*. In his earlier *Citizen of the World* (1762), Goldsmith used the conceit of an educated Chinese man writing letters home to satirise contemporary London society and the follies and absurdities of the British class system. Telling the tale of the urine-drinkers as if his Chinese hero, Lien Chi Altangi, had witnessed it first hand (but referencing von Strahlenberg in a footnote), Goldsmith imagined what would happen were the custom to be introduced in London. He surmised that 'we might have many a toad-eater in England ready to drink from a wooden bowl on these occasions, and to praise the flavour of his lordship's liquor'.[12] He viciously imagined a willing chain of condescension and deference, flattery and obsequiousness, with a lord eagerly drinking from a priest, a knight from a lord, and a 'simple squire drinking it double distilled from the loins of knighthood!' It was, for Goldsmith, a delightfully carnivalesque image that refused to be banished – 'For my part I shall never for the future hear a great man's flatterers haranguing in his praise, that I shall not fancy I behold the wooden bowl . . .'[13]

The *Citizen* was read widely and it remained in print until the early years of the twentieth century. The view of Siberia that Goldsmith presented was further augmented by more travelogues emerging from the region. The *mukhomor* mushroom was positively identified as the fly-agaric, the urine-drinking was confirmed and the antics of the intoxicated described. In his *Description of Kamchatka Land*, published in Russia in 1755, and translated into English, German and French thereafter, the botanist Stephan Krasheninnikov wrote how he had seen dried fly-agaric mushrooms consumed at feasts, rolled and swallowed whole, or drunk with an infusion of berries. The signs of intoxication were announced by the twitching of limbs and the urge to run and jump and dance, followed by the onset of stupor, sleep and strange hallucinatory dreams.

Russian officers, it seems, had tried the local delicacy with wild and debauched results. One officer believed himself to have been commanded by the spirits of the mushroom to perform an act of bravado by strangling himself. He had to be restrained. Another, perceiving a vision of hell, fell to his knees and publicly confessed all his sins, an act he no doubt regretted for he thereafter became the butt of ridicule. We do not know what he confessed, only that he indelicately revealed things that ought to have stayed secret, but his sins were probably sexual in nature. And one final officer was accustomed to consuming fly-agaric before a long march, for he found that it prevented fatigue. However, overdoing things one day and having 'indulged to the point of delirium',[14] he apparently crushed his own testicles and died.

The extent to which these alarming scare-stories were literally true remains moot, for it seems that the Kamchatka locals managed to avoid such debauched excesses of behaviour. They must be contrasted with the account by the Polish Brigadier Joseph Kopec, who in 1797 claimed to have eaten the mushrooms himself. Travelling through Kamchatka, he fell ill with a fever and was received by the inhabitants of a small settlement. Ailing in a yurt, Kopec was persuaded to eat some fly-agaric by the local missionary, who assured him that it would restore his health. This he did, and falling into a deep sleep he experienced vivid dreams 'of the most attractive gardens where only pleasure and beauty seemed to rule. Flowers of different colours and shapes and odours appeared before my eyes; a group of the most beautiful women dressed in white going to and fro seemed to be occupied with the hospitality of this earthly paradise.'[15] Waking from this blissful slumber to discover that the pleasure gardens were illusory proved traumatic for Kopec, so much so that he eagerly doubled his dose the following evening. This time he was presented with long-forgotten images from his childhood and, more worryingly, disturbing precognitive visions of the future which, he later claimed, came to pass exactly as foreseen.

As we shall see, the mushroom's effects more typically lie midway between the poles of pleasure and pain marked out by these two explorers, Kopec and Krasheninnikov, but nevertheless, to an uncritical Western readership, their stories doubtless added to the belief that fly-agaric intoxication was the exotic, yet abominable, practice of barbaric peoples from realms lying off the edge of the map. However, another of the mushroom's genuinely psychoactive effects was to exert a particular influence on the Western literary imagination of the nineteenth century, and ultimately upon the perception and uptake of psilocybin mushrooms in the twentieth.

One of the common consciousness-altering effects of the fly-agaric observed by Krasheninnikov was that users found their perception of scale to be radically altered, a phenomenon now termed macropsia. 'Some might deem a crack to be as wide as a door,' he wrote, 'and a tub of water as deep as the sea.'[16] This peculiar feature was confirmed by Georg von Langsdorf in his *Remarks Concerning the properties of the Kamchadal Fly-Agaric*, published in Russia 1809. 'The nerves are highly stimulated,' he recorded, 'and in this state the slightest effort of will produces very powerful effects. Consequently if one wants to step over a small stick or a straw, he steps and jumps as though the obstacles were tree trunks.'[17] It was this absurd image of a man so out of his wits, so lost to irrationality, that he could no longer see that a straw was just a straw that seems particularly to have caught the attention of Western writers.

At first the story was only to be found in the pages of the early European mycological literature, a curiosity to flesh out the dry botanical descriptions and always to be accompanied by stern admonitions against the temptation to experiment. It became widely read, however, when an English translation was quoted at length in John Lindley's *Vegetable Kingdom* of 1853, and it turned up again in James

Johnston's popular science exposition *The Chemistry of Common Life* of 1855.[18] It was almost certainly there that a young Mordecai Cubitt Cooke encountered it, for he delighted in the knowledge of the Siberian predilection, realising that it was not so very different a habit from the consumption of alcohol, tobacco and opium at home. The moral disdain with which it was usually treated was, he felt, more than a little hypocritical.

Cooke devoted a chapter of his *Seven Sisters of Sleep* to the fly-agaric in Siberia, relating its recorded effects in unusually positive terms.

> At first it generally produces cheerfulness, afterwards giddiness and drunkenness, ending occasionally in entire loss of consciousness. The natural inclinations of the individual become stimulated. The dancer executes a *pas d'extravagance*, the musical indulge in a song, the chatterer divulges all his secrets, the oratorical delivers himself of a philippic, and the mimic indulges in caricature. Erroneous impressions of size and distance are common occurrences . . . a straw lying in the road becomes a formidable object, to overcome which a leap is taken sufficient to clear a barrel of ale, or the prostrate trunk of a British Oak.[19]

He repeated this description, word for word but with a little less enthusiasm, in his *Plain and Easy Account of the British Fungi* of 1862. This charmingly written guide and early example of popular science writing consisted of hard scientific data and botanical description, overlaid with a genial mix of coloured plates, folk tales and anecdote. Unlike other guides published at the time, it was designed for a mass audience and was well-received and widely read: it probably did more than any other book to circulate the fly-agaric story. But later on in his career, when the young and incautious drugs-relativist had given way to the upright and serious mycologist, Cooke performed an about-face, warning that the 'effects which follow on partaking of this fungus have been recorded somewhat in detail, and resemble intoxication, but with dangerous symptoms which result in death'.[20]

With stories about the perception-altering effects of the fly-agaric circulating freely through Victorian society, it is perhaps unsurprising that they stimulated another more famous Victorian academic and writer for whom optical illusions, paradoxes, and other inversions of logic and proportion provided endless sources of literary inspiration:

the Reverend Charles Dodgson, better known by his nom de plume Lewis Carroll.[21] One of the key moments and an essential plot device in his children's classic *Alice's Adventures in Wonderland* occurs when Alice meets a hookah-smoking caterpillar lying recumbent on a large mushroom. One side, the caterpillar lugubriously tells Alice, will make her grow taller; the other, smaller. By judiciously balancing her diet, Alice is able to control her height and so manoeuvre her way through the bizarre juxtapositions of proportion that bedevil Wonderland.

Since the 1960s, *Alice* has been read by members of the psychedelic underground as a knowing pharmacological odyssey, replete with hidden and not-so-hidden drug references and, not least, hinting that the author had a taste for magic mushrooms. The most celebrated instance of this must be that hymn to LSD, the song *White Rabbit* by San Francisco band Jefferson Airplane, a song that played no small part in broadcasting to the world the news about mushrooms. Here, backed by a classic sixties West Coast psychedelic guitar sound, the banshee-like Grace Slick extols the virtues of the psychedelic experience in a thinly disguised code, and in ever more hysterical tones. 'You've just had some kind of mushroom, and you're mind is moving low . . . Go ask Alice, I think she'll know.' But while this particular reading of *Alice* remains fashionable in psychedelic circles, it now seems highly unlikely to be true.

That there is no record of him using mushrooms or other drugs himself suggests that Carroll took only literary, and not literal, inspiration from the Siberian mushrooming antics. Indeed, there would have been little incentive for him to hide any such habits secretively between the lines of a children's fairy tale. An array of psychoactive drugs, including opium, were freely available in Victorian society, purchasable over the counter at most pharmacists. Consuming these for prophylactic or pleasurable ends carried nothing like the stigma nor the legal penalties that it does today: Conan Doyle, for example, saw no reason not to make Sherlock Holmes an occasional user of cocaine.[22] We do know from his records that Carroll was, while never a valetudinarian, not of particularly robust health. He was an insomniac, took homeopathic remedies for persistent migraines, suffered the occasional epileptic fit, was opposed to smoking tobacco and drank only in moderation. With the best will in the world, he makes an improbable pioneer of bohemian drug-taking.[23]

*

The literary impact of the fly-agaric mushroom, however, continued to grow apace, inspiring other works of fiction that have now been largely forgotten. In 1866 Charles Kingsley (1819–1875), better known for his popular fairy tale *The Water Babies* and for his lovingly detailed prose descriptions of the north Devon coast in *Westward Ho!*, published *Hereward the Wake, Last of the English*. Here, alongside the heroic tales of derring-do – replete with the romantic dalliances and dabblings in sorcery so typical of the Victorian adventure story – Kingsley includes an episode of unwitting fly-agaric intoxication.

Hereward's love interest, the beautiful Torfrida, longs to know both her hero's name and the secret of what lies under his ample beard (a tattoo of a cross, it transpires, marked on his chin in the 'English style'). As luck would have it, her nurse, who had been abducted by pirates years before, hails from Lapland and, like all her kind, is well versed in the secrets of magic. She presses the juice from some 'scarlet toadstools' and slips it into the beer drunk by Hereward's men. They grow 'merry-mad' and with their tongues loosened they reveal to her the required information. Consequently, Hereward returns to find his men 'chattering like monkeys', and terrified by the prospect of stepping over a gutter in the road, which they think to be a vast and terrible gulf. No amount of physical persuasion by Hereward will convince them otherwise, and one by one they fall asleep by the roadside. In the morning all conclude they were bewitched for, Kingsley informs us, 'they knew not – and happily the lower orders, both in England and on the Continent, do not yet know – the potent virtues of that strange fungus, with which the Lapps and Samoieds have, it is said, practised wonders for centuries past'.[24] He makes very clear, in other words, the literary origins of this episode.

More extraordinary still must be the short story by H. G. Wells (1866–1946), *The Purple Pileus*, published towards the end of the nineteenth century in 1897. It concerns the hapless and henpecked Mr Coombes, who is so fed up with being bullied and humiliated by his wife that he decides to commit suicide by eating poisonous mushrooms. Far from killing him, however, his feast of purple, yellow, and red-and-white mushrooms has quite the opposite effect, invigorating and empowering him in a wholly life-transforming manner. The ecstatically intoxicated Coombes returns to his wife, who is so shocked and impressed by the change in her husband that she meekly defers to his authority, remaining respectful and obedient even when the effects of the mushrooms have faded away.

By the end of the nineteenth century the putative power of the fly-agaric to inspire writers and artists seems to have taken on a life of its own, quite independently of the more comprehensive Russian and German ethnographies emerging from Siberia. The general and widespread Victorian obsession with all things to do with fairies had given rise to a series of fairy paintings – most notably by the artists Noël Paton, John Anster Fitzgerald, Richard Doyle and the troubled Richard Dadd, who spent much of his life incarcerated in Bedlam asylum. Mushrooms appeared in many of these paintings, not only to suggest the diminutive size of the fairies who pranced and frolicked amongst them, but to act as signifiers for the eldritch netherworld that fairies were supposed to inhabit. Thus, in Fitzgerald's *Fairies' Banquet* of 1859, a group of fairies dine around a mushroom table, while in a Doyle watercolour, painted between 1870 and 1880, we see *Wood Elves Playing Leapfrog over Red Toadstools*. On the Victorian stage, fairies appeared in everything from pantomimes to Shakespeare, and in the Princess Theatre's 1856 production of *A Midsummer Night's Dream*, to the delight and wonder of the audience, Puck rose up from the stage atop a mechanical mushroom.[25]

Not one of these examples of mushroom Victoriana was actually inspired by a first-hand experience with the fly-agaric: as Kingsley had noted, people had yet to acquire the taste. In the case of Fitzgerald, his otherworldly paintings can most definitely be attributed to an imagination bolstered by that true Victorian staple, opium. By the turn of the century, the Victorian love of fairies had dwindled, and they, along with the mushrooms with which they had become inexorably linked, were banished to the domain of the nursery. The fly-agaric made its way into children's books, children's Christmas and good-luck cards (especially in Germany, the Baltic countries, and Eastern Europe), and other ephemera and knick-knacks. Cartoon fly-agarics littered the forest floor in Walt Disney's *Snow White and the Seven Dwarves* in 1937 and performed a memorable 'Danse Chinois' in the Nutcracker Suite of Disney's *Fantasia*. From there they entered the sickly sweet fairy gardens and enchanted forests imagined by Enid Blyton, and became forever conjoined with that embodiment of suburban kitsch, the garden gnome.

More recently, two prominent writers have rejected this prevailing gnomic representation of the fly-agaric in favour of that presented within the Siberian travelogues and ethnographies. In her darkly brilliant fin de siècle Gothic novel *Nights at the Circus* (1984), the late Angela Carter (1940–1992) describes the inspired meeting of a Western journalist, Jack Walser, driven temporarily mad by a blow to the head, and a Siberian shaman.[26] The shaman, understanding none of Walser's incoherent babbling, nevertheless recognises the madman as one of his own, and so plies him with fly-agaric-infused urine. 'Then his eyes began to spin round and round in his head and to send off sparks like Catherine wheels . . . Walser entered an immediate fugue of hallucinations . . . The hallucinogenic urine put the sluggish motor in his skull into overdrive. "Oh!" he declaimed to the oncoming Siberian dawn. "What a piece of work is man!"'[27]

The celebrated children's author Alan Garner also explored the possibilities of fly-agaric shamanism in his novel *Thursbitch* (2004). One half of the story concerns John 'Jack' Turner, an eighteenth-century packman, who presides over a peculiar and local form of paganism in a remote Cheshire village. Jack eats dried fly-agarics, or 'corbel bread', and shares his wonderfully named 'piddlejuice' with fellow villagers at auspicious times of the year. Just like the Siberian tribespeople, they all fall into a stupor before arising again some hours later to perform their rituals, dancing jerkily all the while. At other times Jack strides out over the hills, red-and-white-spotted kerchief round his neck, high on his corbel bread and communing deeply with the landscape. 'The sound of the brook entered him, and he grew to the stone. He waited. The sun was singing, but not loudly, and the small white clouds rang against each other, soft as Jinney's bells.'[28] Beautiful as this description is, Garner is renowned as much for the dark atavistic themes that underpin his work as for his uncompromisingly laconic prose style. Reading the novel, it is disturbing to discover quite what his fictional villagers are eating and drinking, and quite what they get up to in their piddlejuice rituals. Through his mastery of storytelling, Garner has skilfully revived the power of the fly-agaric to shock.

Both Carter's and Garner's 'Jacks' are especially interesting from a cultural point of view for they are presented, if not tacitly as shamans, then as possessing distinctly shaman-like qualities. This is telling, because unlike many of the early travellers to Siberia, and most of the mycological writers like Cooke, we tend to think now that it was exclusively the Siberian *shamans* who used the fly-agaric. In other words, the two predominant stories to have reached us in the eighteenth century from the far north-east, that of the shaman and that of the intoxicating mushroom, have become commingled within the contemporary imagination. It is time now to return to Siberia to consider what truth there is in this; to see exactly who was using fly-agarics, where they were doing it, and for what ends.

When we think about faraway places, distant in time or distant in space, our tendency is to simplify or reduce them in some way. We paper over the complex textures, cracks and folds of other geographies, climates and cultures with the simple stories, stereotypes and other pleasing patterns of our imagining. Beyond hinting at the broadest of contours, these shiny coverings say very little about what might actually lie beneath them; but they reflect the people who placed them there – that is, *our* hopes, fears, prejudices and fantasies. So thought the late Arab-American scholar Edward Said, who bequeathed to the world the concept of 'orientalism'.[29] The 'Orient', he maintained, does not exist. It is a Western fabrication, a mishmash of second-hand notions, half-truths and projections; harems and hashish on the one hand, fanatics with Kalashnikovs on the other. None of these do justice to the civilisations they purport to describe, and their subtle coercive power merely serves the West's ambition to proclaim moral and intellectual superiority.

The same applies when we think about Siberia and shamanism, choked as they are with the accretions of four hundred years of Western speculation and myth-making. Take the very name, Siberia. The small Khanate of Sibir was just the first of many nations to be overcome by the nascent Russian empire after it crossed the Ural Mountains in its economically driven expansion eastwards. Not knowing otherwise, the Russians projected the name out and beyond, until both it and their empire encompassed almost the entire northern half of the continent. The name obliterates all sense of the sheer scale of the region, its geographical diversity, and the complexity of its ethnic, cultural and religious make-up, reducing it to a simplified, homogenised and digestible whole.

But the region we call Siberia is vast. It is as large as Europe and the USA combined, covering seven time zones and several distinct biogeographical regions: tundra in the north gives way to great swathes of forest or taiga in the centre, to be replaced in turn by the monotonous steppes of the south.[30] The indigenous cultures and nations that forged a living out of these harsh conditions were no less diverse. Hunter-gatherers, agriculturists, pastoralists, reindeer herders or whalers: every tribe was keenly adapted to the particularities of its environment. At the end of the sixteenth century the Russians counted approximately 120 distinct linguistic groupings in Siberia, making up a bewildering political landscape of tribal feuds and rivalries, alliances and misalliances. The names by which we now know those tribes represent early Russian attempts to simplify the situation on the ground, inaccurately lumping together peoples that previously had no common identity: the 'Tatars', say, in the south-west, or the 'Samoyeds' in the tundra of the north and west.[31]

Similarly, when we think about Siberian shamans, it tends to be a singular image that comes to mind, an image that has not substantially altered since the earliest Western travellers encountered these enigmatic characters. Take the following, written by Richard Johnson, an English explorer who travelled to Siberia in the mid sixteenth century, which still captures our Western view of the shaman precisely.

> The priest doth begin to play upon a thing like a great sieve, with a skin on the one end like a drum . . . his face is covered with a piece of shirt of mail, with many small ribs and teeth of fishes, and wild beasts hanging on the same mail. Then he singeth as we use here in England to halloo, whoop, or shout at hounds, and . . . in the end he becometh as it were mad, and falling down as if dead . . . I asked them why he lay so, and they answered me, 'Now doth our god tell him what we shall do, and whither we shall go.'[32]

Western travellers, writers and thinkers, from Richard Johnson right up to Garner and Carter, have all tended to portray Siberian shamans in this stereotyped way, painting them as privileged individuals, set apart from ordinary society, who, while in some kind of a trance, travel to a spirit-world and interact with its denizens for the general benefit of their communities, usually imploring spirits to cure illness or reveal the future. But while it is true to say that most, if not all, of the animistic tribal communities in Siberia supported people, typically men, in the role of what we now call shaman, it would be quite wrong to think that they belonged to a static, monolithic, universally adhered-to religion, or even to a pristine ur-religion as some scholars have maintained. There never was a singular 'shamanism' in Siberia.

For one thing, the term itself was particular only to certain Tungusic-speaking societies. It happened to be the first name encountered by seventeenth-century Russian and Dutch explorers, who transcribed it as *schaman*, a form amenable to the large numbers of German-speaking scholars working in Siberia for the Russian empire. Outside Siberia the name stuck, but by far and away the most common term within it was *kam*, followed closely by *bö*, *enenalan*, *tadibei* and *oyun*. For another, 'shamans' everywhere assumed slightly different roles and undertook different duties, coexisting alongside other religious functionaries, priests and magic-workers. Every 'shaman' expressed an individuality of style through distinctive techniques, costumes, performances and attitudes, and each adhered to a particular world view.[33]

For example, one Khant shaman was recorded as saying that in trance he reached an upperworld by means of a rope dropped down from the heavens. Once there, he rode across the sky in a spirit-canoe, casually brushing aside stars that got in his way. But a *kam* from the Altai mountains described how in *his* trance he mounted a spirit-horse, which carried him far across the steppes to an Iron Mountain littered with the bones of dead shamans. A hole at the top of the mountain granted him access to the underworld where the spirits lived. These two, supposedly similar shamans inhabited quite different universes.[34]

Western scholars have laboured long and hard to find some elusive quality, some defining characteristic, that would capture the essence of the shaman – 'the master of ecstasy', 'the wounded healer', 'the primitive psychoanalyst', 'the madman'[35] – missing the obvious point that shamanism has always been characterised by difference and diversity. The term 'shaman' is therefore a Western, orientalist construct, applied willy-nilly to a range of practices around the world that *we* have deemed to be the same, regardless of how the practitioners see themselves. María Sabina provides us with a good case in point, for while she is typically labelled as a 'shaman', she never thought of herself as such: she called herself a *chota chjine*, a 'wise one' or a 'woman who knows'. Indeed, the term 'shaman' has been so widely applied and so overused that postmodern scholars wonder whether it has any analytical value at all.[36] It is, however, far too late for them to try to close this particular stable door, for the horse has long bolted and the term has near universal currency.

The situation is rendered more complicated by the fact that at the time of the travelogues, Siberian shamanism had already been affected by Buddhist and Islamic influences, and by nearly two hundred years of Russian Orthodox Christian rule. An initial degree of tolerance gradually gave way to hostility and repression from the reign of Catherine the Great (1729–1796) onwards, with the result that shamanism still existed, but was very much in decline. It was, however, virtually eradicated by the persecutions of the Stalinist era. Shamans were denounced as primitive anti-Communists and branded as insane, and their drums and equipment were seized and burned. There are reports that shamans were thrown out of helicopters by the KGB as proof to others that their talk of spirit-flight was nonsense; while the last shaman of the Nanais people was pushed through a hole in the ice of a frozen lake, with a weight tied to his foot.[37]

Shamanism was believed to have been dead in Siberia by the 1980s, but there are signs that it is being revived and, where continuity has been lost, reconstructed. We therefore need to make a distinction between pre- and post-Soviet era shamanism, bearing in mind that the shamanism encountered by the Western explorers in the eighteenth and nineteenth centuries was already one that had been affected by Western and other outside influences.

What, then, can be said about the shamanistic use of the fly-agaric within this complex situation? First, the mushroom seems to have been ignored or overlooked in all but two comparatively small regions of Siberia: in the west around the tributaries of the Ob river and including the Taimyr peninsula, amongst the Khanty, the Ket, the Forest Nenets, the Nganasan and the Mansi peoples; and in the far north-east, roughly to the east of the meridian that passes through the mouth of the Kolyma river, and including the great Kamchatka peninsula, amongst the people known as the Even, the Yukagir, the Itelmen, the Inuit, the Chuvanians, the Chukchi and the Koryak. Russian settlers in the coastal regions of the Kamchatka peninsula were also known to have partaken.

Second, the shamans within these regions who employed the mushroom as a tool to access the spirit-realm seem to have been in the minority: it was far from necessary or essential to being a shaman, and most used repetitive drumming or dance as the means to enter trance. For every shaman that enjoyed prestige for being able safely to consume a 'poisonous' mushroom, there were others who were thought the less of for their reliance upon it. Those shamans who used the mushroom did so to assist themselves in performing the activities of their profession: contacting spirits or the spirits of the dead, interpreting dreams, treating disease, finding names for newborn children, reading the past or the future, or travelling to the different worlds of their cosmologies. An important cross-cultural belief within these areas was that the spirits of the mushroom would appear to the shaman, and then impart to him or her the desired information. The shamans' songs – such as the one that opens this chapter – were thought to be sung 'through them' by the mushroom spirits.

The mushrooms were more widely used *outside* of shamanism, however, and to a number of different ends. Ordinary people sometimes took them to try to catch a glimpse of the spirit-worlds that the shamans ordinarily inhabited.[38] Indigenous bards would inspire and fortify themselves with fly-agaric mushrooms before launching into impassioned recitals of epic heroic sagas, lasting through the night. They would then, apparently, fall to the floor exhausted (a situation somewhat reminiscent of Western rock music!). Others would eat the mushroom in low doses when undertaking hard physical labour, such as hauling boats, carrying heavy loads or haymaking, for they found that it alleviated fatigue.[39]

But the mushroom was used most commonly for the age-old purposes of pleasure and intoxication.[40] Apart from the mushroom's immediate psychophysical effects, much of the enjoyment seems to have stemmed from the fact that intoxicated people behaved strangely and ridiculously (although it was claimed that this was all at the behest of the mushroom spirits whose orders could not be gainsaid). The bemushroomed might leap about, blather uncontrollably, or even stand stock still pretending to be, well, a mushroom.

Consequently, no festive occasion was complete without some mushroom-inspired antic or other, and fly-agaric brews were eagerly quaffed at weddings, at feasts to celebrate a particularly successful hunt, or even just to celebrate a neighbourly visit.[41] The similarities between recreational indigenous mushroom use and Russian consumption of brandy and vodka did not go unnoticed, and while in some areas it was forbidden to combine the two, in others spirits of the distilled kind gradually replaced those indwelling in the mushroom as the favoured means of making a fool of oneself. Nevertheless, overall, the fly-agaric remained in great demand, which explains both why the lower echelons were forced to drink urine, and why Langsdorf found that the mushrooms were valuable enough to be traded for reindeer – a high price indeed.

However the mushrooms were used, an average dose seems to have been three to five mushrooms. As many as eleven might be consumed for the purposes of obtaining visions, but figures higher than that are improbable.[42] Fly-agarics were most commonly eaten dried (a choice that, as we shall see, has a sound biochemical basis), but were also taken fresh, raw, cooked, made into tea or infused with berries, and of course distilled via human kidneys. The phase of development of the mushroom was also important: young, unopened mushrooms were thought stronger and more suitable for helping physical exertion; older, flatter mushrooms were better for visions and intoxication. Again, there is an element of biochemical truth in this claim, so the Siberians clearly knew their mushrooms.

With regard to the situation in post-Soviet Siberia, any attempt to make a detailed assessment is hampered by the sheer lack of published research in this area: there are very few ethnographic reports available in the West. However, during the 1980s the Estonian mycologist Maret Saar found that, within the traditional regions, fly-agaric was still being used in much the same way as before.[43] Saar conducted fieldwork amongst the Khanty in western Siberia, and interviewed an Estonian herder who had lived amongst the Chukchi in the far northeast for twenty years. Furtively and away from view, fly-agaric was being taken recreationally, but also to obtain 'second sight', to prognosticate or divine the cause of illness: that is, it was being used 'shamanistically'. Although Saar's informants claimed that they had never used the mushroom themselves, and that they were relaying stories from their fathers, their knowledge of the mushroom pointed to first-hand experience. The extent to which this represented an unbroken lineage stretching back to the pre-Soviet era, or a self-conscious revival of a long-abandoned practice, has yet to be determined.

More colourful confirmation has come from a less formal field trip to the Kamchatka peninsula, undertaken by a group of American mycologists in the early 1990s.[44] There, aside from an interesting and under-explored mycoflora, they discovered a charismatic woman shaman, the eighty-two-year-old Tatiana Urkachan, who, clothed in a tell-tale red-and-white polka-dot dress, lectured them about the correct use of fly-agaric mushrooms for healing and intoxication. Interestingly, she was adamant that she never ate the mushroom herself, for she was already too powerful a shaman to need it. Nevertheless, she claimed to be able to communicate with the mushroom spirits, and thereby to control someone who was bemushroomed.

The eighty-two-year-old Even shaman Tatiana Urkachan, photographed during an American mycological expedition to the Kamchatka peninsula in 1994. She is seen here giving an impromptu lecture on the correct use of the fly-agaric mushroom for both healing and intoxication, though she claimed that she was too powerful a shaman to need the mushroom herself. Photo by Emanuel Salzman.

The initial meeting with Urkachan, though obviously exciting, did not go well, however, for she greeted the party with the ominous news that they had offended the fly-agaric spirits and that until due recompense was made they would never leave Kamchatka! It transpired that, shortly after they arrived, one of the mycologists had placed tufts of cotton wool upon a red Russula mushroom as a joke to try to fool his fellows, and that this was the cause of the offence.[45] Due obeisance was made to the fly-agaric spirits, and the party returned unscathed yet mystified as to how this shaman knew about a jest performed days before and miles away. Even today, Siberian shamanism retains the power to unsettle Western explorers.

One puzzling question is how these Siberians learned to enjoy the effects of the mushroom, when almost everywhere else in the world it has been shunned as worthless and dangerous. And how did they discover that human urine is as potent as the fungus itself? One theory is that humans learned about the peculiar properties of the mushroom by closely observing the behaviour of reindeer. Just like their human herders, reindeer also become intoxicated after eating fly-agaric, and apparently pleasantly so, for they will actively seek it out. Not only that, but during the winter months, when they survive largely by eating lichen, the deer seem compulsively to crave human urine, presumably because it contains essential minerals lacking from their frugal diet: they will even fight over which of them gets to lap it up from the snow. There are reports of people stepping outside to relieve themselves being almost stampeded by their herds and, indeed, the animals are so attuned to the smell of urine that this has long been employed as a convenient means of rounding up straying animals.

It is not far-fetched to suppose that humans discovered the pleasures and pains of the mushroom after copying their animals and then, watching how the effects of the fly-agaric passed from human to deer, progressed to urine-drinking themselves. It is, however, a modern urban myth that shamans or anyone else drank reindeer urine: an intoxicated deer would be slaughtered and eaten, by which means the effects would be passed on.[46]

This leads us to another of the great fly-agaric stories, the idea that the modern-day figure of Father Christmas, or Santa Claus, is an attenuated folk memory of Siberian fly-agaric shamanism. Here, Santa's red-and-white costume represents the parti-coloured mushroom, and his flight from the north on a reindeer-driven sleigh recalls the shamanic spirit-journey. That shamans would enter yurts via the smoke-hole in the ceiling explains why Santa drops into houses down the chimney, and also why he comes bearing gifts, a symbolic memory of presents from the spirit-world.

This rather delightful story has gained a particular currency amongst contemporary members of the psychedelic counterculture – in exactly the same way that *Alice* did for the sixties generation – for whom the idea of a drug-taking shaman lying behind the innocent Christmas figure so loved by adults and children alike has obvious appeal. It has also gained strength by being pitted against another, less palatable version in which Santa acquired his red-and-white coat courtesy of an advert in the 1930s by the Coca-Cola Company, who were keen to promote their sugary brew by clothing Santa in the distinctive colours of the brand. Which, if either, is true?

The history of Father Christmas is complex, and at times obscure. The personification of Christmas began in the early seventeenth century, when the part of Sir, Lord, or Father Christmas was written into English plays, usually as a comic and carnivalesque figure. Santa Claus, on the other hand, has his origins in the medieval St Nicholas, the patron saint of children. He was brought to the United States by the Dutch, who performed the custom of leaving presents, supposedly from the saint, in the shoes and stockings of children on the eve of his feast day. New Amsterdam was eventually captured by the English, who renamed it New York, and Santa Claus and the customs surrounding him were forgotten. However, Santa was revived in the early nineteenth century by writers such as Washington Irving (1783–1859) and, most importantly, Clement Clark Moore (1779–1863) in his hugely popular poem of 1822, 'A Visit from St Nicholas'. It was Moore, with his vivid imagination, who pretty well single-handedly created the modern image of Santa: a fur-clad magic spirit, flying in his reindeer-driven sleigh, and delivering presents to well-behaved children via the chimney. The newly revived Santa had acquired his distinctive red-and-white garb by the end of the nineteenth century, well before Coca-Cola started to employ him in their advertising: that particular suggestion can be discounted. From there, the image crossed the Atlantic and became easily fused with the English Father Christmas.

The notion that 'Santa' was a shaman, however, has a comparatively recent origin. It was first proposed by Robert Graves in one of his typically throwaway remarks about the supposed history of hallucinogenic fungi.[47] The idea was picked up by the American writer Jonathan Ott – presumably via Gordon Wasson, with whom both men were friends – who mentioned it in his popular book about the narcotic plants of North America in 1976.[48] From there, it was seized upon by the young academic Rogan Taylor, who brought it to much wider attention with a colourful article published in the British *Sunday Times*, in 1980.[49] The story was repeated as fact in the weekly science journal *New Scientist* six years later.[50]

Recently, however, the historian Ronald Hutton has poured cold water on the idea, pointing out its inadequacies.[51] For one thing, Siberian shamans did not travel by sleigh and their various cosmologies never included reindeer spirits. For another, they never wore red-and-white clothes, nor did they physically climb out of the smoke-holes of their yurts while in trance, for their spirit-journeys to upperworlds took place entirely in an otherworldly dimension. Hutton reiterated the fact that fly-agaric use was intermittent amongst shamans, and that Americans only began to be aware of Siberian shamanism towards the end of the nineteenth century, long after Moore's poem was composed. Charming and appealing as the fly-agaric Santa story is, it is unlikely to be true, and we can say with great certainty that, in writing his poem, Moore was drawing not upon any shamanistic folk memory, but upon his particular talent for creative writing.

The story of the shaman, then, the man with a foot in two worlds, is one that has so captivated our armchair-bound imaginations that we in the West have spread it around the globe. Whenever and wherever we have encountered something which, on the surface, appears remotely similar, we have named it shamanism, so often, in fact, that many in the West believe that shamanism is some primal human urge that will always push its way to the surface as the ur-religion, provided the primitive conditions are right. These are the stories that we tell about others which say more about ourselves.

When anthropologists, botanists, missionaries and explorers began pushing their way into Central and South America in the nineteenth and twentieth centuries, what struck them most about the spiritual practices they encountered (and labelled 'shamanism') was that they were almost always accompanied by the use of powerful plant hallucinogens: the idea that shamans everywhere always used drugs became one more thread embroidered into the story. No wonder, then, that when we reflected upon Siberia in the light of these discoveries, we interpreted the data to fit our expectations, to fit the very image of the shaman that we had created.[52]

Throughout the first half of the twentieth century, Western scholars came gradually to the mistaken conclusion that, whether a vital component or a late degeneration of a previously drug-free tradition, the mushroom belonged solely to the domain of the shaman.[53] One work in particular did more than all the others to fix this idea in the popular imagination, a book by none other than Robert Gordon Wasson. Although the main focus of his research was always the nature and extent of mushroom use in Mesoamerica, in the mid 1960s he turned his attention to the Old World, and to the fly-agaric. The book he subsequently had published in 1968, *SOMA: Divine Mushroom of Immortality*, proved to be, arguably, the most influential of his oeuvre, for it opened a completely new chapter in the mushroom's cultural history.

Wasson proposed that the mysterious plant Soma, praised throughout the verses of the ancient Indian text known as the Rig Veda, had in fact been the fly-agaric, and his 'revelation' that there had been a mushroom cult practised by Indo-Europeans, as well as by Siberian shamans, sent shockwaves through academia. The date of its publication, 1968, is significant too, for it was received with open arms by the psychedelic counterculture that was both reaching the zenith of its popularity and eager to find historical self-justification in foundational narratives: Wasson's thesis proved just the thing.

8

Soma

Like impetuous winds, the drinks have lifted me up. Have I not drunk Soma?
In my vastness, I surpassed the sky and this vast earth. Have I not drunk Soma?
I am huge, huge! flying to the cloud. Have I not drunk Soma?

Rig Veda, 10.119, verses 2, 8 and 12[1]

Amongst the foundational Indian Hindu texts (such as the *Upanishads*, the *Bhagavad Gita* and the *Ramayana*) are four bodies of sacrificial hymns known as the Vedas, of which the Rig Veda is the oldest, composed somewhere between 1500 and 1200 BCE. The provenance of this earliest text remains a matter of considerable academic debate. The orthodoxy in the late 1960s, the time when Gordon Wasson was writing about the matter, held that it had been composed not by the indigenous Indus Valley civilisation of the subcontinent but by Aryan invaders from the north. Originating somewhere in Europe, perhaps the Caucasus mountains, the Aryans were thought to have been marauding nomadic warriors who rampaged their way down through the treacherous passes of the Hindu Kush, and into the Indus Valley. There they seized control, subduing and violently imposing their patriarchal religious and cultural values upon the local populace. The Rig Veda, consisting of 1,028 sacrificial hymns, was thought therefore to be a record of Aryan religious belief and practice, albeit one that contained traces of the displaced Indian religion, acquired through syncretic absorption.

If its origins are obscure, then the text of the Rig Veda is even more so, for it raises as many questions as it answers. It is an extraordinary work, with moments of poetic brilliance that still have the power to illuminate and electrify; and yet it is replete with idioms and metaphors that elude modern comprehension. Many deities are honoured throughout – notably the thunderbolt-wielding Indra, and the fire god Agni – but one in particular, seemingly at once a god, a plant and an intoxicating drink, is singled out by the ecstatic and rhapsodic nature of the hymns in his honour: the god Soma. In one hymn the author beseeches Soma to inflame him 'like a fire kindled by friction'; '. . . make us see far,' he implores, 'make us richer, better. For when I am intoxicated with you, Soma, I think myself rich.' Elsewhere the poet cries out, 'We have drunk the Soma; we have become immortal; we have gone to the light; we have found the gods.'[2]

Throughout the nineteenth and early twentieth centuries, Western Indologists tussled with the identity of the god-plant that had given rise to such poetic raptures. A body of texts later than the Rig Veda, the *Brahmanas*, composed around 800 BCE, gave descriptions of plants that could, if necessary, be substituted for Soma in times of need. Eliminating these substitutes, scholars were able to advance a number of other possible plant candidates for Soma based upon the evidence of descriptive passages within the Rig Veda. Soma was, for example, apparently crushed between stones, mixed with milk and filtered through wool, making a materialist reading of the text as a sort of recipe book plausible enough. Candidates included various species of climbing plants in the genera *Ephedra*, *Periploca* and *Sarcostemma*, some of which are mild stimulants; the more psychoactive *Perganum harmala*, or Syrian rue; a fermented alcoholic drink of hops, butter and barley; *Cannabis sativa*, taken in the form of a yoghurt drink, *bhang*; and, unlikely as it seems, the rather innocuous rhubarb, *Rheum palmatum*.[3] Forty-three candidates were advanced in the nineteenth century alone, a figure that rose to over a hundred during the twentieth.[4] No consensus was ever reached, and the matter was deemed insoluble.

Quite what inspired Gordon Wasson to wade into this exegetical quagmire is unclear, but it may have been a conversation with Aldous Huxley that prompted him. The brilliant Huxley had become a founding father of psychedelia through his literary investigations into the effects of mescaline, *The Doors of Perception*, published in 1954, the year before Wasson's momentous encounter with Sabina. Huxley, fascinated by mysticism in all its guises, was wholly familiar with the Rig Veda. In his most famous novel, the distopian *Brave New World* (1932), he borrowed the name 'Soma' for the 'perfect drug', which kept his genetically engineered populace happy, placid and in a state of willing political acquiescence.

On learning about Soma, Wasson, like Huxley, immediately suspected that it must have been a hallucinogen of some sort, and so he set about deploying his research talents to determine exactly which one: if it turned out to be a mushroom, so much the better. Originally entertaining several different candidates,[5] he eventually broke with academic orthodoxy by concluding that Soma had not only been a mushroom, but was none other than the fabled fly-agaric. He announced his revolutionary thesis in 1968 with the publication of another of his lavish books, *SOMA: Divine Mushroom of Immortality*, made more widely available than any of his previous works through the production of a trade paperback three years later.

Displaying all of Wasson's hallmarks – panoramic breadth, radical ideas, sweeping arguments and an infectious rhetorical style – *SOMA* makes for a giddy read. The first part of the book is devoted to his fly-agaric thesis, and the second to an essay by the scholar of Sanskrit, Wendy Doniger O'Flaherty, on the history of the Soma question. Another part reprints in translation every reference to the fly-agaric in the Siberian ethnographies, travelogues, collections of folk tales and other secondary sources that Wasson could find (a veritable goldmine for later English-speaking scholars). The remainder details Wasson's speculations about the use of the parti-coloured mushroom in European, Siberian and Chinese history and prehistory, explores his own experiences after consuming it, and debunks some of the myths surrounding it. Take, for example, the popular belief that the legendary Viking warriors or berserkers, renowned throughout the medieval Norse sagas, entered their battle-frenzies by eating fly-agaric. Wasson convincingly demonstrates that this was the imaginative invention of nineteenth-century Swedish scholars: there is no mention of fly-agaric in the sagas, nor any evidence that the Vikings or berserkers required anything more potent than alcohol to access their bellicose rages.

The book's central concern, however, remained the question of Soma's identity, and Wasson employed all of his writerly skills to draw the reader steadily to the conclusion that Soma could only have been that most eye-catching of mushrooms, the fly-agaric. He began his argument by suggesting that certain lines and passages from the Rig Veda were actually poetic descriptions of the mushroom, and illustrated his point with some rather dramatic photos taken by Allan Richardson. For example, a picture of a heavily spotted mushroom is captured with the line 'With his thousand knobs he conquers mighty renown': the mushroom does indeed appear to be resplendently covered with a thousand knobs. Wasson continued by arguing that no one gifted enough to have composed the verses of the Rig Veda would have wasted such an emphatic metaphor as 'the mainstay of the heavens' upon a mere climbing weed like *Ephedra* or *Sarcostemma*. Only a plant as glorious as the fly-agaric merited this kind of praise, as further photos purported to illustrate. Nor could an alcoholic drink have given rise to such magnificent poetry. '[The] difference in tone between the bibulous verse of the West and the holy rapture of the Soma hymns,' he wrote disparagingly, 'will suffice for those of any literary discrimination or psychological insight.'[6]

If Wasson's argument had rested upon second-guessing the motivation of the Vedic poets, which is notoriously opaque, it would doubtless have made little intellectual impact. Indeed, the discovery that he encouraged photographer Allan Richardson to 'tinker' with the photos to make them more persuasive suggests that he was less than confident about the strength of this opening argument himself. For example, two plates in *SOMA* are supposedly illustrative of the line: 'By day he appears *hári*. By night, silvery white' (*hári* meaning red).[7] The same cluster of fly-agarics are shown by day and by night, the latter indeed glowing with a preternatural silvery light. In fact Richardson simply took the same colour photo, reprinted it in black and white, and applied a special darkroom technique that made the mushrooms appear to glow in accordance with the text. All the other photos were similarly cropped and treated so that they would emphasise the chosen qualities.[8]

These exegetical interpretations, and all this photographic jiggery-pokery, however, were merely the hors d'oeuvres to the main course of Wasson's thesis for, he argued, the Rig Veda offered more meaty clues to the identity of Soma. Throughout the text there is no mention of roots, stems or seeds, all of which point to Soma being a mushroom. Soma is repeatedly said to have come from the mountains, which fits with the notion that the invading Aryans brought the fly-agaric cult with them and that they had to import the mushroom from outside the Indus Valley (where it does not grow). Scarcity would also explain why later substitutes had to be found. The text states that the Soma sacrifice took place over a matter of hours, with Soma crushed to produce a tawny-coloured liquid. This, said Wasson, ruled out an alcoholic beverage, which would take much longer to ferment; and he produced photographic evidence showing that, when pressed, fly-agaric does indeed express a tawny liquid.

The centrepiece of his argument, however, was the fact that Soma was described as existing in two forms, one of which appeared to be human urine. A cryptic verse in the Rig Veda states that 'The swollen men piss the flowing Soma'.[9] This was drawn to Wasson's attention in 1963 by O'Flaherty,[10] his translator, and whereas she was nonplussed by the verse, Wasson immediately spotted a connection with the habits of the indigenous Siberians. Soma had to have been the fly-agaric, for no other known drug had ever been recycled in this manner.[11]

This radical idea gained immediate academic support in influential quarters, and not just from within Wasson's inner circle of collaborators: Roger Heim, Richard Evans Schultes and Albert Hofmann. The imposing French founder of structural anthropology, Claude Lévi-Strauss, was won over, as was the great Cambridge Sinologist Joseph Needham, who followed Wasson by suggesting that knowledge of Soma had spread overland into China, to be employed by the ancient and medieval Taoists in their eternal quest for immortality.[12] The American anthropologist Weston La Barre, whose own ideas resonated with Wasson's, wrote enthusiastically about the latter's methods and conclusions.[13] La Barre believed that Siberian shamanism was the original ur-religion, which had crossed into the Americas with the migrations over the frozen Bering Straits during the last ice age. The widespread use of psychoactive plants in the Americas could easily be explained, he argued, for it had originated in Siberia with the shamanic predilection for fly-agaric.[14]

Not all scholars, however, have been so enthused. A rather persuasive objection to the fly-agaric thesis was raised by two Indian scholars, Santosh Kumar Dash and Sachidananda Padhy.[15] Interested in the prescription and proscription of plants throughout Indian history, these ethnobotanists scrutinised a manuscript known as the Book of Manu, the Manusmruti, or the Manava-Dharmasastra. In this book of laws – attributed to Manu, the 'first man', and believed to record the rules and norms that obtained throughout the first millennium BCE – they uncovered some ardent prohibitions against both the eating of mushrooms and the drinking of urine. A 'twice-born' person (someone belonging to the upper two castes) who eats a mushroom falls down a caste, while anyone touching or drinking urine is expected to perform 'an arduous penance'.[16] Wasson knew about the Book of Manu but rather glossed over it in a later chapter of *SOMA*, concluding that its verses probably didn't apply to the Soma plant. On what basis he reached this supposition is unclear. He may well have been correct, for the Book of Manu is of uncertain provenance and could have been written as late as 100–300 CE, long after the Vedic religion had given way to Hinduism.[17] But then again, these prohibitions could easily have been a legacy from Vedic times, in which case the fly-agaric theory would be seriously weakened. Certainly, they presented a serious obstacle for Wasson's supporters to overcome.

The most trenchant critic of the fly-agaric thesis, however, was the Cambridge Vedic scholar John Brough, who condemned it shortly after its publication. At what was a surprisingly genial dinner at Wasson's home in Connecticut, Brough handed over an advance copy of his clause-by-clause rebuttal of the American's elaborate exegesis. The details of his critique (later printed by the School of Oriental and African Studies in London[18]) are lengthy, complicated and dry, but a few will serve to illustrate the magnitude of his objections. He pointed out, for example, that many of Wasson's arguments, not least the supposed existence of Soma in two forms, were overly sensitive to the vagaries of translation; other equally valid or more faithful renditions of the Rig Veda simply eroded many of Wasson's certainties. Only one hymn in the entire corpus appears to describe the effects of Soma – hymn 10.119, cited at the head of this chapter – and the artifice of its structure, Brough thought, precluded its having been written under Soma's influence. A dramatic monologue of this kind could easily have been written by someone with no personal experience of Soma, and without this hymn there is little else in the text that points to Soma being hallucinogenic. 'Exalted language is expected in liturgical utterances,' he wrote, 'and we can hardly suppose that all of these were drug-induced. The point need not be laboured.'[19] Furthermore, he wondered why a mushroom had to have been put through such an elaborate process of crushing, mixing and filtering, when the priests could simply have eaten it, or chewed its dried remains. And finally, and most damning of all, he pointed out that nowhere does the text state that priests actually *drank* the flowing Soma urine.

Obviously bruised by the encounter, Wasson picked himself up and delivered his own, somewhat bombastic counterblast; but in spite of harnessing all his considerable rhetorical skills, his rather puffed-up *Rejoinder to Professor Brough* added little in the way of new evidence that might have overcome the Cambridge man's objections.[20] One of the major disagreements between the two centred on the question of whether fly-agaric consumption in Siberia had any relevance to the identity of a plant used in northern India, some thousands of miles away. Brough maintained that the sheer geographical distance between the two regions – separated as they are by China and the Himalayas – meant that the one had absolutely no bearing upon the other. The identity of Soma could only be determined on the strength of the internal evidence of the Vedic texts and any local archaeological evidence that emerged.[21]

Wasson disagreed profoundly. Implicit in all his work was the belief – which we encountered in Chapter Two – that hallucinogenically inspired gnosis always transcends linguistic and cultural boundaries and is therefore universal. So if Siberian shamans found ecstasy on fly-agaric mushrooms, then the Vedic priests would have done so as well; and if Siberians learnt to recycle the mushroom's active ingredients, and prolong or deepen the intoxication by drinking one another's urine, then so would the Aryan invaders. For Wasson, Siberia formed the template about which the Soma practices could be reconstituted because fly-agaric gnosis was always and everywhere the same. But wedded as he was to the modernist approach of the comparative method, he failed to recognise that religious experience, whether obtained by eating mushrooms or granted by divine grace, is necessarily constrained by culture, and so historically contingent and particular. There is absolutely no reason why a fly-agaric cult should look the same in India as in Siberia, or anywhere else for that matter. Brough was right: the situation in Siberia was interesting, but could have no cultural relevance to Soma whatsoever.

In any case, the evidence from Siberia that Wasson did bring to bear on the Soma question was altogether too partisan, and too rose-tinted. He exaggerated the extent to which the fly-agaric was used within shamanism, and underplayed the extent to which it was used recreationally (thus contributing to the popular confusion that Siberian shamans always used the fly-agaric). He dismissed any account that claimed that the mushroom produced violent, aggressive or stultified behaviour, for this would have weakened his argument (hence also his eagerness to demolish the berserker theory). But it is hard to imagine why the Vedic priests would have written such ecstatic poems about a mushroom that rendered one so insensible, and distorted perception to such a degree, when the majority of Siberian shamans had the discernment to avoid it altogether.

The supposed link between Siberia and Soma mattered so much to Wasson because it formed a crucial piece of evidence in support of his grand theory that the origins of all religions lay in a prehistoric mushroom cult. He felt that his case in Mexico was watertight. Clearly, the use of *teonanacatl* by the Aztecs and the presence of the prehistoric mushroom stones pointed to a long tradition of psilocybin mushroom use in Mesoamerica; it was not so great a leap of faith to imagine this as ubiquitous throughout Central America. In the Old World, however, the evidence for a mushroom cult was not as forthcoming – less charitably put, it was pretty well absent – and Wasson admitted that he had been unable to find 'an umbilical cord linking the Mexican and Siberian cults'.[22] But he knew that this situation would be dramatically overturned if his identification of Soma proved correct.

If, as the orthodoxy of the time maintained, the European Aryans had invaded the subcontinent down from the Caucasus mountains, bringing their Soma religion with them, the use of the fly-agaric, far from being restricted to Siberia, had actually originated within, and been widespread across, Eurasia. Furthermore, if Weston La Barre's thesis that psychedelic shamanism had had its genesis in Eurasia, and had been introduced into the Americas with the eastward migrations that took place at the end of the last ice age, was correct, then Wasson's European mushroom cult acquired a global significance. Spirituality everywhere, from European paganism and the major world religions to First Nation spiritualities in the Americas, could be said to have begun when the religious impulse was triggered in some prehistoric ancestors, munching their way through the magic fairy-tale mushroom. Soma, then, slotted perfectly into this picture, forming the missing piece of the jigsaw that Wasson had been looking for to link the Old and New Worlds.

Let us imagine, just for a moment, that the fly-agaric cult had been carried, as Wasson supposed, from Europe to India, across to Siberia, from there over the Bering Straits to Alaska, and finally down into the Americas. We would surely expect to find evidence for this extraordinary religion scattered along the way, especially in North America. But this is not the case. Evidence for the supposed fly-agaric cult is notable only by its absence, the mushroom having been by and large shunned by Siberian and American indigenes alike, and even by the *curanderos* of psilocybin mushroom-loving Mexico.

There was a flurry of excitement in the late 1970s when Wasson announced that he had, in fact, found evidence of a Native American 'shaman' who used the scarlet mushroom. In 1978, at the third of a series of conferences on psychoactive drugs, organised by Jonathan Ott and others, Wasson introduced a charismatic Ojibway herbalist (and university-trained ethnobotanist) named Keewaydinoquay 'Kee' Peschel.[23] Leading a solitary existence on a remote island in Lake Michigan (at least during the summer months) Kee claimed to have been initiated into a shamanistic tradition of using fly-agaric – known as the *miskwedo*, the 'red-topped mushroom dependent on trees' – which she consumed four or five times a year. To a rapt audience, she retold the 'Legend of *Miskwedo*', explaining how the mushroom had come to be used by the Ojibway people, and how they discovered that its sacred properties could be passed on through urine.

No wonder Wasson was excited, for this, if true, would be just the sort of evidence he needed to link Siberia with North America: he optimistically predicted that other cases of indigenous fly-agaric shamanism would soon come to light. As ever, though, he read far too much into a single uncorroborated piece of evidence. Kee herself admitted that many of the Ojibway were vehemently opposed to the use of the fly-agaric, and that her original teacher forbade her to experiment with it. Other versions of the *miskwedo* story told by the Ojibway differ substantially from Kee's, to the extent that they amount to prohibitions against eating all mushrooms![24]

Further doubt must be cast upon Kee's evidence because of the little-known fact that she and Wasson were lovers. Theirs was a complicated and stormy relationship, and it is not inconceivable that Kee embellished or changed her story to fit with Wasson's expectations as a means of prolonging his less than enthusiastic affections.[25] At best, her testimony demonstrates the existence of a localised fly-agaric-based spirituality, quite possibly of recent origin; at worst, it may have been fabricated. Without further corroborative evidence, the extent of this uncertain fly-agaric use in indigenous North America must be regarded as occasional, sporadic and localised.

As for the foundation stone of Wasson's grand argument, the supposition that the Aryans brought the fly-agaric cult with them, scholars have been unable to agree on exactly who they were or exactly where in Europe they were supposed to have originated. Indeed, the notion that Vedic culture was the product of an Aryan invasion has recently been thrown into question. The lack of any archaeological evidence that would suggest a violent confrontation had taken place with the massive and sophisticated Indus Valley civilisation has led most Indologists to the conclusion that the Aryan immigration must have been an entirely peaceful one, or simply a gradual process of cultural, not tribal, diffusion. Other Indian scholars have even questioned the idea of a cultural influx, arguing that the Aryans *were* the people of the Indus Valley civilisation. They have, quite correctly, pointed out that the invasion theory, first mooted in the nineteenth century by white Europeans whose countries were bent on imperial expansion, was both a product of and supportive of the Western imperialist agenda.[26]

To be fair, much of this is recent scholarship that has occurred since Wasson's death. Nevertheless, he failed to spot the rather glaring circularity in his own argument. 'How astonishing,' he wrote, 'that we can draw parallels with [the use of Soma and] the fly-agaric cult in Siberia.'[27] But having used Siberia to argue his case for the identity of Soma, he then suggested that the Soma cult had, in fact, spread from India to Siberia. In this light, it is Wasson's tortuous connections that appear astonishing.

Like the other great proponents of the comparative method in anthropology (Edward Tylor and James Frazer) and the study of religions (Max Müller and Mircea Eliade), Wasson was an armchair scholar. He based his theory not upon the messy and often conflicting evidence garnered in the field, but upon the evidence of linguistics and texts gathered at one remove, in the library. The picture he generated from afar of a supposed, ancient, ur-religion was altogether too simplistic, too static, too monolithic. He underestimated the speed with which religion and culture change, adapt and evolve, and their sheer complexity and particularity. Like his illustrious forebears, he was too eager to shoehorn the data to fit his ready-formed thesis, when other more parsimonious, but less exotic, explanations would have sufficed. The idea, then, of a fly-agaric-based gnostic religion spreading ever eastwards is too problematic to be accepted. As for Soma, until Brough's objections are met the status of the fly-agaric theory must remain highly questionable. There exists no shortage of plausible alternative candidate plants, for if Wasson had thought to end the Soma guessing game, to stop the merry-go-round once and for all, he was very much mistaken. The debate stimulated by his radical fly-agaric thesis simply spun it round with an extra impetus, and before long a queue of eager pundits were ready to line up with their own theories.

Brough had questioned why the poetically enriched text of the Rig Veda would contain botanical information about the identification of the Soma plant, when it was essentially a collection of hymns: after all, the Christian liturgy does not contain botanical descriptions of the vine. In 1989 two scholars, David Flattery and Martin Schwartz, published their *Haoma and Harmaline*, in which they agreed that the text of the Rig Veda was so vague on botanical details as to be taxonomically worthless.

Instead, they turned their attention to Iran, which had also been invaded by the Aryans, and to the ancient Iranian text known as the Avesta, which is linguistically related to the Rig Veda. Less contaminated by poesy than its Indian counterpart, the Avesta more clearly pointed to another candidate for Soma, or haoma (pronounced 'horma') as it was known in Iran: the Syrian rue, *Perganum harmala*.[28] Its active ingredient is harmaline, which also turns out to be an important chemical ingredient of the psychoactive brew drunk throughout the Amazon basin, known as ayahuasca. Harmaline is a mono-amine oxidase (MAO) inhibitor, which means that it switches off the enzymes in the stomach that ordinarily block the absorption of harmful plant alkaloids. In other words, the mixing of harmaline into the ayahuasca brew is what allows the other principal ingredient, the more potent hallucinogen DMT, to be absorbed orally. Harmaline is, however, a psychoactive in its own right and, argued Flattery and Schwartz, could have prompted the rapturous adulation of the Avesta and the Rig Veda.

Then, in 1992, the classicist Mott Greene published a book which, while agreeing with Flattery and Schwartz that the descriptions of the Soma plant in the Rig Veda were so distorted by poetic fancy and hyperbole as to be botanically unusable, rejected their haoma thesis.[29] Greene saw the *preparation* of Soma, being practical and instructional, as free from exaggeration, and he used this to pinpoint Soma's identity. According to the Rig Veda, Soma was crushed between stones, mixed with milk and filtered through wool. For Greene, this elaborate process would only be necessary if the species concerned required special preparation – if, for example, it were poisonous – and this led him to the conclusion that Soma could only have been the ergot fungus, *Claviceps purpurea*. The preparation process was, he argued, the essential means by which the psychoactive ingredients, dissolved into the fat of the milk, were separated off from the insoluble toxic alkaloids within ergot.[30] It was the means, in other words, by which ergot could be safely consumed.

Conversely, mushroom enthusiast and champion of plant psychedelics Terence McKenna thought that Soma could only have been a psilocybin mushroom, most probably the dung-loving *Psilocybe cubensis*.[31] McKenna centralised neither the description nor the preparation but the *experience* of Soma, as described within the pages of the Rig Veda. His own disappointing personal experiments with fly-agaric had failed to produce the ecstasies promised by the Rig Veda, whereas psilocybin mushrooms had repeatedly fulfilled all his expectations in this regard. He bolstered his argument with some of Wasson's throwaway remarks, in which the latter had wondered whether mushrooms other than the fly-agaric might have been used by the lower castes in India.[32] Both men were convinced that psilocybin mushrooms would be discovered in India (as, indeed, they have been), and that their presence there could explain the sanctity of cattle within Hinduism: by providing the dung upon which sacred mushrooms could grow, cows had become sacred by association.

These, then, are just a few of the dazzling array of suggestions that have been put forward by scholars and amateurs alike, their popularity rising and falling with the horses of the Soma carousel. But the wheels of this ever-spinning ride turn upon two assumptions: that Soma has to have been hallucinogenic, and that the Rig Veda can be read, quite literally, as a recipe book.

It must be remembered that Soma could have carried great symbolic force without actually having been hallucinogenic. As we have seen with the Greek rites of Eleusis, this orientalist assumption, made from afar, may say more about *us* than about the supposed practices of the ancients; about the fact that *we* cannot countenance religious epiphany *without* the use of psychoactives.[33] The following passage, written by an unknown Chinese poet, should serve as a salutary reminder that ecstasies need not be hallucinogenically inspired:

> The first cup moistens my lips and throat, the second breaks my loneliness, the third cup searches my barren entrails but to find

> therein some five thousand volumes of old ideographs. The fourth cup raises a slight perspiration – all the wrong of life passes away through my pores. At the fifth cup I am purified, the sixth cup calls me to the realm of the immortals. The seventh cup – ah, but I could take no more . . . Where is Elysium?[34]

The drug in question here is none other than that innocuous staple of daily life, tea. So while the text of the Rig Veda may very well be the psychedelic cookbook we dearly want it to be, it could equally well be so couched in the idioms and metaphors of the time and culture in which it was written – not to mention poetic hyperbole – that a materialist and literalist reading is both inappropriate and wrong. There are no sure grounds upon which to decide the correct reading.

Soma could have been hallucinogenic, or a stimulant of the mind, or inert but imbued with symbolic force. It could have been a single plant, or a combination of different plants that varied through time according to availability. It could have been extremely rare, a plant as yet unknown to Western science, or even driven to extinction through the voracious appetites of the Vedic priests. Then again, it could have been a metaphor for something else entirely. In the absence of decisive archaeological or textual evidence, there is no way of evaluating the relative merits of the various possibilities to determine which, if any, is the right answer. The plethora of plausible candidates merely demonstrates the elasticity of interpretation that a text like the Rig Veda affords, or indeed demands, and should alert prospective scholars to the improbability of their putting the matter beyond doubt. Like a magic mirror, the Rig Veda has allowed people to see, within its opaque signs and poetic metaphors, whatever it was that they set out to find. The hard truth is that, while the search for Soma is diverting, exciting even, it is ultimately futile.

Outside academia, such abstruse debates have done little to shake Wasson's fly-agaric thesis from its position of near-universal acceptance by mushroom enthusiasts and other psychedelic subcultures around the globe. It is difficult now, nearly forty years later, to appreciate quite what an impact Wasson's radical thesis had when first it appeared.[35] The timing was everything. *SOMA* was published at the height of psychedelia, a year after the summer of love, and just as the wave of inevitable moral backlash was starting to break. LSD, the name if not the drug itself, was on everyone's lips, and by this stage many people – not just hippies, but scientists, academics, intellectuals, musicians, artists and other assorted members of the great and the good – had made the illicit journey into the psychedelic otherworld and returned with shining eyes. The British and American governments, meanwhile, were doing everything they could to stamp out what they feared would be a cataclysmic pandemic of people turning on, tuning in and then dropping out. All this amounted to fertile soil for an idea that proclaimed that European culture and civilisation were actually rooted in psychedelia, steeped in the juice of a red-and-white mushroom familiar to everyone from the accoutrements of childhood. Wasson's idea caught the zeitgeist perfectly. Forty years later, it is now so widely and dearly held that Soma was the fly-agaric that only a major paradigm shift will unseat it from popular belief.

Interestingly, then, Wasson's interpretation of the Rig Veda had the unintentional effect of transforming it into a holy, foundational text for the psychedelic movement. Ironically, the baby-boom generation, who took psychedelics with such glee in the 1960s, and who were so disparaging of received wisdom that they questioned and cast aside scientific and religious orthodoxy alike, still felt the need to base their beliefs and practices upon holy writ. Distance of space and time rendered the Rig Veda quite amenable to the purpose: it was elastic enough for a psychedelic reading to be plausible, and not so far removed as to be unintelligible or irrelevant. Wasson's speculations, however, triggered a spate of other, less cautious minds to wonder whether holy scriptures closer to home might contain veiled references to this supposed magic mushroom cult. Could it be, they wondered, that Judaism and Christianity were actually founded upon the psychedelic fairy-tale mushroom? Crazy and far-fetched as these conspiracy-laden theories might sound, one of them was to scandalise the Anglican Church.

9

Chemistry and Conspiracy

If the great religious systems of India had their beginnings in the cups of the Soma priests, it no longer seemed unreasonable to assume that their Hebrew counterparts and contemporaries may have been doing the same thing . . .

Clark Heinrich[1]

It seems when you start to look, there is mushroom imagery everywhere . . .

James Arthur[2]

At any moment in history, at any particular time, certain ideas or guiding principles can be said to be hastening culture along its way, steering it in a particular direction. We talk of the zeitgeist, the spirit of the times, of surfing the cultural wave. While many are content to swim with the tide, a few taste other possibilities in the water, sensing alternative destinations. We remember and praise the free-swimmers who have changed the way we view ourselves and the world by tugging us in wholly new directions. But history is unforgiving, and forgets or takes little notice of the small minority who, peering into the eddies and pools of culture, sense that all is not as it seems, that there are stronger and more sinister forces at work within its murky depths. These are the conspiracy theorists, the amateur (and occasionally scholarly) sleuths who reach startling conclusions about the way the world actually 'is' on the basis of hitherto unappreciated connections.

The second half of the twentieth century was a plentiful time for conspiracy theories, with the publication of a seemingly endless series of unashamedly popular books on the theme of 'unexplained mysteries'. Speculations about Atlantis, yetis, UFOs, Stonehenge, ley lines, the pyramids and secret Bible codes rubbed shoulders with mildly paranoid conspiracy theories about freemasonry, the Illuminati and other imagined secret rulers of the world. Erich von Daniken's bestseller *The Chariots of the Gods* epitomised this in the 1960s (humanity had been visited by ancient astronauts), just as Dan Brown's publishing phenomenon *The Da Vinci Code* does today (Jesus actually married Mary Magdalene, their children secretly control the world, and Leonardo da Vinci knew about it). But retrospectively, Wasson's book *SOMA* can be seen as a foundational text in what was to become a sub-genre of this literary phenomenon: works that purported to reveal how the world's religions were actually founded upon secret fly-agaric cults.

SOMA was undoubtedly, for all its defects, the most scholarly and serious-minded of these, but it was not quite the first. It was preceded by an extraordinary text that drew at least some of its inspiration from Wasson's earlier discoveries in Mexico: Andrija Puharich's *The Sacred Mushroom: Key to the Door of Eternity*, published in the late 1950s. Puharich (1918–1995) was born in the US, the son of poor Yugoslav immigrants, but trained as a physician, graduating from Northwestern University Medical School in 1947. He chose, however, to devote most of his life not to medicine but to his first love, parapsychology. In *The Sacred Mushroom* he related the story of how, while researching extra-sensory perception (ESP), he met the talented Dutch sculptor Harry Stone, who was also a medium. Stone possessed the ability to enter deep trances in which he channelled the spirit of an ancient Egyptian priest, Ra Ho Tep, who had lived around 4,500 years ago. Enthused in this manner, Stone acquired an ability to speak in the ancient Egyptian tongue and, even more remarkably, to write in hieroglyphs. And when his utterances were professionally translated, Puharich found that they described the ancient Egyptian ritual use of the fly-agaric.

Invigorated by this discovery, and the feeling that greater powers were at work, Puharich found himself rather mysteriously drawn to where the mushrooms, previously unrecorded in Maine, were growing in the local woods. When a specimen was picked and offered to Stone in one of his characteristic trances, the medium proceeded to make ritual gestures with the mushroom before ceremonially rubbing it upon his head and tongue. In the ensuing chaos, with the bemushroomed Stone drunkenly reeling about the room, Puharich managed to test Stone's psychic ability for ESP, and found, to his delight, that it was significantly improved.

The Sacred Mushroom was widely read on both sides of the Atlantic, especially during the 1970s, and many were convinced by Puharich's assertion that the ancient Egyptians had used fly-agaric mushrooms. However, in order to believe this remarkable story, the reader had to make several staggering leaps of faith: that the mushroom would have been available in Egypt, a famously arid region where the mushroom's host-tree species are absent; that a priest by the name of Ra Ho Tep had really existed; that communication from beyond the grave and ESP were possible; that the mushroom really could awaken psychic abilities; and that Stone was not making the whole thing up.

There are, perhaps needless to say, some rather glaring anomalies in Puharich's account. For one thing, in the above scenario, Stone never actually consumed the mushroom but merely rubbed it briefly on his tongue. This would not have allowed significant quantities of the mushroom's active ingredients to enter his bloodstream (the Siberians knew that the mushroom had at least to be chewed), and the fact that the effects came on almost immediately suggests that his reeling intoxication was, if not pretence, then certainly psychosomatic.

For another, Puharich claimed to have undertaken a bioassay of the fly-agaric to determine its active ingredients, but of the three he 'discovered' – muscarine, atropine and bufotenine – only the first is actually present in the mushroom, and in such small amounts as to play an insignificant role in its psychopharmacology. As Jonathan Ott has noted, Puharich was either an inept chemist or a fraud. Most probably he fabricated the analysis and drew his conclusions from the incomplete scientific knowledge of the mushroom at the time.[3] Puharich also claimed to have cultivated the mushroom, a feat that has yet to be achieved owing to the fact that the fungus only grows in a symbiotic or mycorrhizal relationship with certain trees, usually birch, pine and fir.

To be fair, Puharich's interest in telepathy did not mark him out as particularly exceptional for the time. Cold War anxieties led to a flurry of parapsychological research in the late 1950s, with both the US Army and the Westinghouse Electric company seriously pursuing the military potential of ESP.[4] Even Aldous Huxley and Gordon Wasson were moved to participate in Puharich's experiments (though Wasson eventually gave them short shrift).[5] But his methods were, even by the standards of the day, flaky. (According to his second wife, he developed a penchant for hosting mushroom bacchanals during the 1960s – where couples might ostentatiously and volubly make love to the accompaniment of operatic arias sung by appreciative onlookers – justifying them as legitimate scientific research.[6]) The episode with Stone shows that Puharich was as willing to let himself be deceived as he was to deceive his readership in his quest to 'prove' the existence of ESP.

Puharich's *The Sacred Mushroom* and Wasson's *SOMA* were followed by a plethora of academic, and less-than-academic, fly-agaric conspiracy theories, all of which shared two defining features. First, they assumed that anything resembling, however remotely, a mushroom in art, mythology, literature or material remains – that is, anything round, domed or umbrella-shaped, coloured red and white, or for that matter simply unusual enough to have caught their attention – must indeed have been a mushroom and, moreover, 'evidence' of a secret fly-agaric cult. Second, they concluded that absence of any corroborative evidence – plant names or botanical description, accounts of accompanying rituals, modes of preparation, pharmacological effects – simply indicated how sacred, and how secret, the mushroom must have been. In the topsy-turvy world of the conspiracy theorist, absence almost always indicates presence, and a negative is frequently a positive.

For example, Carl Ruck, a classicist at Boston University, was so fired up after reading Wasson that he embarked on a quest to uncover hitherto unappreciated references to the fly-agaric in ancient Greek literature. Thus, he 'found' the fly-agaric in Euripides' play *The Bacchae* (the intoxicated revels of the god Dionysus' devotees supposedly followed fly-agaric feasting); the story of Jason and the Argonauts (the Golden Fleece was actually a mushroom, no less); and the trials of Hercules (as also were those suspicious 'apples' of the Hesperides).[7]

The writer Scott Hajicek-Dobberstein attempted to do the same for Buddhism, principally through his radical interpretation of a collection of Buddhist hagiographies, *The Stories of the Eighty-Four Siddhas*, which were translated from Sanskrit to Tibetan some time in the late eleventh or early twelfth century. In one story, the siddha (roughly, 'saint') Karnaripa is forbidden by his teacher to eat any tasty or delicious alms, and is commanded to accept only as much food as he can place on the point of a needle. The cunning Karnaripa manages to bypass this austerity by balancing a large pancake, laden with sweetmeats, on his pin, much to the annoyance of his teacher. By now the supposed resemblance of this feast to the shape of the fly-agaric, proposed in all seriousness by Hajicek-Dobberstein, should come as no surprise.[8]

But the most influential of all these conspiracies was surely the best-selling *The Sacred Mushroom and the Cross*, written by the British scholar John Allegro (1923–1988) and published in 1970. This sensational book rocked both the British intelligentsia and the Church of England, and triggered a moral outcry within the press by alleging that Christianity was a sham, nothing less than the remains of a fertility cult centred on the consumption of the real sacrament, the fly-agaric.

Allegro contended that ancient humanity, beholden to the awesome powers of nature, had been obsessed with the notion of fertility, both of humans in terms of reproductive success, and of the land in terms of generating successful harvests. Observing how the rain appeared to impregnate the ground, primitive humans everywhere had conceived a sky god – known sometimes as Zeus or Jupiter or Baal or Jehovah – whose holy ejaculations (the rain) maintained the fecundity of the earth goddess, as revealed by the growth of plants. Some plants, however, were thought to contain more of this holy essence, for when eaten they elevated Man to the level of God, and so enabled him to shape his own destiny. The fly-agaric was one such plant, and had been used throughout the Middle East in biblical times; but the knowledge was jealously guarded, lest the secret fall into profane hands.

When Jerusalem was sacked by the Romans in 70 CE, Allegro's narrative continues, this cult was hurriedly forced to conceal its practices, which it did by coding them into a set of stories about a figure called Jesus: initiates would understand them metaphorically, but profane outsiders would wrongly take them as literal descriptions of the man, and his teachings and life. Tragically for the true believers, the actual meaning of the New Testament became lost and the stories were everywhere taken at face value. The entire edifice of the Christian religion was a terrible mistake, the original atavistic fertility cult it stood for forgotten and buried like some mausoleum in the desert sands. That is, until Allegro uncovered it.

Allegro reached this conclusion on the basis of philology, the study of language in its historical and comparative aspects. He argued that ancient Sumerian had formed a linguistic bridge between the Indo-European and Semitic language groups. To decode the Hebrew Bible, one simply had to find the cognate Sumerian word and translate it, and if one took the trouble to do so, one would find that the Bible was replete with references to the fly-agaric mushroom. Take 'John the Baptist', for example. The name John, according to Allegro, comes from the Sumerian GAN-NU, meaning a red dye; while Baptist is from TAB BA-R/LI, meaning mushroom: the fly-agaric. Jesus' supposedly miraculous birth recalled the mysterious and sudden appearance of mushrooms, while his name came from the Sumerian SUM: 'the semen that saves'. 'Christ-crucified' actually meant 'semen-anointed, erect mushroom'.

Had this been proposed by anyone else, it would doubtless have been swiftly discarded as moonshine. But Allegro was a scholar, a lecturer in Old Testament Studies at Manchester University, a fact that gave the book authority in the eyes of the general reading public, unversed as its members were in the finer details of philology. The Church simply could not afford to stand by and ignore this unexpected threat to its integrity. Prior to publication, the book was brought to nationwide attention through its serialisation in the tabloid *Sunday Mirror*, where a drawing of Christ nailed to a mushroom was accompanied by the lurid and arresting headline: 'Worship by orgy turned these women into witches – a startling theory that Christianity is a hoax based on a sex-drug cult'.[9]

A group of outraged clergymen made an attempt to stop the book reaching publication, with an appeal to the Director of Public Prosecutions.[10] They failed, but the publisher, Hodder & Stoughton, was forced to issue a humiliating public statement apologising for the offence caused by the book, adding that they were shortly to publish a scholarly rebuttal written by the Reverend John King (a shrewd move this: it both absolved them of responsibility, and allowed them to cash in on the scandal twice over). There were even mutterings that Allegro might be tried under Britain's creaking blasphemy laws, but nothing came of this.[11]

Almost as soon as it hit the bookshops, the backlash began. Critics in all the broadsheet newspapers berated it with scathing reviews. The Dean of Christ Church College, Oxford, Dr Henry Chadwick, wrote in the *Daily Telegraph* that Allegro's book 'reads like a Semitic philologist's erotic nightmare after consuming a highly indigestible meal of hallucinogenic fungi', before branding it 'a luxuriant farrago of nonsense'.[12] His opinion was not untypical. A group of scholars from Oxbridge and London University, led by the Emeritus Professor of Semitic Philology at Oxford, Sir Godfrey Driver, wrote to *The Times* to inform the editor that Allegro's book contained nothing of scholarly value, and was 'an essay in fantasy rather than philology'.[13]

John King's patient and lengthy rebuttal of the book eventually appeared, and while his belief in the literal truth of the orthodox Christian message may now jar with some contemporary sensibilities, his painstaking and elegant demolition of Allegro's edifice remains essential reading for anyone still convinced of the role of the fly-agaric mushroom in world religions. He pointed out that the Sumerian link between the Indo-European and Semitic language groups, the foundation of Allegro's thesis, was far from proven, and that many of the supposed Sumerian root words were Allegro's own hypothetical fabrications. He noted that the fly-agaric and its host-tree species are entirely absent from the flora of the Middle East. And he pointed out the absurdity of the notion that such a revolutionary idea could have been kept hidden so effectively for so long: if so, it would make it the best-kept secret in the world. 'If a man under the influence of, or out of devotion to, the fly-agaric could dream up what we call the gospels,' he wrote despairingly, 'then we need more men under the influence of the fungus.'[14]

Reading Allegro now, some thirty-five years after the furore, it is hard to see why the book was ever taken seriously enough to merit such a considered, high-level rebuttal. Certainly, the grotesque and pornographic image of Christianity painted by the book would still be offensive to many – hardly a page goes by without some reference to giant or erect penises, ejaculation and torrents of fertilising semen raining down onto the soft and vulvic earth – but the book reads now, not as a cogent academic thesis, but as the outpourings of a troubled mind.

And so it would have been seen then, had not the sensation caused by its serialisation meant that it became an instant best-seller. By luck or judgement, and like Wasson's *SOMA* beforehand, it precisely captured the zeitgeist, the post-sixties preoccupation with sex and drugs, and the growing distrust of scientific and religious orthodoxy. The Church worried, rightly perhaps, that the book would presage an irreversible decline in church attendance, Christian morals, and Christian membership. Wrongly, though, they focussed on the most lurid symptom of that decline and not its cause, which was the Anglican Church's struggle to maintain its relevance amidst rapidly changing social attitudes. As King noted, 'to a large extent . . . we orthodox Christians have only ourselves to blame for making a fertility religion seem desirable'.[15]

Then, as now, the book was taken at face value, as an attempt to reveal the true origins of Christianity; but on reading the book and its far-fetched conclusions, the idea that Allegro actually believed what he was writing seems more and more implausible. Here was no hippy apologist looking to justify a private penchant for LSD, but a respected and serious-minded academic. Earlier, he had been singled out as exceptionally talented when he was elected to be the sole British member of the team sent to examine the newly discovered Dead Sea Scrolls (found in 1947). One repeatedly wonders what possessed him to write it.

It has been suggested that his motivation was financial, and that the book was a cynical attempt to cash in on Wasson's *SOMA*, published two years earlier.[16] It is certainly true that Allegro made a lot of money out of the book and was apparently paid £30,000 in total for the serialisation rights,[17] but the few references to Wasson in the book make it probable that Allegro was broadly ignorant of the man and his work. If he was trying to piggyback on anyone's publishing success, it was surely Erich von Daniken's. I think there is another, more convincing, explanation, however.

It was while working on the Scrolls that Allegro first crossed swords with the Church. Impatient at the slow pace of work, he urged his colleagues to make their findings available to an ever more curious public. The Church, however, urged caution as they were worried that the Scrolls – believed at the time to be a 'missing link' between Judaism and Christianity – would undermine the Christian message. For if Christianity had merely *evolved* out of Judaism, that weakened its whole philosophical foundation, namely that Christ was the incarnation of God, bringing with him the New Covenant that caused a sudden and decisive rupture with the Old.

These worries eventually proved unfounded – the Scrolls did not form a missing link as such – but the Church's demurral prompted Allegro to publish his own book – *The Dead Sea Scrolls* (1956) – on the discoveries made. This became a best-seller in its own right, selling some 250,000 copies. It was the one book that the average lay person read on the subject, and it too triggered a flurry of outraged letters to *The Times*. Increasingly seen as a loose cannon, Allegro was steadily marginalised within the academy until eventually he was barred from working on the Scrolls altogether.

At the same time, it seems that he suffered a profound and personally catastrophic loss of faith. Not only was he unable to square the supposed inviolate truth of the Bible with the newly emerging textual and archaeological findings, but he also found the Church's resistance to those same findings unbearable – they denied him an easy way of reconciling his own inner conflict. Something in him snapped, and he developed an almost pathological hatred of Christianity.[18] 'It is difficult,' he wrote of the Clergy, 'not to feel contempt for people who have thus brought unnecessary suffering upon millions of simple folk who trusted their spiritual mentors too well.'[19] Here then was a man who, having lost both the means and the reason to pursue his work, ended up embittered and resentful.

The companion volume to *The Sacred Mushroom*, published almost unnoticed in the same year and named appropriately *The End of a Road*, gives further clues to Allegro's intentions. Throughout he refers to himself as a dispassionate scholar 'owing no allegiance to any religion',[20] concerned only with the truth. But this is not how the book reads. It reads as a sustained and vitriolic attack on a once-treasured relationship, now irreparably sundered. It reads as an anguished jeremiad by a man who feels utterly betrayed. And it reads as a call for vengeance. Allegro wanted nothing less than to demolish Christianity and erase it utterly from modern life. 'Throw it out,' he cried, 'with the rest of the discredited cult of the sacred mushroom.'[21] He saw himself as a Samson-like figure, bringing down the edifice to which he had devoted his life. He chose to do this by using all his masterful scholastic skills to fabricate a lie so convincing that people would abandon Christianity, and religion more generally, in droves. And of the whole sorry mushroom-cult story, the only part he actually believed was that Christianity was a sham.

It need hardly be said that the years of resentment had, at the very least, clouded his judgement, for his plan backfired spectacularly. To his horror, one imagines, the public found the idea of his supposed sex and drug mushroom cult titillating and rather appealing, and while his own reputation lay in tatters, the integrity of the Church was actually strengthened by the scandal. Sidelined even further, he quietly stepped down from his position at Manchester University. Quite what it was about the man, his make-up and his biography that caused such a calamitous breakdown remains a mystery, but the whole episode still has an air of tragedy about it, ending as it did with his academic career in ruins.

*

While *The Sacred Mushroom* sold in huge numbers, its dense and uncompromising style meant that it was left by and large unread. The one idea it popularised, however, was that an unusual Romanesque fresco, tucked away in the centre of France in the Abbaye de Plaincourault, depicts Eve being tempted by a mushroom, not an apple: the serpent of the Garden of Eden appears, at first sight, to be coiled round a fly-agaric mushroom. A colour photo of the fresco was included in the book, and a stylised graphic rendition of it was eye-catchingly placed on the inside cover. For the general public, this was the only piece of evidence they needed to persuade them to accept Allegro's theory, for surely here was the incontrovertible proof?

Wasson had, rather regretfully one presumes, already discarded this notion when art historians convinced him that the fresco depicted nothing more exotic than a common, if bizarre, Romanesque stylisation of a tree.[22] Allegro dropped the fresco from the abridged paperback version of the book, printed in 1973, so one assumes that he too was eventually brought round, but the myth lingers on in popular consciousness, resurfacing in field guides and coffee-table mushroom books. Interestingly, Wasson had actually presaged Allegro by contending that the stories of the temptation and of the tree of knowledge were a memory of fly-agaric-based spirituality. But by playing down the suggestion, and by placing it at the very end of his *SOMA* book, he managed to avoid the fuss that Allegro wittingly or unwittingly caused.

Allegro's book, its radical ideas and the moral panic it triggered have been largely forgotten, but every so often new books are issued that reiterate the conspiracy. Using the same 'creative' methodology outlined above, the American James Arthur – 'Ethnomycologist, Author, Lecturer, Theological Researcher, Shaman, Teacher, Soul Healer' – suggests that the fly-agaric 'can be found at the roots of most of the religious writings our planet has to offer'.[23] Clark Heinrich, who reconstitutes Allegro with the addition of his own exegetical insights, writes of the bread (white) and wine (red) consumed ritually at the Last Supper that 'if this quite specific and literal-sounding exhortation is not about the fly-agaric mushroom then it is one of the most bizarre passages in the history of religion'.[24] But beyond feeding the appetites of conspiracy-hungry mushroom enthusiasts, these sorts of unfounded speculations have made no religious or cultural impact, consigned as they are to the lunatic fringe.

*

To each of the fly-agaric conspiracies specific and convincing objections may be raised, but a fundamental problem common to all is that the documented psychopharmacological effects of the mushroom do not seem to match the rhapsodic ecstasies that the theorists insist it produces:[25] it is hard to imagine anyone founding a religion on the basis of such a giddy, unpredictable and often unpleasant experience.

To be fair, the effects, as we have seen, can include feelings of overall stimulation, excitement and euphoria, with a perceived increase in strength or stamina; but they are typically accompanied by nausea, muscle twitches, general loss of coordination and stupor. Visions or hallucinations, of the sort recorded by the Polish brigadier Joseph Kopec, are rare, but visual distortion and macropsia – the alteration of perception of scale and proportion, seized upon so profitably by Charles Kingsley and Lewis Carroll – are common. A period of intense agitation lasting several hours – in which subjects may behave strangely or ridiculously, but rarely as self-destructively as the Russian officers witnessed by Stephan Krasheninnikov – is usually followed by a coma-like sleep accompanied by strange but lucid dreams. Then again, as Wasson found to his chagrin, the mushroom may do nothing at all beyond making one violently sick. 'I am at a loss,' he wrote, after his unsuccessful attempt to replicate his psilocybin ecstasies, 'to explain the . . . failure of my own experiments with the fly-agaric.'[26] Unlike the comparatively dependable action of psilocybin, there is simply no way of knowing which way the fly-agaric experience will go.

This is due to the mushroom's spectacularly complex biochemistry. The long-held idea that muscarine was the principal psychoactive agent of the fly-agaric proved to be misplaced (muscarine was isolated in 1869[27]). The symptoms of muscarine poisoning, the excessive and uncontrollable production of sweat, saliva and tears, are rarely observed in fly-agaric intoxication, unsurprisingly given that we now know muscarine to occur in such low concentrations (0.0003 per cent) as to have a negligible effect.

In the mid 1960s, nearly a hundred years after muscarine was first discovered, two new alkaloids were identified almost simultaneously by three teams of scientists operating independently in Japan, England and Switzerland.[28] After some negotiation, it was agreed that the new chemicals should be labelled ibotenic acid (α-amino-3-hydroxy-5-isoxazoleacetic acid) and muscimol (5-(aminomethyl)-3-hydroxyisoxazole), and various tests, including self-experimentation by some of the scientists, demonstrated that these were indeed the chemical culprits.[29]

These two closely related substances belong to a class of chemicals known as the isoxazoles, but even though they may occur in roughly similar concentrations within any given mushroom (0.03–1 per cent for muscimol, and 0.03–0.05 per cent for ibotenic acid[30]), muscimol is about five to ten times more potent and seems largely responsible for the species' pleasurable psychoactive effects. Like psilocybin, muscimol works by mimicking the action of a neurotransmitter, but unlike the former, which binds to serotonin receptor sites in the brain, muscimol interferes with another transmitter, gamma-aminobutyric acid or GABA, which works lower down in the central nervous system (CNS). Ibotenic acid mimics glutamate, a slightly different neurotransmitter, but both are excitatory amino acids involved with the control of neuronal activity within the CNS. Quite why stimulation of these receptor sites produces the observed effects remains unclear.

Interestingly, ibotenic acid readily transforms itself into muscimol through a process known as decarboxylation.[31] Drying, cooking or even the digestion process itself will, to varying degrees, facilitate this process. The Siberian practice of drying the mushroom to lessen its toxic effects turns out, therefore, to have a sound biochemical basis, for it maximises the concentration of the more desirable muscimol within any particular mushroom. Nevertheless, unmetabolised ibotenic acid has been found to be excreted in large quantities via the urine, which explains why the Siberians were able to recycle the mushroom's properties so effectively.[32]

Experimental evidence has shown that the proportions of these two related ingredients vary according to the age of a mushroom, and anecdotal evidence suggests that there may be seasonal variations as well, with mushrooms picked later in the year proving more toxic.[33] Greater concentrations of isoxazoles are found in the peelable skin and the cap than in the stem.[34] Other psychoactive ingredients may also be present, but scientists have been unable to agree upon exactly which chemicals and in what concentrations. Up to twenty further alkaloids are suspected – with ever more arcane names – but the presence and significance of muscazone, muscaridin, choline, uracile, hercynin and β-D-n-butylglycopyranoside, to name a few, remains hotly debated. To confuse the picture further, the mushroom also contains a dizzying mix of pigments and amino acids, and even heavy metals that it readily absorbs from the soil. None of these have any psychoactive effect, but might contribute to its toxicity.[35]

As yet, no systematic bioassay has been undertaken to determine how these constituents vary according to the substrate upon which the mushrooms grow, the host-tree species, the variety or sub-species of mushroom, the country of origin and so on; no two specimens picked in different parts of the world will have exactly the same chemical constitution. In New Zealand, for example, fly-agaric mushrooms are largely avoided by mushroom aficionados, because they induce a violent sickness with no pleasurable effects whatsoever.[36] Presumably, Siberian fly-agaric mushrooms have a more favourable constitution. Compared to say the Liberty Cap, in which psilocybin content averages at a steady 1 per cent wherever it is found in the world, the fly-agaric is like a chemical cocktail shaker into which a blindfolded bartender has poured whatever ingredients come to hand – vodka, juice or carpet cleaner. It is this huge variability in its relative chemical composition that makes the fly-agaric so capriciously psychoactive, so unpredictable in its effects – and so shaky, and hence unlikely, a foundation upon which to build a religion.

There is, of course, one other overarching objection to all the fly-agaric conspiracies, which is that if the effects of the mushroom are so desirable, so inspiring, and so religiously elevating, why haven't we seen the emergence of modern fly-agaric sects? Why hasn't the habit become more than a little widespread or acceptable? In the three hundred years that we have known about its properties, it is a struggle to find anyone in the West who has ever intentionally sought the mushroom out. There are a few, but compared to psilocybin mushrooms, which have gone global in a mere fifty years, the fly-agaric is consistently seen as a poor way to get high. Dutch smart shops have recently put dried fly-agaric on the market, riding on the back of the psilocybin mushroom trade, but sales are poor, and the market has by and large decided against this substitute.

Of course, human inquisitiveness is such that there have been a few notable outbreaks of intentional fly-agaric consumption throughout the last three hundred years, but always on a small scale, and always localised.[37] The knowledge either came from serendipitous discovery or slipped out from the academy, for a few scientists experimented with the mushroom if only to see what all the fuss was about. We have already encountered the pioneer of American mycology, Captain Charles McIlvaine, who at the beginning of the twentieth century tested the toxicity of hundreds of mushrooms by eating them.[38] The fly-agaric, he discovered, did little more than give him a headache, and almost regretfully he concluded that intemperance, in whatever form, was always accompanied by similar penalties. But his experiments were not the first.

The earliest recorded Western encounter with the mushroom took place in London, towards the end of the eighteenth century, not long before Everard Brande was called out to administer to the poor family poisoned by Liberty Caps. William Curtis (1746–1799) was one of the early pioneers of botany, the director of London's Chelsea Physic Garden from 1773 to 1777, who helped popularise the techniques of ornamental flower cultivation through his *Botanical Magazine*. It was while working on his magnum opus, the *Flora Londinensis* – a guide to all the plants and fungi growing within ten miles of the capital, which took him eleven years to complete – that he encountered the stories about the mushroom emanating from Kamchatka. The realisation that the resplendent fly-agaric was either inebriating, 'making some joyous, others melancholy', or that it drove you 'raving mad', merely piqued his curiosity and encouraged him to try some himself. On chewing the mushroom, though, he found that it produced a terrible burning sensation in his mouth and stomach. Just to be sure that he hadn't been mistaken, he prevailed upon his reluctant and, one presumes, long-suffering gardener to repeat the experiment. The poor man gingerly broke off a piece, placed it in his mouth, and immediately complained of similar symptoms. They both concluded that the mushroom was very probably poisonous, and that they were lucky to have escaped unharmed.[39]

In truth, they probably ate such small amounts that their symptoms were psychosomatic. The same cannot be said for Dr B. Grassi,[40] a young doctor from the small village of Rovellasca, in the north of Italy, not far from Milan, who in 1880 published an enthusiastic paper on the effects of the fly-agaric.[41] Grassi had been inspired to investigate the matter after treating a peasant who had wittingly or unwittingly – we cannot be certain – eaten a meal of the red-and-white mushrooms, and had had a rather splendid time of it. During this period, European vineyards were being threatened by a particularly virulent parasitic insect accidentally imported from America, and the price of wine was rising accordingly. Worried about the effect this would have on the peasant population, who after all needed a glass of wine as much as the next man, Grassi started to wonder whether the fly-agaric might not prove a useful substitute for alcohol during the lean months ahead. He set about investigating the matter with all the enthusiasm of youth.

First he tested the mushroom himself by eating dried portions of it in small doses, repeated at regular intervals throughout the day. Having found its toxicity to be negligible, and the inebriating effects rather pleasant, he offered the mushroom to a series of healthy volunteers, most of whom, it must be said, were young and pretty women. While the single male experimental subject experienced nothing at all, even after consuming massive doses, the women responded much more favourably. In almost all cases they became flushed, agitated, twitchy and talkative. Experiencing the usual macropsia, they rushed about maniacally, expressing delight at their feelings of euphoria and melancholy; indecorous flirtation and the incautious babbling of secrets preceded a deep slumber. The following day, some expressed guilt at their immodest behaviour, but most thought the experience pleasurable and worth repeating. The handsome Grassi was, unsurprisingly, delighted with the results, and arranged to have the mushroom sold at the local apothecary so that all could avail themselves of its pleasures. To what extent the local townsfolk took up his seductive offer has yet to be discovered.

Grassi's medical endorsement of fly-agaric intoxication marks him as exceptional. More typically, episodes of amanita insobriety, where they have occurred, have emerged following serendipitous discovery and have been met with medical disapproval. Take the following, from the case notes of two Scottish doctors, C. H. W. Horne and J. A. W. McCluskie of Glasgow's Western Infirmary, who were taken aback in the autumn of 1962 when two 'professional salmon poachers', one of whom was distressingly confused, reeled into the waiting room: the afflicted man apparently saw everyone's skin peeling off. It transpired that the chancers had been habitual fly-agaric users for about two years, but that they had overindulged that night by drinking an infusion of twenty mushrooms soaked in beer: a considerable dose. The doctors seem to have found the whole episode a cause for levity, and having safely applied the usual emetics, wrote the episode up in a wry article for the *Scottish Medical Journal*.[42]

The two poachers, who made an excellent recovery, were apparently evasive when asked whether others shared their predilection. (Theirs must be set against the admittedly anecdotal story recounted to me by Edward Pope, who spent the tail end of the 1960s living in Scotland with members of The Incredible String Band. There he was informed by a local shepherd that the fly-agaric, thought poisonous by the locals, was known as the gozinta: 'the mushroom gozinta you, and you gozinta the ground'!) The poachers did claim that in certain London bars it was possible to buy a mixture of vodka and amanita juice, called a 'Catherine', after Catherine the Great of Russia who was supposed to have been partial to the brew. Certainly, Catherine the Great's loathing of indigenous shamanistic practices makes this last claim implausible,[43] although the Lithuanian archaeologist Marija Gimbutas claimed that fly-agaric vodka was widely used in her homeland in the years leading up to the First World War, and had even been used to drug prison officers during several daring escapes.[44] But as no other sources have come to light to corroborate the poachers' story, we cannot say whether there really ever was a fly-agaric craze in the British capital.

By the end of the 1960s, when both psychedelia and London were in full swing, speculation about the use of the fly-agaric in the past was rife, with one writer for the infamous underground paper the *International Times* claiming that the ancient Druids had used the mushroom in their initiatory potions.[45] But it was the brouhaha over Allegro's book that prompted a few to try the mushroom themselves. A review in *Friends* magazine, by Colin Moorcraft, gave instructions for its preparation[46] – Moorcraft recommended drying, powdering and mixing several mushrooms to obviate the problem of variable potency – while a man called Lynn Darnton, who was to play a key part in the rediscovery of the Liberty Cap, wrote up what were presumably his own personal experiences for the more controversial *Oz* magazine. Under the influence of the mushroom, Darnton, having first overcome his nausea with a concoction of aromatic oils, heard 'a single, pure, flute-like note played inside [his] head', then 'choirs of angels and deities singing from the tops of hills' before he plunged headlong into the now familiar deep sleep. In his dreams, he was taken deep into the earth, where he witnessed 'a Grand Meeting of Gnomes' who communicated with him telepathically.[47]

The poet Heathcote Williams – later famous for his elegiac poem

Whale Nation – wrote eulogising the mushroom, but the overblown language of the time makes it difficult to determine whether he meant his lines to be taken literally, or as a thinly veiled reference to LSD:

> Take it, eat it,
> Cut away the fronds and smoke it,
> And with your third eye Contact Lens
> Turn into a Planisphere.[48]

Certainly when Mick Jagger administered fly-agaric to James Fox in Nicholas Roeg's 1970 cult film *Performance*, this would have been read as a code for Acid, at least by those in the know.[49]

While experimentation with fly-agaric remained rare in Britain, the preserve of a few plucky inner explorers, the biggest fly-agaric craze outside pre-Soviet Siberia occurred in the Pacific Northwest area of America. During the 1970s, the mushroom, along with its more potent cousin the Panther Cap, *Amanita pantherina*, began to be used by hippies inspired after reading Puharich, Allegro and, principally, Wasson.[50] The medic Andrew Weil, commenting at the time, noted that there seemed to be two groups: those preferring psilocybe mushrooms and those preferring fly-agaric (both of which grow abundantly in the region). Each group was broadly ignorant of the other, the former deriving their mycological knowledge from academic training, the latter possessing a more folksy general knowledge about edible fungi. Steven Pollock went to Alaska on an amanita quest in the early 1970s. He not only found that the mushrooms were being regularly used by young enthusiasts, but was able to find an abundant supply growing in the woods around Anchorage. His first experiment produced what he called a pseudo-delirium, interspersed with moments of nausea and euphoria; his second, with dried mushrooms, gave him a more pleasant feeling of intoxication. Jonathan Ott, however, reflecting back on this period, thought the amanita users a minority, even though fly-agarics were being traded on the black market.[51]

The fly-agaric is still championed by a few enthusiasts on both sides of the Atlantic. Take Runic John, for example, a stalwart of the British festival scene, and a practitioner of a form of reconstructed Anglo-Saxon neo-shamanism. He makes his living running a stall selling a cornucopia of exotic, yet legal, psychedelic plants from around the world. Pots containing strange roots and dried leaves jostle with elixirs, potions, snuffs and smoking mixtures and, of course, with pre-

Runic John's festival stall selling a cornucopia of exotic, but legal, hallucinogenic plants and potions. The jar second from right, front row, contains dried fly-agaric mushrooms.

serving jars filled with specially dried fly-agarics, available in pill form for the weak of stomach. John told me that he found the amanita experience positive and powerful, especially in low doses, useful for providing extra stamina to complete his lengthy tramps across the Lancashire hills. Once he even met a fly-agaric spirit in the form of a woman, who taught him how to make a shamanic headdress; but he confessed that he made most of his living selling cultivated psilocybin mushrooms (that is, while they were still legal).

In the Pacific Northwest, Hawk and Venus, the so-called Red Angels or Soma Shamans, have dedicated their lives to the use of the fly-agaric.[52] Taking Wasson's identification of Soma at face value, they regard the Rig Veda as an instruction manual for the use of the mushroom. They have eaten it every day for nigh on twenty-seven years, to their spiritual betterment, they maintain. Through books, videos and websites, they promote the mushroom, teach the 'correct' spiritual preparation required – fasting and other austerities – and warn against the ever-present dangers of overdose. That they have been able to use the mushroom safely for so long seems beyond question, but the revelations they attribute to the mushroom – which they say has not only revealed to Hawk the exact time and manner of his death, but also the moment when the inevitably approaching day of Armageddon will come – may prove not to be its greatest selling point.

We have already encountered Clark Heinrich, who has attempted to resurrect and elaborate upon Allegro's theory, in part by attempting to promote the mushroom. His experiences with it are, perhaps, absolutely typical, and quite explain why it is treated with such ambivalence. He records that, during the 1970s, he determined with a friend to take the mushroom every day for thirty-one days (with not a little inconvenience to their respective partners). Saving the biggest mushroom till last, they enhanced the trip by recycling their own urine, which seems to have been the catalyst for what followed. Heinrich found himself ascending through ever-increasing levels of bliss, each more magnificent than the last, until he achieved union with the clear white light, the source, the Godhead. Understandably excited, he attempted to repeat this experience a few days later, but this time the experience was hellish. Losing all sense of time, space and identity, he struggled to remember who he was. Coming to, and remembering his name (at last!), he saw grotesque rope-like columns stretching upwards to the sky – which turned out to be threads on the handkerchief upon which he was slumped – and promptly vomited. The whole cycle, loss of identity, coming to, and being sick, was repeated again and again for what seemed like an eternity, until the effects eventually wore off.[53]

And even Keewaydinoquay, the fly-agaric shaman, had this to say after one particularly bad bout with her beloved *miskwedo*: 'Sunday I wanted to die and thought I might, Monday I knew I was going to die and was glad of it, Tuesday I became aware that I hadn't died but had flunked some tests I'd taken somewhere, Wednesday I was still nauseated but I didn't want to die, and Thursday I was glad I hadn't. By Sunday again I was really smarting and wondering how I could have been such a damn fool.'[54]

For every fly-agaric enthusiast, then, there are twenty, or perhaps fifty or more, who read these accounts and politely decline, preferring the more trustworthy psilocybin experience. Amanita-eaters suffer the curse of Cassandra: their panegyrics, their promises of the ecstasies to be found within the red-and-white-spotted mushroom, may very well be true, but they inevitably fall upon deaf ears. No one ever quite believes them.

Before leaving the fly-agaric behind to continue with the story of the rise of the psilocybin magic mushroom, I want to end this section by returning to the question of why a mushroom that most people go to assiduous lengths to avoid has become the focus of so many myths and conspiracy theories. Why, in other words, has style triumphed over substance?

It is not too problematic to see why some people find conspiracy-hunting per se so appealing and all-consuming. The conviction that you have, say, unmasked the 'secret truth' behind Christianity is paradoxically, and in spite of its implicit paranoia, empowering. In a bizarre way, one feels elevated from the ranks of mediocrity for having uncovered the liberating truth. It turns one into a movie hero, a sleuth, an outsider able to spot the hidden connections that those in epistemological authority refuse to entertain.

But this in itself is not enough to explain why the fly-agaric has attracted so much attention. Rather than look for the explanation in any inherent biochemical qualities the mushroom possesses, as the conspiracy theorists usually do, I prefer to seek it in the way that the mushroom's extraordinary form has engaged the human imagination. It is surely here that the mushroom has worked its most powerful effects. The one psychoactive property for which it is justly famous is the way it alters perception of scale, rendering straws into tree-trunks and cracks into chasms. Yet merely bringing the mushroom to mind seems catalyst enough to magnify our thoughts and ideas, to distort and stretch them way beyond the elastic limits of evidence and reason.

Nothing in the natural world looks quite like the fly-agaric. There are other red mushrooms, red-leaved plants, red insects, red frogs, mammals and birds with flashes and dashes of red, but nothing else (at least on dry land) is red with white spots. The fact that these spots aren't markings at all, but the broken remains of the volva, the egg-like sac from which the mushroom bursts forth, does not diminish the point that the mushroom's coloured form is so rare, and so simple, as already to strike us as alien, unusual or peculiar. Anyone, even a child with no knowledge of natural history, could identify one – they are that distinguishable.

Furthermore, red is a colour that excites us. Culturally, it is the colour of anger, lust, passion, energy and heat, while naturally it is the colour of blood. In the animal kingdom, it is a universal sign of poison and danger, a common language reached after millennia of convergent evolution that warns potential predators to stay clear.

This alone would have ensured that the daringly coloured mushroom stirred our imaginations, but there is a second reason, which is to do with the way that we habitually categorise the natural world. It is ironic that as our knowledge of the physical, chemical, biological and ecological workings of the environment has massively increased, so our actual contact with it has fallen away almost to nothing. Most of us spend our lives cocooned in human-created worlds, cities and suburbs where the other species with which we would ordinarily cohabit cannot prosper, or from which they are actively removed. We have become quite divorced from the rhythms and other inhabitants of the natural world. As we have made the transition from agriculturists to industrialists to consumerists, we have retained a body of folklore, legend and urban myth about certain plants, certain animals: the powerful, colourful and distinctive species that linger in the imagination as easy markers or symbols for the human qualities we admire, and project. We speak of being as brave as a lion, as wise as an owl, as cunning as a fox, as noble as an eagle. Most of us have never come face to face with these or any other animals or plants of the wilderness (excepting the urban fox, of course), and yet they intrude into our unconscious lives in ways that the more indistinct, forgotten, species – say, wheatears, shrews and sulphur tufts – do not.

Apart from any of its psychoactive properties, then, the power of the fly-agaric to enchant and bewitch begins with its striking and memorable colour and form. The discovery that the fly-agaric is psychoactive merely confirmed what we knew already: that it is dangerous, powerful, mysterious and odd. As the one mushroom we all recognise, it has become the representative of all of its kind. From the safe distance that urban living affords, we have endowed the fly-agaric, the archetypal mushroom, with all the peculiar qualities that the fungi in general have come to represent for us. Like them, it seems to invite speculation and wonder in equal measure.

Only one final myth about the fly-agaric has yet to be mentioned, and that is the myth that gave it its name: the idea that, crushed in milk, its flesh attracts and kills flies. Repeated tests have shown that a few insects are killed, while most, like their human observers, are merely stupefied by the mushroom and go on to make a full recovery.[55] In the topsy-turvy world of the fly-agaric, a world where nothing is quite what it seems, even its foundational myth proves to be a half-truth. Truly, this mushroom is the greatest of dissemblers.

PART THREE: PSILOCYBE

Psilocybe!
Makes you high-be!
Psilocybe makes you fly!

Boris and his Bolshy Balalaika, 'Toadstool Soup'

10

Academic Exercise

Experiment 1 (10 mg Psilocybin on 6 April, 10.20)
10.50 – 'strong! Dizziness, can no longer concentrate . . .'
10.55 – 'excited; intensity of colours: everything pink to red . . .'
11.05 – 'The world concentrates itself there on the centre of the table. Colours very intense.'

Rudolf Gelpke[1]

The implications of Gordon Wasson's Mexican discoveries – María Sabina, the magic mushrooms, the myco-gnostic revelations to be had south of the border – took time to be felt by the wider culture in America and beyond. But they were immediately seized upon by a number of scientists who were not only keen to find answers to the many questions thrown up by Wasson's researches – legitimate academic questions about chemistry, biology, anthropology and psychology – but were also impatient to try the mushrooms themselves.

From the perspective of our time – in which prohibition reigns, the 'war on drugs' lumbers on, and any open academic self-experimentation with psychedelics would amount to career suicide – this does seem rather strange. However, the political situation was very different in the late 1950s. Although cannabis, heroin and cocaine were all illegal throughout the Western world, what we would call the hallucinogenic or psychedelic drugs were not. The latter, including LSD and mescaline (the active chemical ingredient of the Central American peyote cactus), were being eagerly employed on both sides of the Atlantic in the rapidly expanding discipline of psychology.

In an almost utopian spirit, it was hoped that these psychedelics would usher in a new era of understanding of the mind – as important a tool for psychology as the telescope and microscope had been for astronomy and biology. They would strip bare and magnify the contents of the unconscious; they would allow psychiatrists to find the chemical basis of mental illness; and, at the very least, they would allow scientists to empathise with the world of the schizophrenic by enabling them temporarily to inhabit it themselves. The discovery of psilocybin mushrooms provided yet another window into this mysterious domain of the mind, and the way was clear for scientists of any discipline to step through this particular looking glass themselves.

Perhaps the most cogent advocate for the prescribed use of psychedelics at this time was a young English psychiatrist working in Canada, Humphrey Osmond (1917–2004). It was Osmond who, one spring morning in 1953, administered mescaline to Aldous Huxley in the experiment that formed the basis of *The Doors of Perception*; and indeed it was Osmond who, in a poetic sparring match with Huxley, coined the term 'psychedelic', meaning 'mind-manifesting'.[2]

For Osmond, convinced by the structural similarities between mescaline and adrenaline that there must be a naturally occurring chemical trigger for schizophrenia, his elusive 'M-factor', the principal role of the psychedelic drugs was to provide a model or experimental psychosis.[3] But he also believed passionately that psychiatrists should experience psychedelics first hand, and many times, or else how could they ever begin to empathise with their patients? 'These [psychedelic] states deserve thought and pondering,' he argued, 'because until we understand them, no account of the mind can be accurate' – adding that 'the mind can not be explored by proxy'. Gently and without sensation, he went further, arguing that psychedelics had intrinsic religious or philosophical value, and might be used profitably by people with no immediate interest in mental illness. 'For myself, my experiences with these substances have been the most strange, most awesome, and among the most beautiful things in a varied and fortunate life,' he wrote. 'These are not escapes from but enlargements, burgeonings, of reality.'[4]

Although some of the psychedelics Osmond was using were comparatively new, his attitudes were not, for the history of psychology is replete with scientists exploring their own minds through the consciousness-altering effects of drugs. Perhaps the earliest was the French psychiatrist Jacques-Joseph Moreau de Tours (1804–1884), who ate lumps of hashish in the hope of finding a preventative for mental illness (needless to say, his ideas about the merits of cannabis are now quite unfashionable).[5] The London physician Francis Anstie (1833–1874) seems to have been able to categorise the action of various drugs in his *Stimulants and Narcotics* of 1864 because he had tested many of them on himself. The American philosopher and pioneer psychologist of religion William James (1842–1910) felt able to comment on the nature of mysticism after he experienced the classic numinous feelings of awe under the influence of laughing gas (nitrous oxide). Sigmund Freud (1856–1939) was partial to the odd syringe of cocaine, while countless psychiatrists – most famously Weir Mitchell (1829–1914) in America, Heinrich Klüver (1897–1979) in Germany and Havelock Ellis (1859–1939) in Britain – experimented with peyote.

Scientific endeavour is one thing, but public and political opinion is quite another, and it does not necessarily follow that these personal scientific preferences reflected the popular mores of the time. Nevertheless, that a respected scientist like Osmond could publish his pro-psychedelic opinions in a peer-reviewed American scientific journal without fear of censure or castigation says much about the climate into which Gordon Wasson announced his discoveries. Not surprisingly then, given the general air of excitement that prevailed in the late 1950s as to what psychedelics might reveal about the mind, Wasson's team of chemists, mycologists and anthropologists had few reservations about trying the mushrooms themselves.

For one, at least, self-experimentation was simply a matter of logic and expediency. The Swiss chemist Albert Hofmann was approached by Roger Heim in the late 1950s with a request to identify the active chemical ingredients of the newly identified Mexican mushrooms. Heim, you will recall, was Professor of Mycology at the prestigious Muséum National d'Histoire Naturelle in Paris, and had accompanied Wasson to Mexico in 1956 in order to classify and describe the hallucinogenic mushroom species. Attempts made by his own laboratory at the Muséum, and by various American pharmaceutical companies (Merck; Smith, Kline & French), had failed in the task, so Heim approached Albert Hofmann as the leading authority in the field of plant alkaloids: Hofmann, after all, had discovered LSD some fourteen years earlier.

By this stage, LSD was already showing signs of becoming, as Hofmann would later christen it, his 'problem child', and he found that his co-workers at the Basle-based pharmaceutical giant, Sandoz, were reluctant to take on the commission lest they became tarnished by association, to the detriment of their careers.[6] He therefore led the investigation into the secrets of *teonanacatl* himself, but immediately hit a problem, which was that none of the chemical extracts isolated had unequivocal effects upon the animals that new compounds were usually tested upon. Without clear, emphatic reactions, how was he to tell which alkaloid was the right one?

Thinking the problem through logically, he began to question whether the dried mushrooms provided by Heim – who was the first person ever to have cultured them in the lab – were still active: perhaps the drying process rendered them impotent? There was only one way to find out, and that was for a human volunteer to eat some. Thus it was that one afternoon in July 1957, after having carefully assessed the risks, Albert Hofmann consumed thirty-two dried mushrooms of *Psilocybe mexicana* in a clinical setting under the supervision of a physician, to see whether they had any effect. He was not disappointed.

Of course, Hofmann was quite used to self-experimentation, having famously taken the first ever LSD trip. He had isolated lysergic acid diethylamide-25, to give it its full name, in 1938 while looking for analeptic drugs that might alleviate migraine, deriving it from the cocktail of alkaloids that make up the ergot fungus (the sclerotia of which, as we have seen, have been used in folk medicine for centuries for their vasoconstricting effects). Initial tests proved disappointing, however, and LSD-25 was quietly shelved, but acting on an intuitive hunch, and the feeling that he had overlooked something of importance, Hofmann went back to re-examine it in the spring of 1943.

It was while handling the material that he began to feel rather unwell and was forced to go home, overcome by a not altogether unpleasant dazed and drunken feeling. Lying in bed, he noticed that his imagination seemed preternaturally stimulated. Cleverly surmising that the cause of this malaise was chemical not biological, and that some microscopic trace of LSD must have entered his bloodstream, perhaps via the skin, Hofmann returned to the lab three days later on 19 April to test whether this was indeed the case. He deliberately took what he considered would be a microscopic dose, 0.25 mg dissolved in a little water. What he could not have known in advance was that LSD is many thousands of times more potent than any of the other hallucinogens, and that, by the standards of the later psychedelic sixties, he had taken a massive dose (today, a standard dose on the street is between 0.06 mg and 0.08 mg). Rather tellingly, his laboratory notes break off abruptly, for he was suddenly plunged into an overwhelming maelstrom of colours and hallucinations. Accompanied by a concerned colleague, he made what must be the most famous bicycle ride in history, weaving his way home as the road, and his Swiss sense of ordered reality, undulated before him.

Sandoz were, understandably, more than a little perplexed at what to make of Hofmann's quite literally mind-blowing discovery but, sensing a marketing opportunity, they conducted the necessary toxicity tests, and put the drug on the market anyway, exactly as Osmond would later stipulate: for the purposes of inducing model psychoses, and for enabling the researcher to enter, and so empathetically understand, the world of the schizophrenic. From there, it was readily taken up by researchers on both sides of the Atlantic.[7]

By the time he ingested the dried Mexican mushrooms, then, Albert Hofmann was a seasoned psychedelic traveller, having made several personally illuminating LSD voyages. Nevertheless, like many of the psychedelic experiences he diligently recorded, which are characterised by dark and sinister undertones, it was not altogether pleasant. Things began to appear unsettling and strange to him, so he asked to be driven home to more familiar surroundings, but he found that the streets had been 'demonically transformed'.[8] Arriving home at last, he lay on a couch and closed his eyes, whereupon he saw visions of landscapes and strange, exotic, architecture. Opening his eyes again, he was quite taken aback when his room became suddenly encrusted with Mexican designs and motifs, and the doctor, leaning over him to check his pulse, was transformed into a rather menacing Aztec priest. 'I would not have been astonished,' Hofmann later wrote, 'if he had drawn an obsidian knife.' By the time the experience peaked, the 'rush of interior pictures' in his mind's eye had become so intense that he felt as if he had been possessed by a demon, and feared he 'would be torn into this whirlpool of form and colour and would dissolve'.[9] Luckily for science he did not, and he returned both unharmed and certain that nothing was at fault with the dried specimens.

From then on, animal experiments were abandoned as quite ineffectual, and all extracts were tested on a now increasingly willing field of human volunteers drawn from Sandoz's employees.[10] In this way Hofmann's team swiftly isolated the active ingredients, the chemicals psilocybin and psilocin, which he named after the mushrooms, and Sandoz hurriedly placed their branded *Indocybin* on the market as one more in their stable of mind-altering drugs for use in psychological research.

*

As for Roger Heim, his interest in the hallucinogenic mushrooms was first and foremost mycological, and it was intellectual curiosity that prompted him to accede to Wasson's request that he come to Mexico in 1956 to make the first systematic identification of the mushroom species used by the *curanderos*. The discovery of previously unrecorded species always brings justified rewards for the taxonomist, but for Heim these were doubled when he realised that it was a nondescript and hitherto overlooked genus, *Psilocybe*, that contained most of the hallucinogenic species.

However, his interest was not quite as professional as it might seem. As luck would have it, this expert mycologist, chosen by Gordon Wasson because of his glittering academic credentials, had dabbled with the fly-agaric as a young man way back in 1923. Whether this had been an accident, the upshot of youthful exuberance, or part of a calculated experiment remains unclear, but years later, reflecting on his own experiences, he wrote that 'consumed fresh and in small quantities, [the fly-agaric] can produce an agreeable sensation which limits itself to a feeling of contemplative well-being'.[11] In Mexico, therefore, he had no qualms about taking part in a Sabina-led *velada*, and then, having brought fungal samples home and cultivated them in the lab, undertaking a series of self-experiments (the results of which he eventually published in the prestigious *Annals of the French Academy of Sciences*).

Heim's meticulous writing style and his mastery of classical French prose, for which he was justly recognised, have the unfortunate effect of rendering his mushroom trips somewhat dull, especially when compared to Wasson's purple rhapsodising. No detail is trivial or insignificant enough to go unreported. Thus, we know that when on 14 April 1957 Heim consumed thirty-two fresh specimens of *Psilocybe mexicana* – the same dose as Hofmann – he was forced to fight off a cold feeling with a hot-water bottle. And that when, on 18 May 1956, he took five fresh mushrooms of *Stropharia [Psilocybe] cubensis*, they tasted strongly of radish.[12]

In none of his experiments did he allow himself to become overwhelmed by sensation, as Wasson and Hofmann had, for he strove to record and analyse what was happening to him in real time, as the experience unfolded, and as accurately as possible. He struggled to find the exact letters and words (his handwriting becoming strangely bunched up, like the teeth of a saw) to describe the ever-changing 'fairy spectacle' of visions fleeting across his mind's eye, the exact hue of the coloured lights that danced about his study, the precise manner in which ordinary things had become portentous and meaningful. In spite of his analytic approach, however, and anticipating one of the effects that would eventually make the magic mushroom so popular, this dry and serious man found that he always got the giggles, and that he would laugh and joke hysterically until the tears rolled down his face.

One by one, members of Heim's and Hofmann's research teams in France and Switzerland made their own voyages of discovery into the strange realm of the mushroom, recording every visual mote, every perceptual nuance and every physical sensation as diligently and as faithfully as their superiors. Mushrooms were eaten fresh or dried, in low or high doses, in the clinic or at home, during the day or at night, supervised or alone: every combination was attempted, every difference noted.[13] A certain naivety pervades the records of these early explorers, who were new to the terrain and uncertain of what features might later prove significant or, even, quite what it was they were looking for. One is reminded of the pioneering mountain climbers, hurling themselves up the unscaled Alps with nothing more substantial than tweed jackets, stout boots and rucksacks-full of enthusiasm to serve in place of expertise and technical know-how. In both cases, mishaps were prevented only by luck, not judgement.

Take, for instance, Heim's laboratory assistant, Roger Cailleux, who seems to have been unusually sensitive to the effects of the mushrooms. Even on barely active doses he saw visions of oriental-style writing and great threshing tapestries upon which strange motionless people had been embroidered in exotic tableaux. Intrepidly, he took a large dose of *Psilocybe semperviva* mushrooms very early one September morning in 1958. In the darkness before dawn, he watched as swathes of colours drifted across his room, arranging themselves into a cone of light standing freely before him and engraved with abstract designs. Closing his eyes, he felt as though he was witnessing the processes of evolution itself, forms twisting and writhing and begetting ever more forms.

However, the perceived presence of what he called 'silent and invisible people' began to unsettle and irritate him, and thereafter the trip took a downturn. He felt as if his personality was disintegrating, and that he was no longer inhabiting his body (which was permanently shivery and cold), and became terrified that he might stop breathing. Only the coming of dawn, and the beautiful sight of the tree in his garden with its unusually intense greenery, kept him clinging onto his sense of self. Eventually, and to his great relief, the effects of the mushrooms began to wear off. Never experiencing Heim's 'happy clarity of spirit and exceptional well-being', Cailleux was left with an unsettling aftertaste, like the remembered presence of a bad dream, that lingered through the day and beyond.[14]

Others, intellectuals and artists not immediately within the Hofmann–Heim scientific circle, volunteered to try this new mushroom-derived drug and to reflect their experiences through the more considered lens of their literary prose: most famously, the Swiss Islamic scholar Rudolf Gelpke (1928–1972), and the French artist and writer Henri Michaux (1899–1984). But though their experiences were often profound and occasionally overwhelming, and though they were often beautifully rendered – 'the orchestra of the immense magnified inner life is now prodigious' thundered Michaux – it became clear, to the scientists at least, that the addition of yet more varied accounts of scintillating colours, indescribable visions and numinous feelings were doing little to further scientific understanding. A more controlled and clinical approach was called for, one in which the occluding variables of setting and personality could be factored out and the true effects of psilocybin upon the mind could be determined. In other words, scientists on the continent stopped experimenting on themselves and began the more usual empirical process of testing the new drug upon others.

Much of the early, unsung research into the pharmacology of psilocybin was conducted by the psychiatrist Professor Jean Delay (1907–1987) and his team at the Sainte-Anne Hospital in Paris.[15] In addition to performing tests on rabbits and mice, Delay administered psilocybin to healthy volunteers and, less ethically perhaps (but by no means unusually for the time), to patients sectioned in asylums, afflicted with a variety of mental illnesses. He investigated psilocybin's effects upon the body and mind, and defined no fewer than thirty-one 'psychic' effects. The most commonly experienced were, for example, attention difficulties, euphoria (and/or disphoria) and extraversion, but the list also included auditory hallucinations and the delightfully named 'bizarrerie de l'ambiance' (peculiar moods or atmospheres).[16] The recovery of buried memories, at least with some patients, offered Delay the hope that psilocybin might prove useful in psychotherapy: at least one woman with an eating disorder achieved a partial cure under this regime.[17]

But perhaps the most interesting research conducted in Paris throughout the 1960s – by and large unreported – was that investigating the effects of psilocybin upon creativity. The question as to whether drugs unleashed or provoked artistic inventiveness was not new, for artists, poets and musicians had been experimenting with drugs since at least the early days of Romanticism. But for every Théophile Gautier (1811–1872) – the French novelist who claimed, in 1843, that hashish induced in him an irresistible urge to write[18] – there has been a Charles Baudelaire (1821–1867), whose creative powers were quite clearly stymied by drugs (in this case his opium addiction).[19] For the prodigiously talented writer Aldous Huxley, who at times experimented with mescaline, mushrooms and LSD, there was no easy relationship between psychedelics and creativity. Quizzed on the matter, he replied that a drug like LSD can only help the creative process *indirectly*. 'I don't think one can sit down and say, "I want to write a magnificent poem, and so I'm going to take lysergic acid." I don't think it's by any means certain that you would get the result you wanted – you might get almost any result.'[20]

In the 1940s, the London psychiatrists Walter Maclay and Erich Guttman, who were interested in understanding the hallucinations of their mentally ill patients, gave mescaline to artists and asked them to paint what they saw. Believing mescaline to create a 'model psychosis', the researchers hoped to find what in German is called the *Stilwandel*, a sudden or radical change in drawing or painting style that sometimes accompanies the onset of mental illness: at the time, different pathologies were thought to produce different stylistic changes, the one symptomatic of the other. Maclay and Guttman's results were disappointing, in that they found no discernible 'mescaline style', but what the drug did do was give artists the confidence to explore more fully the stylistic innovations with which they were already experimenting.[21] In that sense, creativity was definitely enhanced.

The Parisian psilocybin experiments were a more sustained attempt to investigate the matter, and at their simplest they involved scientists giving psilocybin to artists and then observing what happened to their painting style both during and after the session. But the rationale of the experiments was particularly French, for it was shaped by a school of thought that dominated continental thinking throughout the second half of the twentieth century, namely structuralism.

Structuralism posits that what is important about a cultural system is the *relationship* between its fundamental units. So, for example, language only acquires meaning because of the way words are positioned in relation to other words, rather than because of any fundamental or essential meaning residing within the words. A word such as 'asylum' can refer to a place where the mentally ill are incarcerated, or to the shelter offered to political exiles, but its exact sense depends on the way it is deployed around the other words of a sentence. Words, dress codes, popular music, adverts, myths, the content of dreams: structuralists have read all of these as 'signs', the meanings of which can be found in relation to the network of other signs that make up each particular system. Though not a French invention, the idea took particular hold in France. Claude Lévi-Strauss (1908–) used it to revolutionise anthropology; Roland Barthes (1915–1980), cultural and literature studies; Jacques Derrida (1930–2004) and Michel Foucault (1926–1984), philosophy; and Jacques Lacan (1901–1981), psychology and the study of the unconscious.

This humanistic tendency informed the Parisian experimenters who, following Freud and Jung, saw the cause of mental illness as originating in the dynamics of the unconscious. In their structuralist view, the elements of the unconscious were thought to fit together like words in a language, but the question was how to shed light on this shadowy aspect of the psyche and so be able to untangle this foreign tongue. Artists were chosen for the experiments because they were believed already to possess the gift of expressing the unconscious. Certainly all art was culturally mediated, and had to be rendered skilfully in culturally sanctioned forms (as in, say, an allegorical painting). But the experimenters held onto the view that 'great' pieces of art had originated from deep within the artist's unconscious mind. The Parisians hoped that psilocybin would roll back the reducing filter of the conscious mind a little further and so allow the unconscious an unprecedented freedom of expression. And by studying the elements of this raw art, drawn up fresh from the depths, they hoped to lay bare the very workings of the unconscious, to piece together the elements and structural relationships that gave it form and coherence.

And, after all, who knew what powers and forces moved in the deep recesses of the mind? Perhaps psilocybin-induced encounters with these fundamental forces would refresh and reinvigorate artists, sweep them along in wholly new aesthetic directions? Perhaps psilocybin might clear away conscious and unconscious blocks to activate a latent but unexpressed artistic ability, transforming the mediocre into masters? Perhaps the drug might reveal the origins of the primitive shamanic impulse itself, or even provide access to unusual psychic powers, and the ability that shamans so often claimed, to divine the future?

The first experiments were conducted by one of Delay's students, Robert Volmat, and his colleague at the Sainte-Anne Hospital, René Robert, who together supervised some thirty-five sessions in which psilocybin was administered to twenty-nine willing artists. The results were superficially dramatic. One artist abandoned his usual brushes and gouache for pastel and watercolours, which he smeared across the canvas with his hands, painting a new picture every three minutes for nearly an hour and a half.[22] Another revelled in what he called the 'dissolution of forms' that psilocybin occasioned.

Volmat and Robert were certain that they could distinguish an emerging psilocybin style, characterised, they said, by elongated brush strokes, parallel lines and repetition, that came not from some impairment of coordination but from 'a compelling rhythm that replaces the normal experience of time'.[23] They found that, as the drug took hold, the artists were faced with a challenge: to try to hold onto their ordinary style, their ordinary way of working, or to succumb to the drug and experiment freely with this new way of seeing and creating. Those that opted for the former seem to have gained little from the experience; those opting for the latter, a great deal.

For the researchers, this was all evidence in support of their structuralist theory of mind. They likened the psilocybin experience to a 'fertile moment' in which conditioned barriers were stripped away, and in which there was no thought for the future, only for expressing the unconscious in the moment. And they were pleased to find that many of their subjects felt liberated by the experience, and attempted to incorporate their creative discoveries back into the day-to-day: in other words, the insights of the psilocybin *Stilwandel* seemed not to be entirely forgotten.

Later, Roger Heim, of all people, together with Pierre Thévenard, took an interest in this line of research and, in particular, the question of whether mushrooms could unlock any latent artistic potential hidden in the depths of the mind. Throughout the 1960s, they conducted a series of similar experiments, giving mushrooms (*Psilocybe cubensis*) to volunteers. They not only published their findings but also filmed the experiments, recording the reactions of their subjects and the process of artistic creation as it occurred.[24] Every nuance of every picture produced during the experiments was rigorously analysed, not only for 'artistic merit', but also for what it revealed about the inner workings of the artist's mind.

Take Maitre Breitling, for example, who, unusually for Heim's volunteers, had a disphoric encounter with the mushrooms. He spent the entirety of the trip drawing an apocalyptic dragon, its mouth filled with ivory fangs, its head covered in crenulated scales and repeated eyes. Obsessed with analysing this creation, he addressed it in distinctly Freudian overtones: 'filthy beast! dirty beast!' he kept repeating.[25] When he declared that the dragon stood in the middle ground between the eagle and the lion – a structuralist statement if ever there was one – surely he had stumbled into some dark but archetypal region of the unconscious from which the Western mythological wellstream had sprung? And when he drew a plumed serpent, surely this was the figure that appeared throughout Central American indigenous mythology, and surely this meant that much of Aztec art and mythology was derived ultimately from mushroom visions? So thought Heim and Thévenard, at least.

The mushrooms had a profound effect upon another volunteer, the young Mademoiselle Michaux, who was by profession a graphic designer. Heim had no qualms about rather patronisingly pointing out the inadequacies of her previous artistic endeavours, which he felt were mediocre, kitsch and without merit. He was delighted to find, however, that under the influence of mushrooms her drawing style was radically transformed, her pictures becoming almost cubist or primitivist in appearance. The young woman was enthralled: '. . . there is not one detail [of my drawing] that isn't significant. I'm inspired. I've been a genius for three hours.'[26] Thinking that she had somewhat overstated the case, Heim nevertheless wrote of one of her portraits that 'it is without doubt unconscious, but it is self-explanatory, luminously comprehensible. We are far from the abstract, and perhaps quite close to the totemic origins of magic.'[27] And indeed Mlle Michaux maintained that after the experiment she had had several flashes of precognition in which she accurately predicted the future. Heim declined to pass judgement upon this, but behind the Gallic insouciance he was clearly excited by this encounter with what he saw as the buried unconscious wellspring of shamanism, mythology and the creative impulse itself.

Did the mushrooms really reveal the workings of the artists' unconscious minds? Such a view is now unfashionable, Freud and 'the unconscious' having fallen from favour within the academy (except, perhaps, within the discipline of literary criticism). Nor is art – especially modern art – thought to be an unmediated expression of the unconscious: it is far too self-consciously done. A more current view would be that this mushroomic art was simply the product of a distorted perception, no more or less revealing than if the artists had been given spectacles of frosted glass to wear.

And exciting as Heim and Thévenard's and Volmat and Robert's results appeared to be in terms of making people more creative and revealing a 'psilocybin style', other experiments elsewhere produced more ambivalent results. Those conducted in Germany using LSD, for example, had the opposite effect: artists felt inhibited and impaired by the experience, and were terrified that the very qualities that made their art unique would be stripped away. The determining factor seems to have been not the different drugs, but the way in which the experiments were conducted. The French seem to have been able to create a particularly safe atmosphere and a conducive environment for artistic experimentation, in a way that the Germans were not.[28] What would become clear, however, was that whether or not psychedelics made one *more* creative or artistically talented, they opened up extraordinary new realms of visionary experience, which countless artists, musicians and poets felt moved to try to represent: they gave artists something new to paint, in other words. Psychedelia was about to explode into the world as a major cultural force.

Before turning to the story of how the mushrooms made their way from the cloistered world of the academy to the psychedelic underground, however, we need to return to America, where academic interest in Wasson's discoveries was also keen but was not always motivated by such humanitarian ideals. James Moore was an American

James Moore, the CIA operative sent to Mexico in 1956 with orders to bring back magic mushroom specimens for testing as a potential 'truth drug'. Gordon Wasson was unaware of Moore's duplicity until many years later. Courtesy R. Gordon Wasson Archive, Harvard University, Cambridge, Massachusetts.

chemist at Parke, Davies and Company, who in 1955 wrote expressing an interest in accompanying Wasson on his travels, ostensibly for 'scientific reasons': he wanted to identify the mushrooms' active ingredients. Wasson was persuaded, not least because Moore had access to a funding body, the Geschickter Fund for Medical Research, which had generously offered to underwrite the entirety of his next expedition. So it was that Moore accompanied Wasson, Heim, the photographer Allan Richardson and the French anthropologist Guy Stresser-Péan to Mexico in 1956. What none of them knew, until many years later, was that Moore was working for the CIA, and that the Geschickter fund was a government front for a secret project called ARTICHOKE (later MK-ULTRA), the aim of which was to find 'mind-bending' drugs for use in the Cold War.

Throughout the 1950s the CIA tested a variety of psychoactives, including cannabis and LSD, in the hope of finding the elusive 'truth drug', or at least a chemical that might weaken an opponent's will. They did this through a series of highly questionable experiments: giving LSD to unwitting soldiers; running a brothel and secretly spraying LSD into clients' faces; paying psychologists around the country to pull in students who might volunteer to try exotic drugs.[29] No story about an unusual plant or drug was left unexplored, and the CIA were somewhat ahead of the game when they sent an agent to Mexico in search of the reputed mushrooms, long before Wasson had heard of them. Their heavy-handed tactics, however, had failed to get past the reticence of the indigenous locals, and their agent – whose name we do not know – returned empty-handed. Somewhat piqued that a middle-aged banker had succeeded where they had so spectacularly failed, the CIA attempted to bring Wasson on board, approaching him in 1955. After some consideration, Wasson politely demurred, believing that intelligence work would compromise his own researches.[30] Undeterred, the CIA sent Moore to infiltrate Wasson's team and to bring back the strange hallucinogenic mushrooms himself.[31]

Moore was a disaster from the start. He very quickly managed to get himself ostracised by the rest of the group for his continual whingeing and complaining, for whereas the others revelled in the challenges of travelling in a developing country, Moore detested it. He got diarrhoea, was mercilessly bitten by insects, and slept badly on the hard earth floors of Mexican homes. And while the others enjoyed a night of ecstatic brilliance in the capable hands of María Sabina, Moore just felt miserable and disorientated. Almost certainly, these reactions were the result of the inner distress caused by his own duplicity: a chemist, not a spy, Moore was ill equipped to handle the conflicting demands of subterfuge. As Allan Richardson noted, '. . . all we knew was that we didn't like Jim. Something was wrong with him.'[32]

Despite his inept handling of the situation, Moore returned, albeit nearly a stone lighter, with a bagful of mushrooms. But before he could extract the active ingredients (animal tests having once again proved ineffectual), Heim's team managed to cultivate the mushrooms and Hofmann's to isolate psilocybin. The CIA were thwarted for a second time, and were obliged to obtain their supplies of psilocybin from Sandoz, just as they had their LSD. In the end, neither drug proved to have any potential for military use.

Perhaps more unwelcome, from the point of view of Wasson's scientific circle, was the arrival in Mexico in 1957 of a rival mycological team, under the leadership of the American Rolf Singer (1906–1994), intent on collecting and identifying the remaining hallucinogenic species. Singer had been a student of Heim's, but was by then working for a university in Argentina, and was commissioned by the Chicago psychiatrist Sam Stein to undertake the expedition. Stein, like many of his profession at the time, was interested in the therapeutic possibilities of psychedelics, and was eager to conduct experiments with these newly discovered mushrooms.

The rivalry between the Wasson and the Singer group was intense, and came to a head when Singer published his treatise on the Mexican mushrooms just months ahead of Heim.[33] Heim had intended to honour his friend by naming one particular species *Psilocybe wassonii*, but as a result of his swiftly published paper, Singer got priority for his *Psilocybe muliercula*. A rather undignified spat ensued – both privately and in the press, with much name-calling from both sides, and which rumbled on into the 1980s.[34] The situation was partly resolved, and Wasson's dented pride partially restored, when *Psilocybe wassoniorum* was later placed into the scientific textbooks.

Singer, by his own confession, was nervous about the effects of the mushrooms, and avoided eating them, but his boss Sam Stein was itching to try them: he had once suffered an accidental intoxication after eating mushrooms, and he wanted to know whether the effects would be the same. Two mushrooms of *Psilocybe mexicana* disappointingly had no effect (unsurprisingly, given that Hofmann and Heim had both taken thirty-two), but just before Christmas, on 22 December 1957, Stein ate two specimens of *Psilocybe cubensis*, fried in butter, which had been grown in the lab from those brought back by Singer. Things did not go well.[35]

As the mushrooms began to work, Stein found himself feeling angry and uneasy. Already worried about having a 'bad trip', he attempted to neutralise the effects with an antipsychotic drug, reserpine, but to no avail. Panicking, and unable to hold down a coherent thought, he got into his car and drove to his doctor's house, luckily avoiding an accident that would have made him the first ever psilocybin-related casualty. The doctor barely knew what to do – coffee, alcohol and more reserpine merely aggravated Stein's feeling of being 'distressed in the head' – and so attempted to calm him by reading out choice extracts from classic works on brain biochemistry. Hardly surprisingly, this failed, and so Stein got into his car (again!), drove home, and sat out the course of the trip, nursed by his concerned wife.

What is extraordinary about Stein's account – published once again quite legitimately in a peer-reviewed scientific journal – is his sense of indignation that the mushrooms failed to live up to expectations. Previously, he had had happy times with atropine and mescaline, and found, to his evident suburban delight, that LSD improved his tennis game.[36] He was forced to the regretful conclusion that mushrooms were just not as good.

Of Stein's original team, however, special mention must be made of the Mexican mycologist Gastón Guzmán. Along with the German Jochen Gartz, and the American John Allen, Guzmán has dedicated much of his working life to discovering, describing and naming every species of hallucinogenic fungus in the world. In 1983, he published his magnum opus, *The Genus Psilocybe*, followed by a lengthy and detailed paper on the world's hallucinogenic species, written with these two collaborators.[37] His passion for the subject seems to have been stimulated when, while working for Stein, he took part in a *velada* in the small town of Rancho El Cura, not far from Huautla, in the rainy season of 1958. In this pivotal experience, friends and relatives mysteriously appeared to him, and apart from the surprising colours that played before him, he was particularly struck by the sight of an enchanted castle that had materialised unexpectedly in a corner of the room. Only the next day did he realise that the object he had been looking at with such rapt attention was his mushroom-drying unit.

In the halcyon days of the early 1960s, then, increasing numbers of academics and scientists in France and America were happily trying the magic mushrooms themselves – or were testing them on volunteers – and were eagerly comparing notes with their academic colleagues. Hofmann's isolation of psilocybin, and its subsequent marketing by Sandoz, meant that the quintessential mushroom experience was now readily available – at least for those licensed to obtain the drug, researchers around the world. No longer did they have to make the difficult trek to Mexico, nor try to master the difficult art of laboratory cultivation. However, one American scientist, a brilliant Harvard clinical psychologist, did make the trip to Mexico, where, on the recommendation of a colleague, he managed to procure some mushrooms, which he ate with friends one sunny day in August 1960. His name was Timothy Leary, and afterwards nothing was ever quite the same.

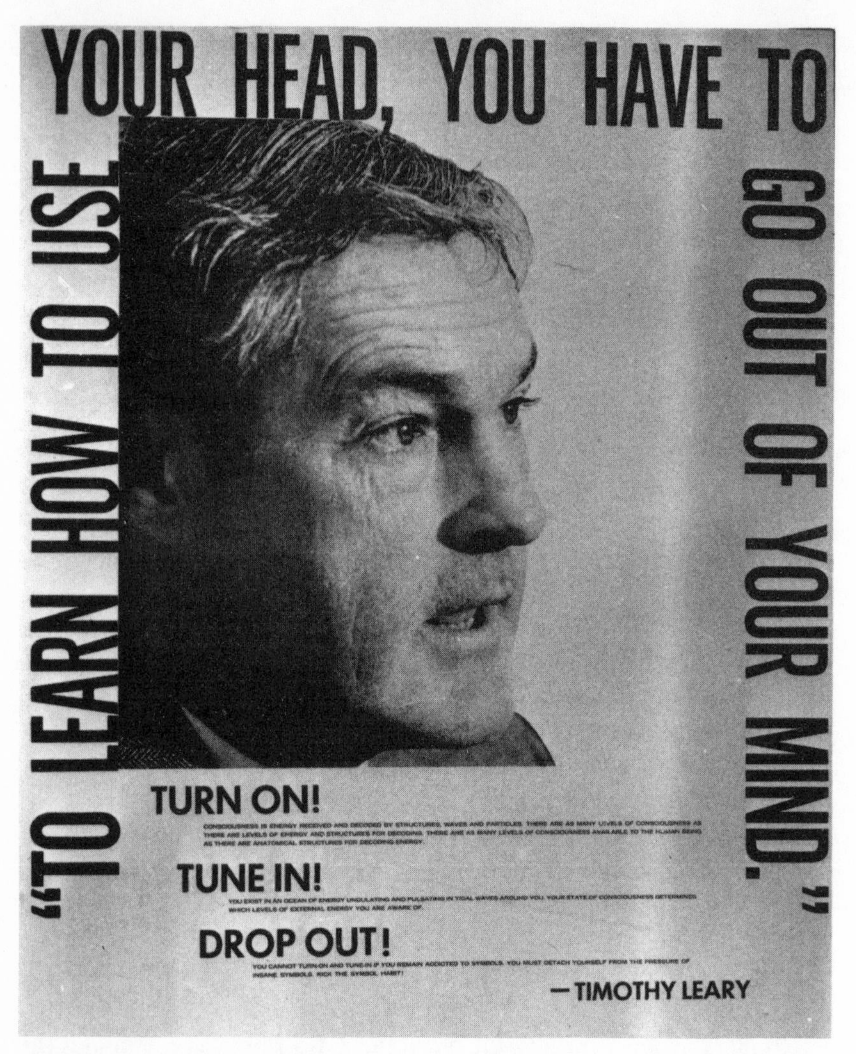

Timothy Leary photographed during the height of his fame as High Priest of LSD. © Hulton Archive / Getty Images

11

High Priests

Then God spoke to me . . . I saw in a quick glimpse the design of the universe. The blueprint of evolution. The impersonal, staggering grandeur of the game . . .

Timothy Leary, recalling his first LSD trip[1]

You take the blue pill and the story ends . . . You take the red pill and you stay in Wonderland and I show you how deep the rabbit-hole goes.

'Morpheus', *The Matrix*

If the psychedelic sixties could be viewed as a Technicolor movie, then its star would undoubtedly have been LSD. When Jimi Hendrix nonchalantly asked, 'Are you experienced?' he meant 'Have you dropped Acid?' When Grace Slick sang about Alice, mushrooms and Wonderland in her masterpiece *White Rabbit*, her meaning was quite obviously allegorical. And when, amidst a psychedelic melange of twisted sounds and backwards guitars, John Lennon shattered the anodyne pop of The Beatles by imploring us to turn off our minds, relax and float downstream, it was clear just what he was referring to.

Magic mushrooms, like mescaline, played a supporting role in this movie, vital to the direction and development of the plot, but lurking in the wings and only taking centre stage in the events that came afterwards. Had Wasson never made his magic mushroom discoveries and broadcast them to the world, had Humphrey Osmond never given Aldous Huxley mescaline that spring morning in Los Angeles, LSD would still have found its way into popular culture as surely as any acid eats its way through matter and metal. The plot wheels were set in motion as soon as Albert Hofmann mounted his bicycle way back in 1943, were propelled forward when Sandoz put LSD on the market, and were unstoppable once the drug was taken up by research psychologists across the Western world. The rather elastic boundaries of their experiments ensured that Acid would never stay bottled up in the clinic, for its psychoactive effects were simply too dramatic and too profound to remain secret for long.

But the exact manner in which the story unfolded was determined by these early luminaries, Wasson and Huxley, and their favoured drugs. For had one man in particular not found out about Wasson's mushrooms and been persuaded to try them himself, the movie would have been quite different. Married with kids, a model tenured academic at Harvard, and a brilliant clinical psychologist who had revolutionised the field of personality testing, Dr Timothy Leary made an unlikely candidate for the position of Acid guru. Nevertheless, within the space of five years that is exactly what he had become, a role that turned him into a pariah and a jailbird, made his name a byword for degeneracy, and led to him being branded by one federal judge 'the most dangerous man in America'.[2]

The story of Acid, with its rainbow trajectory and its highs and lows, has been told so many times that the pivotal moments and key characters have become etched into the popular imagination.[3] To list a few, in America: San Francisco and the Haight-Ashbury, with its love-ins and be-ins; Ken Kesey, The Merry Pranksters, and their 'Acid Tests'; The Grateful Dead; the preposterously named underground Acid chemist Augustus Owsley Stanley III; the gay, naked, Beat poet Allen Ginsberg; the highs of Monterey and Woodstock, and the corresponding lows of Altamont; and, more bleakly still, the psychotic killer Charles Manson. In Britain: London with its 'happenings' and the 'Fourteen Hour Technicolor Dream'; the Notting Hill squat scene; clubs like UFO and Middle Earth and their resident band, Pink Floyd; the maudlin psychiatrist R. D. Laing; underground rags and mags like *Oz* and *International Times*; the hippy trail to India; *Sergeant Pepper*; the Rolling Stones in Hyde Park, and Brian Jones, dead, in a swimming pool. But of all these, Leary stands out as perhaps the most iconic, thanks to the carefully crafted image he presented of himself as the kaftan-wearing academic with flowers in his hair, urging middle America to 'turn on, tune in, and drop out'.

The story, as commonly told, of how Leary went from respected academic to Acid priest, began in August 1960 when he ate magic mushrooms in a rented villa in the Mexican resort town of Cuernavaca. He first heard about mushrooms from his friend and colleague Frank Barron – who naturally had read Wasson's *Life* article –

but, initially sceptical, Leary had to be convinced that this was something worth trying. Judicious inquiries by another colleague, anthropologist Gerhart Braun, led him to Señora Juana, an elderly Mexican woman who willingly parted with a bundle of freshly picked mushrooms for just a few pesos. But whereas Wasson had tramped through the hinterlands in search of a primitive encounter with Sabina and her holy mushroom *veladas*, Leary, rather tellingly, took his mushrooms by the pool, with cold beers and dry martinis close at hand. He was joined by an indolent circle of academics – and their bikini-clad wives – who were moved to try anything to while away the long, hot summer days. The playboy environment did not make the experience any less profound, however, for Leary oscillated between having intense sensual encounters with the women folk, and the feeling that he was being dragged backwards through his own evolutionary history: he regressed until he was no more than a single amoebic cell. 'It was the classic visionary voyage and I came back a changed man,' he later wrote. 'You are never the same after you've had that one flash glimpse down the cellular time tunnel. You are never the same after you've had the veil drawn.'[4]

Later, he wrote that until that moment he had been 'a middle-aged man involved in the middle-aged process of dying'.[5] He returned to Harvard both reinvigorated and determined to abandon all his previous research in favour of further explorations into the 'strange deep realms' that the mushrooms opened up. He quickly learned, following a judicious meeting with Aldous Huxley, that the mushrooms' chemical ingredients were readily available from Sandoz – thus solving problems of supply – and so he initiated what must surely rank as one of the most unusual episodes in American intellectual history: the Harvard Psilocybin Project.

Still at this point a rationalist, atheistic psychologist, Leary's principal concern in initiating the project was to see how people might be beneficially changed through psychedelic experiences: personality assessment was, after all, his speciality. But in what was to prove a crucial decision, Leary rejected the clinical, quantitative approach – that of statistically comparing personality traits before and after the experience – in favour of one that was subjective and qualitative. 'Our experiments would not follow the medical model of giving drugs to others and then observing only external results,' he proclaimed.[6] Instead, he assembled an ever-expanding team of willing volunteers –

academics, poets, artists, musicians, graduate students, friends of friends – and, breaking all the academic rules, took psilocybin with them in a 'supportive environment'.[7] This typically meant a relaxed home setting where volunteers could listen to music, browse through books on art, or make love even. Subjects were asked to record their experiences in whatever way they felt appropriate, be that in a painting, a poem, or an annotated report. In other words, Leary moved in exactly the opposite direction to the Parisian researchers, who had abandoned subjective accounts as too woolly, too imprecise, too vague on crucial details.

Over the years that the Psilocybin Project ran its chaotic course – from 1960 to 1963, during which time some two hundred doses were administered – the project began to look less like a scientific experiment and more like a psychedelic tea party, or worse, like a religious cult with Leary its ebullient leader. For, quite surprisingly in this rationalistic academic environment, increasing numbers of subjects were returning from their 'mushroom' trips with reports of religious, spiritual or mystical experiences. Take infamous Beat poet Allen Ginsberg, for example, who, halfway through his first psilocybin session, descended naked from his bedroom, declaring that he was the Messiah. He intended to walk the streets of Cambridge instructing people to stop hating one another. Careful redirection persuaded him against this somewhat inadvisable course of action.[8] Other equally starry-eyed explorers gathered in Leary's cramped office, muttering about the other world, cosmic consciousness, and the need to have taken psilocybin personally in order to 'understand'.

The project became cliquey, with furtive nods and winks dividing those that 'had' from those that 'hadn't'. Unsurprisingly, faculty divisions started to open up over Leary's steadfast refusal to rein in the project and play by the accepted academic rules. In the end, he attracted too much unwelcome press attention and ruffled too many feathers, particularly over the issue of giving psilocybin to graduate students, and after some wrangling the project was forcibly closed down. Harvard, worried by the stain on its reputation, gave Leary his marching orders, and in the very same year, almost as if to erase all trace of his ever being there, his controversial office on Divinity Avenue was removed to make way for a new and larger psychology building, in which psychedelic research would have no place.[9]

Leary was to prove irrepressible, however, for he positively revelled in his outlaw status. During the latter days of the Psilocybin Project, he had finally taken LSD, a drug that, though at first he was reluctant to try, proved even more astonishing in its effects than psilocybin. 'I became initiated into an ancient company of illumined seers,' he later wrote of his first Acid trip. 'I had taken the God-step.'[10] His colleagues feared for his sanity, but after a day or so he returned to normal – or so it seemed – and thereafter he became perhaps the world's greatest proselytiser for the drug: everyone should take it, he said, everyone should turn on.

Together with Richard Alpert and Ralph Metzner, two other Harvard colleagues fired during the psilocybin furore, he founded IFIF, the International Foundation for Internal Freedom, and then the Castalian Foundation, both with the aim of promoting cultural revolution and spiritual enlightenment through mass consumption of LSD. The trio authored *The Psychedelic Experience*, a tripper's manual loosely based on the Tibetan Buddhist text *The Book of the Dead*, and designed to steer Acid initiates towards a peak ecstatic experience (it was this book that inspired John Lennon to write 'Tomorrow Never Knows', referred to above).

Sponsored by wealthy patrons from the Hitchcock dynasty, and with his growing and well-to-do entourage, Leary moved into Millbrook, a rambling mansion set in a four-thousand-acre estate in upstate New York. It became the prototypical hippy commune, centred, of course, around the taking of LSD. He published two autobiographies, gave an infamous interview for *Playboy* – in which he claimed women could expect to have a thousand orgasms on an average Acid trip – and appeared on radio and TV extolling the chemical's benefits. His notoriety at a peak, he stood as a candidate for Governor of California against the ultimately victorious Ronald Reagan: Leary the academic had made the transition to Leary the prophet or, as he was punningly dubbed, Leary the 'High Priest'.[11]

But by the mid 1960s, the backlash was in full swing. Not everyone found heaven on LSD: some found hell, and a small proportion lost their minds completely. The politicians and media latched onto these horror stories and stirred up a moral panic about 'the drug that made you mad'. In truth, they were more concerned about increasing numbers of people who had taken LSD and lost not their minds but their motivation to work: many, especially good middle-class kids, were taking Leary's advice and dropping out. As Acid's most outspoken advocate, Leary made an obvious target. Prior to the California elections, he was arrested for possessing cannabis, and by the end of the decade he was in jail. Possession of LSD was made a federal offence in America in 1966, and overnight many hundreds of research projects across the States were shut down. Thereafter, research into the effects of psychedelics became impossible and unthinkable, a situation that only now, forty years later, is beginning to change.

Leary's adventures did not end there, however. He escaped prison, and lived a fugitive lifestyle in Tangiers and Switzerland during the 1970s, before eventually returning to America to serve out his time. During this period, his influence waned steadily, only reviving slightly when he found a new following amongst the Rave generation of the 1980s and 1990s. He died in 1996, and at his bequest his ashes were fired into space: in death he orbits the earth, just as, one presumes, he did so often in life. He managed the last laugh in the end.

This, then, is the story as it is usually told: the likely plotline of any *Timothy Leary: The Movie* biopic waiting to be made. But, while none of the facts are in question, it is very much the story as Leary wanted it remembered, an uncritical narrative that reinforces his view of psychedelics as inherently benevolent revolutionary agents. If Leary is to be believed, all the events that later befell him were ineluctably set in motion the moment he took those mushrooms by the pool in Cuernavaca.

It is interesting, though, to compare Leary's reading with that of the researchers working contemporaneously in Paris, who never highlighted the supposedly revolutionary aspects of psychedelics in quite the same way. Indeed, they stuck rigidly to empirical research methods and the scientifically sanctioned way of doing things that Leary eventually rejected as passé. For whereas Leary thought psychedelics would produce a sudden and irreparable transformation in society, the Parisians made no greater claim than that psilocybin might produce a shift in painting style.

That the same drugs could be considered so differently by two teams working in different countries rather suggests that their contrasting ideas were the products of their time and place, and not, as Leary maintained, of some inherent or essential quality of the psychedelic experience. To unpick the Leary myth a little, then, we need to examine the ideas that informed his thinking, and that were seemingly confirmed by his psychedelic experiences. In fact he expressed these ideas very clearly in the six syllables of his famous motto: 'turn on, tune in, drop out'. Catchphrase, mantra and advertising slogan, all rolled into one, it was also a statement of scientific belief.

The psychological school of thought to which Leary remained wedded all his life, and which indeed dominated psychological thinking in post-war America, was called behaviourism. As we have seen, the Parisians adhered to a Freudian-influenced structuralist model that prioritised the role of unconscious elements and forces in determining personality. Behaviourism, on the other hand, regarded external, environmental factors as the principal factor influencing character and behaviour. Behaviourism's most ardent advocate was Frederic Skinner (1904–1990), inventor of the 'Skinner Box' in which rats or pigeons could be 'conditioned' to behave in predictable ways by a regime of punishments and rewards. Skinner's belief in behaviourism was so extreme that once, when asked about the significance of a student's mushroom trip, he replied that manipulation of the correct environmental stimuli would produce identical effects, and that mushrooms, in themselves, were irrelevant.[12]

Most behaviourists would not go as far as that. For Leary, behaviourism meant simply that we come into the world as blank canvases, and that the way in which our personalities subsequently develop is determined primarily by our physical and social environment. Just as geese 'imprint' the first moving object they see as 'mother' – whether that object is indeed mother goose, human researcher, or even, as has been recorded, a passing car – so, Leary thought, we are all conditioned to respond in certain ways because of the various imprints hard-wired into our brains at key moments in our development.[13] From then on, we are compelled into acting out socially sanctioned 'games' – another idea popular in America in the 1950s and 1960s – such as husband, father, breadwinner, housewife, mother, hostess and so on. Society was, in other words, one giant, human-sized Skinner Box.

What psychedelics do, Leary maintained, is allow us momentarily to see these games for what they are, meaningless socially sanctioned interactions that imprison us and prevent us from fulfilling our own unlimited potential: psychedelics let us think outside the box. At the same time, they soften the soldering of our neural wiring and give us the power to re-imprint ourselves in the manner of our choosing, before we return to normal consciousness. 'The Timothy Leary game now existed only as a memory,' he wrote of his first Acid trip. 'I was liberated. Free to do anything I chose . . . Go back down and wander through the planet as anyone I chose to be. Pick a role. Select a costume.'[14] With psychedelics, then, he felt that he had found the holy grail of psychology: an agent that would deliver permanent and demonstrable personality change.

This is why Leary rejected what he called the 'psychology game' of tests and measurements. What mattered to him was whether someone had had that transcendent moment, whether they had stood outside of their conditioning to glimpse the essential nature of selfhood. He came to believe this was something that could never be measured, but could only be recognised by a look or a glint in the eye, by a turn of phrase, a lightness of step. Very early on he started using the language of 'turning on' – have you turned on? have you been there? are you experienced? – for if enough people turned on and saw the petty trivialities of society's games, then these restricting institutions would crumble and a new harmonious psychedelic era would be ushered in. It was to be nothing less than a bloodless revolution. No wonder that he refused to play the Harvard game, for he had bigger issues on his mind.

By contrast, in France, the psychoanalytic tradition meant therapy was regarded less as a sudden transformation and more as an ongoing and continually unfolding process. Through therapy, psychedelic or otherwise, the various repressed elements of the unconscious would slowly and steadfastly rise up from the depths to be integrated into the psyche: but there was always more work to be done. Furthermore, as the traditional modes of resistance were still very much alive, there was little incentive to use alternative means of raising consciousness. The Left was still a vital force in politics, the barricades were never very far away, and students were still able to rock the government, as they did in May 1968. And, of course, the French intellectual tradition, with its roots going back to the pre-Revolution Enlightenment *philosophes*, had too much faith invested in the academic project for it to be so easily abandoned. If it was a game, then it was a good game, and one that was a vital part of the national character and of civilised life.

Political dissent and the struggle for Civil Rights were, of course, crucially important in sixties America too, but many in the counterculture abandoned politics in favour of Leary's spiritual revolution. Even if the Left had 'full control of society', said Leary, 'they would still be involved in an anthill social system unless they opened themselves up first'.[15]

The idea, then, of the 'Acid flash' as causing an instant and irreversible personality change was a particularly American one. The CIA hoped fervently that LSD would turn people into compliant zombies, while politicians worried that it already had. Leary insisted that we were compliant zombies already, and that Acid would set us free, while on the West Coast, Ken Kesey and The Merry Pranksters made the flash a trial by fire with their provocative challenge: 'Have you passed the Acid Test?'

Without wishing in any sense to undermine the enormity and profundity of anyone's psychedelic experiences, this view is problematic. There is nothing intrinsic to an Acid or mushroom trip that makes someone, say, decide to grow their hair, wear beads and kaftans, experiment with non-nuclear living arrangements, or listen to The Grateful Dead. These are cultural choices, matters of lifestyle and identity, determined through the interplay of peer pressure and personal choice. There are countless examples of Leary's contemporaries who, having 'turned on', most certainly did not 'drop out'. Gordon Wasson provides an obvious example, but take also the troubled author Arthur Koestler (1905–1983), who found Leary's solipsistic musings so offensive that he actually walked out of his psilocybin session. 'Chemically induced raptures may be frightening or wonderfully gratifying,' he later wrote in an article for the *Sunday Telegraph*, but 'in either case they are in the nature of confidence tricks played on one's own nervous system.'[16] No sudden change there.

But the best example of all must be Leary himself, who possessed an indelible rebellious streak – bound up with his somewhat self-fulfilling identity as a downtrodden Irish-American – that surfaced time and again throughout his life.[17] In spite of his many hundreds of trips, the Timothy Leary 'game' stayed remarkably constant. Contrary to the myth, he was not turned instantly and magically from a stuffy academic into a laid-back hippy messiah by a single touch of the Mexican mushrooms. Rather, his transformation was a gradual process of messianic drift, occurring across a period of years (during which time he continued to perform scientific research), which was also the product of social and cultural forces. His was a conscious choice, in other words, and psychedelics simply provided him with a platform upon which this rebellious side could find expression. The trouble was that, having dropped out, it was very difficult for him to do anything else, and having chosen the 'outlaw game' he was forced to play it for the rest of his life without respite, even as the world moved on, tuned out and eventually turned off. No wonder he invested so much energy in promoting his own personal mythology.

The gradual way in which Leary let go of the 'academic game' in favour of the 'guru game' is illustrated by the fact that, in spite of his supposed rejection of the 'psychology game', he continued to undertake some statistical research. The results are widely cited by mushroom enthusiasts as evidence for the transformational power of psychedelics, and hence form an important episode in the history of the magic mushroom. But tellingly, with a stake in both games, Leary did not conduct the research well.

Leary's Concord Prison experiment, which ran from 1961 to 1963, had its genesis in the profound and supposedly life-changing experiences of volunteers in the Psilocybin Project. Why not offer psilocybin to prison inmates due for parole, for surely a peak psilocybin experience would reduce their likelihood of re-offending? Leary worked hard to overcome the initial mistrust of the prison authorities, which is perhaps why he actually used quantitative methods here – it was more or less the last time that he did – but, as ever, he and the other psychologists took psilocybin along with the volunteer prisoners. The very first session started badly, as Leary experienced anxiety about tripping behind bars and in the company of convicted criminals. He admitted this, and an inmate confessed that he was equally afraid of the mad scientist who was about to meddle with his mind! Formalities over, the session thereafter ran smoothly.[18]

The results of the Concord experiment have been cited, time and again, as proof of the transformative power of psilocybin in particular, and psychedelics in general: ten months following release, recidivism rates amongst those that had taken psilocybin were roughly half those of a similar control group. After twenty-six months the rates had crept up again, but Leary maintained that this was because of the excessive scrutiny to which the experimental group were being subjected: they were being pulled in on minor parole infractions that would ordinarily pass unnoticed.

The prison volunteers seemed almost universally enthusiastic about the drug and their experiences. One was overcome with remorse when he was confronted with the stark reality of a life of crime. 'I saw what a life of crime was, hated it, fought it . . . I saw that crime was foolish, a coward's way out, a ridiculous flaunt in a child's game.'[19] This was promising stuff.

In a brilliant long-term follow-up study conducted by Dr Rick Doblin and his team from the Multidisciplinary Association for Psychedelic Studies (MAPS), however, in which Leary's original data was painstakingly recovered and revisited, it was found that Leary had used some unorthodox statistical methods.[20] When the rather glaring distortions were properly ironed out, Doblin and his team found that psilocybin had had no statistically significant effect upon recidivism rates, and that Leary had wittingly or unwittingly steered his results in a direction that would endorse his belief in psilocybin as a magic cure-all. Just like Wasson, Leary was too eager to fit the facts to his theory, rather than vice versa. 'The failure of the Concord Prison experiment,' concluded Doblin, 'should finally put to rest the myth of psychedelic drugs as magic bullets, the ingestion of which will automatically confer wisdom and create lasting change after just one or even a few experiences.'[21]

Another experiment with which Leary was tangentially involved, however, proved less ambiguous: Walter Pahnke's 'Good Friday Experiment'. Walter Pahnke was both a physician and a minister, and he devised his experiment as the major part of his PhD in Religion and Society at Harvard, under Leary's supervision. The idea was simple: he wanted to find out whether psilocybin would enhance the experience of attending a religious ceremony, or perhaps even induce mystical experiences, for religious believers. Matched pairs of volunteers, all theology graduate students, were randomly given either psilocybin or a placebo, in a double-blind (the placebo contained nicotinic acid, which produces a tingling sensation in the skin, so that controls would be fooled into thinking that they had been given psilocybin). The volunteers spent the session in Boston University's Marsh Chapel, into which a Good Friday service, taking place upstairs, was relayed. Good Friday was chosen as being the most important or emotionally significant moment of the Christian calendar, the festival most likely to induce mystical feelings.

Measuring the results through a series of questionnaires after the event, Pahnke found that volunteers were statistically more likely to have had mystical experiences (such as feelings of unity, or of transcending time and space) if they had taken psilocybin. Again, the experiment has been endlessly cited as proof of the inherent sacramental quality of psychedelics and the experiences they occasion, which is why Doblin and his team chose to investigate the phenomenon through another long-term follow-up.[22] This time, however, while they found significant flaws with the experimental design, the results held, and thirty years after the experiment, when Pahnke's original questionnaires were given to as many of the volunteers as could be found, time had done little to alter the statistics.

Even so, the experiment was at times chaotic. One volunteer had something of a psychotic reaction and forced his way out of the building, where his erratic behaviour proved problematic. He apparently felt that he had been told by God to go forth and proclaim the dawning messianic age, and was intent on doing just that – beginning by confronting the Dean of the University. He had to be restrained and injected with thorazine to sedate him, a fact that Pahnke curiously omitted to include in his thesis.

Another problem was that it very quickly became clear to all present exactly who had received the psilocybin and who hadn't, for one half of the congregation were enraptured and muttering mystical pronouncements, while the other half, understandably feeling left out, were bored, frustrated, and somewhat disruptive. One volunteer, Huston Smith, now a respected scholar of religion, recalls that as the psilocybin began to work he turned to his companion and ominously declared that 'it was all true'. His companion, sober and left out, later admitted that he had had no idea what Smith was talking about.[23]

Doblin concluded that, despite these flaws, the experiment strongly supported the hypothesis that psychedelics can help facilitate mystical experiences when used by religiously inclined people in a religious setting. In follow-up interviews, many of the experimental subjects recalled that, in spite of occasional difficult moments, the experiences were by and large extremely positive. For Huston Smith, it was 'the most powerful cosmic homecoming I have ever experienced'.[24] And while the majority of the volunteers never tried psychedelics again, many went into the priesthood where their participation in the experiment remained a lasting influence on their outlook and pastoral approach. 'We took such an infinitesimal amount of psilocybin,' said one, 'and yet it connected me to infinity.'[25]

Though Leary was wrong about the power of psychedelics to change personality, LSD nevertheless gave birth to a cultural movement. The great swirling outpouring of fashion, art and especially music that characterised the era all attempted in some way to reflect the psychedelic LSD experience and, of course, to advertise membership of that elect club. Interestingly, however, the number of people actually using psychedelics in the 1960s was a fraction of those doing so today.

Psychedelia was as much a fashion or a craze as it was a direct response to personal drug experiences. 'You didn't have to take the drug to pick up what would have been termed "the vibes",' reminisced the writer Jonathan Green, 'and the LSD culture spread far wider and faster than the drug itself.'[26] But what psychedelia, the cultural phenomenon, did was open up and broadcast a new and favourable way of understanding the strange effects produced by magic mushrooms, and so it was only a matter of time before the freaks and hippies started seeking out the strange hallucinogenic fungi for themselves.

12

Ripples and Waves

'Mushrooming sounds to me like a risky proposition.'
'A bit like life itself,' said Amanda.

Tom Robbins, *Another Roadside Attraction*[1]

What is this madness, I thought. Why are those people dragging themselves through this ungodly drizzle?

Timothy Egan, *University of Washington Daily*, 29 October 1973[2]

How does a new idea come to prominence? How do fashions, crazes, urban myths, conspiracy theories or even new illicit pleasures spread and take hold? The millions invested in marketing and advertising by political parties and corporations alike suggest that the process is not a straightforward one of top-down product placement, for the populace is notoriously fickle. Nor can it be said that new ideas rise effortlessly from the bottom up, for there are often conservative forces working against the spread of novelty and innovation.[3] Then again, those conservative forces are often quite unable to stop the spread of a forbidden or frowned-upon practice, such as a new-found drug pleasure. There is no simple formula or calculation, no easy way to predict whether an idea, product, policy initiative or practice will catch on or take hold. We do at least have a name for that elusive moment when it does, for when a critical mass has been reached and universality is assured. We call it 'the tipping point', a snappy phrase that rather nicely crossed its own tipping point thanks to the best-selling book of the same name.[4]

The magic mushroom 'tipped' on both sides of the Atlantic during the 1970s, having arrived in the wake of LSD. Psychedelia meant that for the first time in Western history the effects of magic mushrooms – the colours and hallucinations, the bodily perturbations, the emotional excitation – had become *desirable*. Consequently, as the news spread out in waves, mushrooms very quickly went from shunned poison, to academic curiosity, to popular bohemian drug choice. And just as with

Acid before it, the first reverberations of this newly discovered delight, sprouting everywhere with impunity, were to be felt in America.

The headlong expansion of what, at the time, was called 'psilocybian consciousness'[5] in the States can be attributed to several factors. It was passed on as a form of 'folk knowledge' by intrepid experimenters who had returned from the Mexican mushroom pilgrimage and discovered hallucinogenic species at home. Hostile press attention surrounding this newfound habit – which largely bemoaned the unwelcome influx of mushroom pickers onto farmers' land – succeeded in bringing it to even wider attention (as moral panics are so often wont to do[6]). But mainly, continued academic interest in the matter of mushrooms ensured that the latest information concerning taxonomy, identification, pharmacology and dosage filtered down into, and circulated freely within, popular culture.

This last point should, at the very least, raise an eyebrow, for it reveals that, although clinical psychedelic research had become impossible after the Leary debacle, other disciplines were able to carry on investigating these drugs quietly and away from view. There was nothing to stop mycologists, for example, describing and identifying new psilocybin-containing species as and when they were discovered. Nor was there an injunction preventing anthropologists, ethnobotanists and pharmacologists from studying the indigenous use of plant hallucinogens in other, distant cultures. And what relevance could the situation at home possibly have if local convention required the scholar in the field to sup upon psychoactive brews? Remarkably, though papers published in serious-minded journals like the *Journal of Psychedelic Drugs*, the *Journal of Altered States of Consciousness* and the *Journal of Ethnopharmacology* at the time read like extended global drugs jaunts, this type of research flourished without censure. Consequently, the 1960s and 1970s saw a series of academic conferences and publications on the matter of indigenous drug use, which worked to keep the latest discoveries about plant hallucinogens very much in the public domain.

The first of these conferences took place in San Francisco in January 1967, was organised by the National Institute of Mental Health, and was attended principally by ethnobotanists, pharmacologists, chemists and psychiatrists. The proceedings were published in a large and rather dry academic tome, *Ethnopharmacologic Search for Psychoactive Drugs*, which included chapters on the history of the discipline, together with pharmacological analyses of drugs as varied as nutmeg (*Myristica fragrans*), Kava (*Piper methysticum*), ayahuasca and the fly-agaric (Gordon Wasson was, needless to say, a contributor). The tone of the book was rigorous, if not severe. Just two weeks prior to the conference, twenty thousand hippies, or 'freaks' as they preferred, had attended the 'Human Be-In' in San Francisco's Golden Gate Park, a great, free LSD festival, attended by Leary, Ginsberg and the other important figures of the Acid underground. No wonder, then, that a chapter was included in the book about the modern problems of psychedelic consumption: it distanced the work from the illicit psychedelic cultures that were blossoming all around it.

A public lecture series about indigenous hallucinogens, given three years later at the University of California, *was* attended by hippies and gave rise to a more popular book, *Flesh of the Gods*, edited by the young anthropologist Peter Furst. This included accessible chapters on a similar range of indigenous plants and compounds – tobacco, ayahuasca, iboga, peyote and the fly-agaric – as well as an overview of New World psychoactives written by Richard Evans Schultes. This last chapter helpfully provided photos of the Mexican mushrooms, and was one of the first accurate, accessible and readily available books to do so. Nevertheless, Furst was well aware of the potential political difficulties surrounding this kind of research, and his introduction included a passionate argument for the importance of stringent ethnobotanical inquiry. The tone was sober throughout, the target audience academic.

In 1973, however, a more provocative volume was edited by another young anthropologist, Michael Harner, *Hallucinogens and Shamanism*, published by Oxford University Press. If this had a distinctly countercultural attitude then that was because Harner was fresh from the jungles of Ecuador, with the taste of ayahuasca still upon his lips.

The book was typical of those written by young, brilliant, up-and-coming academics who are keen to show off their mastery of the subject while, at the same time, wanting to shake up the discipline with new and radical ideas. Harner berated an older generation of anthropologists for failing to apprehend the importance of psychoactive plants within their studied cultures. More controversially, he argued strongly that Western scholars would never grasp the nuts and bolts of indigenous world views without partaking of these drugs themselves.

The implication was that all the young contributors to his book had done so. A lengthy, elliptical and often rambling chapter by Henry Munn on the use of the 'mushrooms of language' in Mexico, for example, gave the distinct impression of actually having been written under their influence. Another contributor, Marlene Dobkin de Rios, admitted to having used both LSD and ayahuasca. Given the political climate of the times, this was potentially explosive stuff, but it was held somewhat in check because the book was still a serious anthropological text.

Harner not only edited the book but contributed several chapters, including his influential if erroneous piece on the supposed use of psychoactive flying ointments by European 'witches'. Successful as the book was, however, it was to be Harner's farewell to academic life, for during his fieldwork amongst the Shuar Indians of Ecuador he had drunk ayahuasca, 'gone native'[7] and begun training as a shaman.[8] His visionary experiences were so bizarre and overwhelming[9] that, after returning to the States and finishing his PhD, he felt an academic career was untenable.

For example, the first time he drank ayahuasca he felt his soul being transported away in a ship crewed by bird-headed deities.[10] Then he was shown the secrets of life on earth by great black whale-like entities with pterodactyl wings, who were, they said, refugees from another planet in distant outer space. They brutally informed him that he was only being given these secrets because he was dying. Afraid that he really was standing on the precipice, Harner called for an antidote, which his shaman guides hurriedly gave him, and thereafter the menacing black creatures disappeared, to be replaced by steadily diminishing coruscating visions.

Awakening the next day feeling surprisingly relaxed and refreshed after his ordeal, Harner went to see another shaman – blind, but reputedly powerful – to discuss what he had seen. Harner was shocked to find that the shaman was not only familiar with the black whale-creatures, but also warned that they were not to be trusted because they were inveterate liars! Harner felt that, through ayahuasca, he had indeed entered into the cosmological universe of the shaman. The shaman was certainly impressed, and straightaway suggested that Harner might himself become a great shaman. The anthropologist needed no further encouragement to go native, and at once set about the arduous training.

On his return to the States, then, he finally left academia to set up his Foundation for Shamanic Studies, and to write several books on shamanism for a popular audience. He began teaching what he called 'core-shamanism', a set of techniques that he abstracted from the many different 'shamanisms' of the world. In truth, this is a rather stripped-down shamanism compared to the one he had encountered in the Amazon. But if it is a shamanism-lite, a shamanism denuded of the drugs and sorcery, the terror and magic, then it is arguably one that is perfectly tailored to Western needs and expectations.[11]

It was another book, however, published a few years earlier, that really kept the idea of plant hallucinogens and magic mushrooms uppermost in the public imagination. It told the supposedly true story of a young, rationalist anthropology student having his ontological foundations shaken by a hallucinogen-wielding, trickster-like Mexican shaman. Espousing exactly the kind of tales that a generation of Acid-heads wanted to hear, it very quickly became a bestseller. It was called *The Teachings of Don Juan: A Yaqui Way of Knowledge*, and its author was Carlos Castaneda.

According to the story, which continued through many sequels, Castaneda met the shaman Don Juan Matus during the early 1960s while researching the indigenous use of peyote for his anthropology Masters at UCLA. Don Juan proved reticent, obfuscating even, on the matter until Castaneda agreed to become his pupil. Thereafter, Castaneda entered an alien world of witches and sorcery, of crows that could talk, lizards that could prognosticate, and spirits that would kill you in an instant if you transgressed their complex and bewildering strictures. Castaneda played the ignorant fool to Don Juan's wise old man, always asking silly questions, never fully understanding what was happening. Don Juan berated, cajoled and generally bamboozled 'little Carlos' in varyingly successful attempts to get him to step out of his shackled, rational, way of thinking so as to 'see' the world in an entirely new way and to become 'a man of knowledge'. Of course it was possible to be in two places at once, to assume the form of an animal, or to step off a cliff and fly. Castaneda was expected to do all these things and more.

Understandably, Castaneda found much of this challenging to his belief system, and so Don Juan employed a variety of hallucinogenic plants in a desperate attempt to shake him loose of his logical, empirical world view. Castaneda chewed peyote buttons, took Jimson Weed (datura) and half-smoked, half-ingested a crumbly magic mushroom mixture, which Don Juan called 'humito', or his 'little smoke'.

The experiences Castaneda went through were usually terrifying, occasionally beautiful, but always remarkable. The first time he used humito, for example, his body became so insubstantial that he fell backwards through the wall he was leaning upon. Another time he came face to face with a gigantic and hideous beast, which Don Juan had warned him was 'guardian' to the other world. It turned out to be nothing more than a hallucinogenically magnified gnat.

In many ways Castaneda's experiences were similar to Michael Harner's: both had entered a radically unfamiliar shamanic world through the use of powerful plant hallucinogens. Both were anthropologists who were impressed enough with what they discovered to have abandoned their careers and gone native. And both chose to write about their experiences in popular and accessible books. The only difference was that while Harner had actually undergone the training he described, Castaneda had not. In fact, it now seems likely that Castaneda made up the entire adventure in the library at UCLA.

Close reading of the text by various scholars has shown it to be replete with factual errors, inconsistancies and narrative blunders.[12] For example, Don Juan, supposedly an illiterate Indian, possesses a remarkably Westernised vocabulary, especially when compared with, say, Sabina. Similarly, despite Castaneda's allegedly impeccable note-keeping, certain events described across the first three books happen in an inconsistent order. Gordon Wasson, though initially excited by the book, became increasingly frustrated at Castaneda's inability and unwillingness to produce a mushroom specimen for formal identification. Smoking mushrooms was unheard of – it produces mixed results at best[13] – and if true would certainly have merited further investigation. Castaneda's failure to deliver led Wasson to become suspicious,[14] and rightly so, for scholars now believe that Castaneda concocted the 'little smoke' after reading *Mushrooms, Russia and History*. Wasson concluded that *The Teachings* were 'science fiction badly written'.[15]

The whole thing – Don Juan, his teachings and Castaneda's terrifying initiation – was an elaborate fantasy. That it was so widely countenanced says much about the spirit of the times. It was, at its core, a version of the 'magical apprenticeship' story, as relevant to people then as *Harry Potter* is today, for it appealed to the same longing for enchantment. What raised it above the ranks of Tolkienesque fantasy fiction was the supposed use of hallucinogenic plants which, in the eyes of the Acid generation, gave the whole thing believability. For, whether readers had actually taken Acid or not, they knew that the psychedelic experience was supposed to unsettle one's faith in the Newtonian, Cartesian universe. Thus, it was entirely credible that we in the West had seriously underestimated how weird reality is, that Castaneda could just conceivably have jumped off a high cliff and landed unharmed, or fallen asleep in one place and dreamt himself awake in another. If Harner, who had had the real training, was taken in by Castaneda's masterful storytelling, it is not surprising that so many others were as well.[16] But when a glut of would-be shamans journeyed south to find their own Don Juan in the Sonora deserts, they were met by bemused disdain, for the locals had never heard of such a person nor of his preposterous doings.

In very different ways, both Harner and Castaneda rejected academic life to pursue their own ideas of shamanism. But another group of their contemporaries, likewise turned on to psychedelics but looking for legitimate ways to pursue their hallucinogenic interests, chose to take the 'Great White Shamans' – Schultes, Hofmann, and Wasson – as their role-models. All of these young psychedelicists strove to emulate the intellectual and publishing successes of their illustrious forbears. They were Andrew Weil, Jeremy Bigwood, Paul Stamets, Jonathan Ott and Stephen Pollock.

The luxuriantly bearded Andrew Weil is now known as America's most vocal advocate of complementary medicine and 'natural' healing, but he began his career promoting other more psychoactive plants and substances. It was while he was studying medicine at Harvard that he encountered both LSD and the Economic Botany course run by Schultes at the Botanical Museum. Engaged and transformed by both, he thereafter became one of the leading writers and thinkers of the psychedelic underground. His articles appeared in the long-running American dope-mag *High Times*, and his first book, *The Natural Mind* – an impassioned defence of the judicious use of natural (that is, plant-based) psychedelics – published in 1972, was a best-seller.

Jeremy Bigwood is now a journalist, but during the 1970s and early 1980s he worked as a research chemist at Evergreen State College in Olympia, Washington. He conducted important research into the biosynthesis of psilocybin by magic mushrooms, and also into perfecting methods of cultivation.[17]

Paul Stamets is now America's leading expert on mushroom cultivation, and devotes his energies to growing gourmet and medicinal mushrooms, and developing ways to use fungi in ecological restoration. After spending his formative years working as a logger he went on to study Microbiology at Evergreen State University where he eventually taught as an Adjunct Professor. His reliable magic mushroom guide, *Psilocybe Mushrooms and Their Allies*, was published in 1976.

Jonathan Ott began life as a chemist, but transferred across to Ethnobotany when he realised that there was a way to combine his love of chemistry with psychedelics. It was while he was a research fellow in Mexico City that he met Schultes, who introduced him to Wasson. The two got on well, and thereafter Ott became something of a protégé of Wasson's.[18] Ott subsequently wrote a number of expository reference books about the history, chemistry, pharmacology, biology, availability and preparation of old and novel plant hallucinogens, marked by his trademark laconic, and occasionally pedantic, style. His continuing opposition to America's drug laws has made him a hero of the psychedelic underground, which he still regularly addresses at conferences, seminars and other similar events.

But of all these thrusting young psychedelicists, however, the late Stephen Hayden Pollock (1948–1981) was probably the most colourful. A physician at the University of Texas Health Centre in San Antonio, Pollock devoted much of his energy to identifying new psilocybin species, and tracing the spread of what he called 'psilocybian consciousness' throughout the world.[19] He worked on refining cultivation techniques, and even found and developed a new species, *Psilocybe tampanensis*, which is one of the commercial varieties cultivated today. At one time he was alleged to have possessed the world's largest collection of hallucinogenic mushrooms. He sailed close to the wind, however, and his liberal attitude towards selling prescriptions to drug addicts is probably what led to his early and tragic demise. He was found murdered in his home in 1981, most probably for crossing powerful drug barons. At the time of his murder he was under investigation by at least five different government agencies, and had he not been killed, it is likely that his mushrooming interests would have landed him in serious trouble.[20]

All of these young men, in their own ways, made significant contributions to the study of psilocybin mushrooms, but more importantly from a historical point of view, they all wrote enthusiastic academic-style books and papers detailing the dissemination of shrooming throughout America as it occurred. It is largely thanks to them that we know how it happened.[21] Of course, the subtext of these works was thinly disguised and would have been very apparent to their intended readership. By including taxonomic, chemical and pharmacological information about the mushrooms, they hoped to spread the word.

At times, the academic veneer of these papers seems very thin indeed: Pollock's justification of a mushroom-inspired skinny-dip in a secluded lake in Colombia as a 'psychomycological investigation' pushes the limits of scholarly inquiry. 'It became a major decision whether to sit in the sun or the shade,' he wrote of this particular trip. 'The phantasmagoria of color flashes superimposed on a panorama of solar diffraction produced by clouds blowing over the mighty Amazon was awesomely beautiful.'[22] Quite so, but hardly the great scholarly insight it is dressed up to be.

Though, as you might expect, there was often intense competition and rivalry amongst the members of this psychedelic 'brat pack', they came together to organise the *First International Conference on Hallucinogenic Mushrooms*, held in Maytown, Washington, in 1976. Unlike the previous conferences and lecture series already described, this was explicitly aimed at magic mushroom enthusiasts, and was designed to disseminate accurate information about mushroom identification. The mycological establishment had rather buried its head in the sand over the burgeoning recreational use of magic mushrooms in America during the 1970s. Indeed, some had even stated publicly that it would be better for people to die from mistakenly eating poisonous mushrooms than it would be for experts to produce a reliable guide to the hallucinogenic varieties![23] Consequently, in the absence of any proper works, a rash of poorly researched guidebooks were circulating, gleaned from the literature rather than from knowledgeable field experience, and offering potentially dangerous and inaccurate advice. Two such were Enos's *A Key to the American Psilocybin Mushroom* (1970), and Ghouled's *Field Guide to the Psilocybin Mushroom Species Common to North America* (1972). The conference was intended to transform this worrisome situation. Speakers included Wasson, of course, and the Mexican mycologist Gastón Guzmán, who taught people the key techniques of mushroom identification.

The conference was a great success in spite of the fact that rumours of an impending drugs bust led some of the organisers to flee prematurely: embarrassingly for them, the bust never happened. A second, larger conference was organised for the following year. Held in Port Warden, Washington, it brought together Schultes, Hofmann and Wasson, along with other up-and-coming writers and researchers of the psychedelic movement: Andrew Weil, Scott Chilton, Jeremy Bigwood and Carl Ruck. The proceedings were eventually published in 1978 as *Teonanacatl: Hallucinogenic Mushrooms of North America*, edited by Ott and Bigwood. The book quite naturally included detailed descriptions of the region's known psychoactive species, along with accounts of Wasson's rediscovery of the mushrooms and how the knowledge had subsequently been disseminated. A chapter by Bigwood explained the ins and outs of home cultivation. The book sold well, but is now so rare, and so coveted for the information it contains, that in the British Library at least the cover must be disguised lest anyone attempt to steal it.

Once again, the success of this conference led to a third in 1978, this time in San Francisco, on the broader subject of hallucinogenic drugs and attended by the usual suspects.[24] The net result of all these papers, books and conferences was that accurate information about identification, preparation and even cultivation of magic mushrooms was freely available and continually circulating throughout the public domain, where it was put to extremely good use by hippies and other psychedelic enthusiasts.

The earliest record of illicit magic mushroom use in North America is from Canada in 1965. A handful of college students were arrested in Vancouver and found to be in possession of Liberty Caps. The records do not tell us who they were, how they made their fortuitous discovery, or whether they had the opportunity to sample the mushrooms,[25] but in all likelihood one or all of the students had made the pilgrimage to Mexico and discovered the properties of Liberty Caps on their return.

From these early ripples, the first proper waves of psilocybian consciousness began washing up in the beach State of Florida. In 1972, students discovered that *Psilocybe cubensis* grows abundantly there in the summer months, as it does throughout the Gulf States.[26] Anecdotal reports at the time suggested that fraternity parties were being livened up with mushroom omelettes and tea, with perhaps hundreds of people tripping at a time. Two years later, the use of *cubensis* was being reported from Mississippi, which was being hailed as the 'mushroom capital' of the States.[27]

It was another region, though, that could have more legitimately laid claim to this title, for the Pacific Northwest proved to be absolutely replete with psilocybin-containing species. In 1973, a young hack, Timothy Egan, wrote a satirical piece for the *University of Washington Daily* detailing the new craze for magic mushroom-picking. Rising before dawn, and braving inclement weather, irate farmers and truculent cattle, Egan accompanied the eager pickers hunting the plentiful Liberty Caps. 'After a while some of the magic mushroom harvesters stopped picking,' he observed. 'They wandered around, stared at the sky, rolled on the wet grass floor and laughed, some of them hysterically, at cloud formations . . . Apparently they'd been sampling their harvest.'[28]

People may already have been sampling the mushrooms for a few years before Egan swelled their ranks, however. The popular author Tom Robbins claimed to have first tried mushrooms in the 1960s, and subsequently wove them into his debut novel, *Another Roadside Attraction*, published in 1971.[29] This hallucinogenic, erotically charged tale centres around the activities of a bohemian, pantheistic couple who accidentally find themselves in possession of the preserved body of Christ. The gypsy-like heroine, Amanda, forages for mushrooms in the Skagit Valley and, when not entering trances herself, visits a mushroom-eating guru on Bow-Wow mountain. It is even hinted that she uses the fly-agaric, but in truth magic mushrooms play a small part in this cleverly crafted critique of contemporary Christian mores. Even so, Egan reported that Robbins was considered the 'godfather of the mushroom cult', and had inadvertently inspired many of the pickers he met that year.

That said, mushrooming remained very much an underground activity during the first half of the 1970s. For example, having heard rumours that Liberty Caps were being used in Oregon, Andrew Weil took upon himself the arduous task of reporting whether this was indeed the case, setting out in the autumn of 1973. But he only succeeded in obtaining a bag of dodgy chopped-up mushrooms for the princely sum of $15, and these later turned out to be tinned field mushrooms laced with both LSD and another compound, the veterinary anaesthetic PCP.

He returned again in 1975, but this time the visit was successful. Together with some local friends, he collected large numbers of Liberty Caps from a farmer's field. The locals had apparently learned about the mushrooms from mycology students at Oregon State University at Corvallis, although one or two seem to have made the discovery serendipitously. Anecdotal accounts, recorded at the time, stating that people had been using the mushrooms for twenty years or so, were almost certainly untrue. Of Weil's companions, 'Susan' had developed something of a taste for the mushrooms, and would often consume two or three each morning so as to put her in the 'right' state of mind to face the day; 'Greg', meanwhile, usually ate them less frequently, in doses of twenty or so, but occasionally as many as a hundred at a time.

The year 1975 proved particularly abundant for Liberty Caps, so much so that collectors gathered sufficient numbers to market them. They were found to be on sale in Eugene for the then exorbitant price of $75 to $100 per pound, wet weight.[30] Local farmers, like their fellows everywhere, were typically hostile to this seasonal inundation of mushroom hunters in their fields, and some complained vociferously to the press.[31] Sensing a marketing opportunity, however, others started charging for access, with prices ranging from $1 to $25 for a day's picking. One enterprising farmer issued pickers with official blue buckets to show that they had registered.[32]

Of the many psilocybin-containing mushrooms discovered in the Pacific Northwest, the next most important species after the Liberty Cap was the newly identified *Psilocybe stuntzii*.[33] It was found growing in the landscaped gardens of the campus of the University of Washington, on imported bark mulch. In 1973, unknown but intrepid students noticed that, when bruised, the mushroom turned blue, a sign that can indicate the presence of psilocybin (but is most certainly *not* a reliable indicator of edibility). Luckily for them, the mushroom turned out to be both non-poisonous (though it is extremely similar to the deadly poisonous *Galerina autumnalis*), and replete with psilocybin.

Almost immediately thereafter, a mushroom craze spread through the University and on to other colleges in the region (such as the Evergreen State University in Olympia). An article in a student newspaper warning against the practice merely accelerated its spread.[34] The mushroom acquired the name Washington Blue Veil and was usually eaten in doses of about twenty or so. Steven Pollock reported that immediately after Ott's first conference in 1976, a bumper crop of Blue Veils was found growing on the playing fields of Turnwater High School. 'Hundreds of students and other magic mushroom fanciers, neophytes and veterans alike, picked and sampled to their hearts' content, apparently with neither significant mishaps nor coercive administrative intervention.'[35] In fact this craze caused considerable indignation amongst the police and school authorities. The High School Principal, Gordon Prehm, found the whole situation 'disgusting'.[36]

Psilocybin and psilocin had been classified as controlled substances under Federal Law in 1970. The Comprehensive Drug Abuse Prevention and Control Act (Public Law 91–513) made the unauthorised possession, sale or use of a psilocybin mushroom a crime punishable by fine or imprisonment.[37] But the fact that all these activities were illegal did little to halt their spread. Arrests were few and far between, and were more often for trespassing than for possession of a prohibited substance.[38] 'On weekends,' Prehm complained, 'the lawn is covered with mushroom freaks.'[39]

A combination of word of mouth, academic dissemination and press scare stories meant that the magic mushroom finally 'tipped' in America in 1976. The Oregonian newspaper called the autumn months of that year 'mushroom madness time'.[40] Paul Stamets recalls that hunting for magic mushrooms approached the status of a 'national sport'. He estimated that on any particular autumnal day there were probably thousands of people out collecting. 'The wave of interest soon became an invasion,' he later wrote, 'a pandemic and a cause célèbre for an entire generation.'[41]

And apparently, while the growing use of wood chips in garden landscaping contributed to the natural spread of psilocybin mushrooms through suburbia, members of the mushrooming underground did what they could to help the process with 'guerrilla inoculations'. Spawning blocks – wood chips imbued with mycelium – changed hands and were freely shared, most notably at Grateful Dead gigs – and thereafter psilocybin mushrooms started appearing in parks, zoos, arboretums and nurseries across the Northwest. It was, Stamets writes, one 'continually unfolding, exponential wave of mycelial mass'.[42]

The spreading waves of mushroom consciousness were not confined to North America. Magic mushrooms were found and used, typically for the first time, in Jamaica, Hawaii, Guatemala, Venezuela, Argentina and Peru.[43] Andrew Weil continued his investigations by dropping in to an idyllic American hippy commune in Colombia, known as 'La Miel' or 'Honey', where he was entertained with *hongos Colombianos*. A few years later, in the mid 1970s, Pollock made a similar trip and found that mushroom use there was rife.[44]

It was more orthodox waves, however, that carried psilocybian consciousness beyond the Americas, for it seems that it was a surfer who took the knowledge from Hawaii to Australia. The initial discovery of hallucinogenic mushrooms in Hawaii may have something to do with the maverick researcher Andrija Puharich. Puharich, you will recall, had investigated the supposed psychic-enhancing properties of the fly-agaric, and written up his results in his popular book *The Sacred Mushroom*. Perhaps hoping to emulate Wasson, he travelled to Hawaii in 1961 to try to 'uncover' evidence for an ancient mushroom cult there. The *Honolulu Star-Bulletin* reported that he had yet to find the conclusive proof he needed to support his theory (unsurprisingly, given that there was none to be found),[45] but it is possible that the publicity surrounding his visit spurred someone to identify the five or so hallucinogenic species now known to grow in the islands.[46]

However the news reached Hawaii, the unknown surfer carried it to Australia in 1969, where he or she recognised the distinctive form of *Psilocybe cubensis* sprouting plentifully from cowpats. That summer proved to be a bumper year for the mushroom in Queensland, which resulted in a craze amongst the young, and a corresponding moral panic amongst the older generation. Psilocybe mushrooms were made illegal in Queensland in 1971. Again, such legislation, and alarming newspaper headlines – 'Children at a suburban school are getting high on mushrooms' – did little to halt the spread.[47] Underground supply lines opened up to meet the demand in the cities, so that in Hobart in 1972, sales of LSD dropped away almost to nothing because of the abundant availability of mushrooms. The habit amongst surfers was apparently to consume these Golden Tops or Golden Caps in 'smoothies', before heading out to sea to ride the waves.[48]

Several other psilocybin mushrooms were found in Australia, the most important of which were the extremely potent *Copelandia cyanescens*, known as 'blue meanies' (a reference to the fact that they stain blue when bruised, to their strength, and to the villains of The Beatles' cartoon adventure, *Yellow Submarine*). Just a handful of these mushrooms produces an intensely strong trip, and in 1971 a seventeen-year-old girl from Adelaide called for medical assistance after having eaten two or three. She was apparently convinced that she had been turned into a banana, and feared that at any moment someone might try to peel her. That she called for assistance is proof that the poor unfortunate was having a horrible time, but the banana story sounds so like the kinds of fabricated LSD scare stories circulated during the 1960s that it is almost certainly an embellishment.

It took longer for indigenous magic mushrooms to be discovered in New Zealand. The phenomenon was thought to be entirely an Australian one, and until the early 1980s the only mushrooms consumed in New Zealand were those posted across the Tasman Sea. Once again, it was surfers who recognised blue meanies and another species, *Psilocybe tasmaniana*, this time growing in the sand dunes at Khomenii beach in New Plymouth. An unknown British botanist apparently found Liberty Caps growing on the Otago peninsula near the city of Dunedin on the South Island, and very soon the knowledge spread by word of mouth. From 1982 onwards lurid headlines appeared in the press, but these were accompanied by few prosecutions.[49]

It is tempting to suppose that surfers took the knowledge with them up the coast of South East Asia but, satisfying as this would be, there is no evidence for it. What is clear is that by the early 1980s psilocybin mushrooms were being used in Bali, Indonesia, Samoa, the Philippines and Thailand.[50] In Samoa, an 'unknown foreigner' introduced the locals to the properties of *Copelandia cyanescens*, which was previously known as the Ghost Hat and avoided.[51] Thereafter the mushroom became popular amongst the younger indigenous population. More often than not, however, mushroom use in South East Asia was restricted to remote, coastal tourist areas.

Though illegal in all of these countries, it was, until very recently, possible to see mushrooms (usually *Psilocybe cubensis*) openly advertised for the benefit of young European travellers in search of paradisiacal psychedelic full-moon parties. In Thailand there was even rudimentary cultivation of mushrooms, with dung collected and watered until the mushrooms sprouted obligingly. They were sold cooked in omelettes, or blended up with coca-cola in what one can only imagine to be a particularly foul brew.[52] Clearly the black market in magic mushrooms formed an important part of the local economy, but the use of the past tense is necessary because it is unclear how far this has been affected by the 2004 tsunami.

What we have, then, is a remarkable spread of an illicit practice taking place in a mere thirty years or so. Propelled by top-down academic inquiry on the one hand, and by gutsy, bottom-up folk experimentation on the other, mushroom-consciousness expanded out from Mexico, to North and South America, and then right round the Pacific rim. The question we must now ask is how the news reached Europe. To answer that, we need to go to Britain where, in the late 1960s, spiritually inclined hippies started reading books, made the connection and struck gold.

13

Underground, Overground

They looked like a couple of dry seeds, but they were the tiny mushrooms that grow so rarely near Avalon . . . this was a gift more precious than gold.

Marion Bradley, *The Mists of Avalon*[1]

The key to the door of fairyland is now within anyone's reach . . .

George Andrews[2]

Of all the sights, spectacles and commotions that beset the visitor to the modern city of London, one image is guaranteed to arrest attention. The Tube map, the schematic plan of the London Underground railway system designed by Harry Beck in 1931, was created to enable travellers to navigate their way across the capital with a minimum of inconvenience. Such is the success of this 'design classic' that hardly a journey is undertaken without at least a quick perusal of its reassuring lines. But it possesses more than elegant functionality. Its bold colours mark out the tunnels that, spreading like a mycelium beneath the urban expanse, keep the city in constant motion, keep it alive. Like some alchemical glyph, it has acquired its own presence, its own aura, for it has come to represent not just the layout but the many-hued, living spirit of London itself.

As with any abstraction, its seductive simplicity belies a more complex reality. For instance, during the years stretching roughly from 1965 to 1972, London was home to another underground entirely. Though never mapped out in any formal sense, this subterranean movement had its own cartography, its own mycelium of connections, junctions and nodes that briefly intersected with the capital's more familiar superstructure. This underground had clubs and venues, galleries and theatres, squats and crash-pads. It had clothes shops, bookshops, cafés and dope dens. It had advice centres, alternative clinics, schools and even a university. It had its own sacred and significant sites. Within these spaces, politicos and journalists hammered out opinions in eagerly devoured papers and magazines, or harangued people on the streets. Artists, dancers and musicians chipped away at the boundaries of taste and form, while mystics, priests and visionaries watched the movement of planets through Acid-widened eyes, or traced out mythical ley lines across the English landscape.

For those few short years this young, radical, disparate, mostly white and male-dominated movement was united by the utopian belief that the establishment was in its death throes. The old way of doing things, choked and stifled as it was by class, propriety, hypocrisy and the other vestiges of empire, was at an end. A new psychedelic Albion was about to rise from the ashes of this political, cultural, sexual and spiritual revolution. Or so they thought. But if the promised Blakean new dawn never quite emerged from all this foment, dissent and hubris, then something else rather more unexpected did.

As far as it is possible to tell, the first ever intentional psilocybin mushroom trip in Britain took place in or around 1970.[3] Appropriately, it was recorded in *Oz*, the magazine that, along with the fortnightly newspaper the *International Times* (or *IT*), was the most vocal champion of the radical political values and psychedelic aesthetics of the day.[4] As we shall see, it is not entirely clear how the momentous discovery was made, but what is striking is that it took someone such a long time. Just as in America, Acid was being used illicitly from the early 1960s onwards, and with it came news of the mushrooms, the shamans of Siberia and the pilgrimage to Mexico. All the pieces of the puzzle were in place, in other words, and by rights, the Liberty Cap should have been discovered a decade earlier than it was.

In March 1961, for example, Arthur Koestler published his essay 'Return Trip to Nirvana', detailing his psilocybin experience with Timothy Leary. Though hardly a ringing endorsement of psychedelia – Koestler lambasted Leary's 'pressure cooker mysticism'[5] – it nevertheless appeared in the *Sunday Telegraph* and brought the concept of mushrooms to a wide readership.

It was followed a month later by a prime-time BBC television documentary, *Eye on Research*, devoted entirely to the subject of 'The Sacred Mushrooms'.[6] Presented by Raymond Baxter, this precursor to the long-running science series *Tomorrow's World* contained interviews with Gordon Wasson about his discovery and consumption of the Mexican mushrooms, and with Albert Hofmann about the successful isolation of psilocybin. Slides and recordings of María Sabina

were juxtaposed with footage of volunteers undergoing psychedelic therapy at Worcester's Powick Hospital, under the guidance of psychiatrist Ronald Sandison.[7] The tone of the documentary was, in keeping with the times, stuffy. The Mazatecs were portrayed as primitive drug addicts, while the scientists appeared as white-coated and angelic, using purified psilocybin to cure 'previously hopeless mental cases'.[8] Perhaps this explains why a striking picture of a Mexican mushroom, *Psilocybe mexicana* – the species most similar in appearance to the Liberty Cap – sprouting from a Petri dish seems not to have inspired anyone to consider the psychedelic possibilities of our own mycoflora. Though the image was reprinted in the British TV guide *Radio Times*, and delivered straight into the homes of Middle England, no one seems to have grasped its significance.

As the decade wore on, however, news about Sabina and the Huautlan mushroom pilgrimage – not to mention the Mexican government's attempts to stamp it out – filtered steadily into the British media. *The Times* reported the expulsion of hippies by the Mexican authorities in September 1967,[9] while earlier that year the *International Times* enthusiastically described the existence of Mexican mushrooms. This piece, written by Bradley Martin as part of a regular column detailing the latest groovy ways to 'get high', announced that psilocybin and

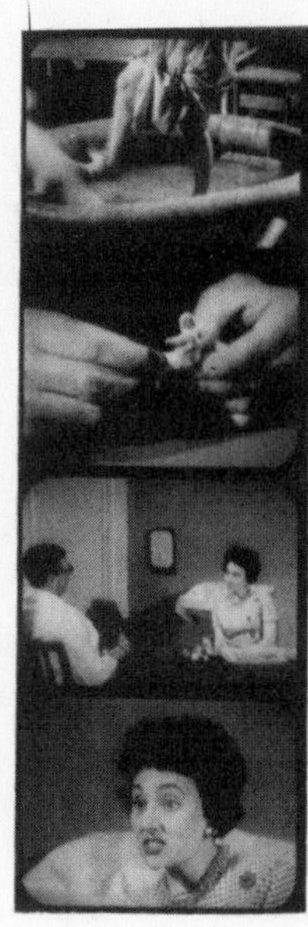

Screen shots from the BBC documentary, *Eye on Research: The Sacred Mushroom*, 1961, showing: María Sabina; presenter Raymond Baxter; Gordon Wasson; photographer Allan Richardson eating mushrooms in Huautla, 1955; chemist Albert Hofmann; examples of art drawn under the influence of LSD; and a patient regressing to childhood during psychedelic therapy. Courtesy R. Gordon Wasson Archive, Harvard University, Cambridge, Massachusetts.

psilocin were 'two beautiful hallucinogens found in the Mexican Sacred Mushroom'. Martin added that 'they grow in this country too – any number of the genus *Psilocybe* will do'.[10] But as he also recommended smoking banana skins (that great sixties myth) and stuffing toad skins 'up your arse', it was not altogether clear whether he meant his advice to be taken seriously: in any case, it all sounded a bit too much like hearsay.

One reason why psilocybin mushrooms were overlooked in Britain for so long was that popular belief still equated the 'magic mushroom' with the fly-agaric, a mushroom that, as we have seen, was not eaten with any degree of enthusiasm. Take the late playwright Jeremy Sandford, for example. Sandford (1934–2003) rose to fame on the back of his shocking TV drama *Cathy Come Home* (1966), which, in one of those defining sixties moments, raised the plight of the homeless to the top of the political agenda. Less well known is Sandford's travelogue detailing his meanderings through Mexico, *In Search of the Magic Mushroom* (1972),[11] in which he swore that the fly-agaric was the only equivalent British hallucinogen. He seems not to have tried it, though, and if Sandford, a true psychedelic initiate, was unaware of the Liberty Cap, then that suggests that most other people were as well.

Sandford was not the first playwright to have taken an interest in the plot possibilities afforded by magic mushrooms, however. Another dramatist had put them on the London stage eight years earlier, but once again clothed them in red and white. Henry Livings (1929–1998) was a British working-class playwright famous for his farces and gritty social commentaries. His *eh?* debuted at the Aldwych Theatre in 1964, and was performed by the newly formed Royal Shakespeare Company no less, under the direction of Peter Hall.[12] The plot revolved around the inauspicious meeting of company boss Mr Price (played by that oleaginous stalwart of British theatre Donald Sinden) and countercultural dandy Valentine Brose (played by the ever-menacing David Warner). In the play, Brose pretends to be looking for work, when in fact he is searching for somewhere to grow his mushrooms. Price's dank factory turns out to offer the perfect conditions. The climax of the play comes when the whole cast accidentally eat Brose's crop of mushrooms and the factory is destroyed in the ensuing chaos: a somewhat clumsy metaphor for the dismantling of capitalism.

Four years later the play was made into a film, *Work is a Four Letter Word* (1968), again directed by Peter Hall and a strong contender for the title of 'worst film ever made'. It is now hard to come by,[13] but one of its few highlights occurs when a young Cilla Black – Liverpudlian pop singer at the time, later doyenne of Saturday evening light entertainment, and now national treasure – clutches a mushroom and with delirious eyes declares that 'God is a circle whose centre is everywhere and its circumference nowhere.'

Livings had clearly never taken mushrooms himself, but had a well-formed idea of the sorts of pronouncements that the psychedelically befuddled were supposed to make. While he was hazy on the details of exactly which mushrooms Brose used – they are described in the script as mauve and prepared by cooking in milk – he almost certainly had the fly-agaric in mind. In the film version, Brose's interest had most definitely turned to Mexican mushrooms, for we first see him seated in front of a poster with *Teonanacatl* written across it in large letters, but nowhere is the idea of a British mushroom countenanced.

The tenacity with which the improbable fly-agaric hogged the psychedelic limelight is explicable given its lurid cultural history, but is still surprising when one considers that scientists had found Liberty Caps to be hallucinogenic as early as 1963. Flushed with success from their Mexican discoveries, Roger Heim and Albert Hofmann (together with their lab assistant Hans Tscherter) wondered whether any European species might not also prove psychoactive. Chromatographic techniques allowed them to test for the presence of psilocybin in a variety of species (without having to eat the mushrooms personally this time), but only the Liberty Cap, *Psilocybe semilanceata*, turned up trumps. Rather satisfyingly, however, it proved to be the closest taxonomical relative to the Mexican species.[14] The finding was replicated in 1969 at London's Imperial College by a young biochemist, Peter Mantle, whose dry paper concluded that 'ingestion of more than about 3g of air-dry sporophores of *P. semilanceata* might be expected to be hallucinogenic'.[15]

Had such a portentous line appeared in an American journal it would, almost certainly, have triggered an immediate response, especially if the research had been conducted in a city that was, like London, in the grip of psychedelia. But a greater gulf between academia and the underground existed in Britain than in the States, at least within this subject area. Mantle's interest in psilocybin was purely biochemical; he was not a hippy, nor did he ever try the mushrooms himself.[16] He obtained specimens from the experimental station at Rothamsted in Hertfordshire, and undertook the assay alone. The final sentence of the paper was mind-expanding only in the sense that it was intended to further scientific knowledge, while the journal in which it appeared, *Transactions of the British Mycological Society*, was obscure enough to ensure that no one stumbled across it accidentally. As for Heim and Hofmann's paper, its publication in the voluminous, multidisciplinary French journal *Comptes Rendus Hebdomadaires des Séances de l'Académie des Sciences* meant that it too went undiscovered in Britain.

If members of the London underground can be excused for overlooking the latest scientific findings, the accusation that they were slow off the mark nevertheless still holds. For one man – a prodigious and charismatic writer and ultimately Professor of Poetry at Oxford – had been volubly proselytising about magic mushrooms since the late 1950s, and not just in high-brow books and essays, but face to face with hippies at his home in Mallorca. He was a friend and correspondent of Gordon Wasson, and he played a vital, if largely forgotten, role in the story of the discovery of the magic mushroom. His name was Robert Graves.

*

The eccentric poet and champion of the magic mushroom, Robert Graves. During his first mushroom trip he felt transported to paradise. © Hulton Archive / Getty Images

Robert Graves (1895–1985) occupies an idiosyncratic position within twentieth-century English literature. Primarily a poet, he was also a prolific writer of essays and historical novels, the most famous of which is probably *I, Claudius* (1934). Educated at Charterhouse and then Oxford, and a veteran of the First World War, he had a lifelong passion for the ancient world. Always destined, perhaps, to become an establishment figure, he nevertheless lived what was by any standards a bohemian lifestyle. He spent most of his adult days in a sort of self-imposed exile amongst the olive groves of his beloved Deià in Mallorca, where he adhered to an eccentric personal mythology. For Graves, poetic inspiration was a gift from the Muse, a once ubiquitous, matriarchal moon goddess long abandoned by patriarchy because of its deification of reason. His ceaseless quest for the Muse – usually in the form of beautiful but cold and domineering women – led him to have a complex and often fraught emotional life (not least his tempestuous liaison with the American poet Laura Riding (1901–1991)). This creative tension, or as he phrased it, the dilemma of 'how to live by intuition and always keep on the beam',[17] suffused his work, while his search for inspiration, in whatever guise, made him receptive to unusual ideas. When he encountered Gordon Wasson's speculative notions about the role of mushrooms in ancient religion he was delighted, for they chimed absolutely with his own way of thinking.

On paper, the two men appeared to have very little in common, the one a measured and exact banker, the other an extravagant and impulsive poet. But they were both amateurs, operating outside the confines of academia, and were equally enthused by the enigmas of the ancient past. They shared the belief that myths were attenuated memories of actual historical events, not psychological processes, and agreed that the comparative method was the means by which to reconstitute these origins. They were enthralled by the idea of trance, whether the poetic frenzy granted by the Muse, or the divinely sanctioned grace of the mystic. And both, publicly at least, expressed an unshakeable self-belief in their oft-times bizarre ideas.

It was Valentina Wasson who first wrote to Graves in January 1949 concerning the murder of the Roman Emperor Claudius, poisoned by mushrooms, but this initiated a long and lively correspondence between the two men. Graves relished their 'queer exchange of letters'[18] and thought Wasson 'one of the very few people whose mind works along the same channels as mine'. Wasson likened their sparky relationship to that of a flint and steel.[19] The banker was a little in awe of the great poet at first, but enjoyed the fact that their friendship helped secure his own reputation.[20] Grateful for this, he even helped the profligate and impecunious poet manage his financial affairs towards a position of solvency.[21] But as Wasson's own reputation grew independently of Graves, there began a definite cooling of the relationship, initiated from the American's side. The exact reason is not clear from their correspondence, but may have had something to do with Graves's wilful habit of misreading, misquoting and embroidering Wasson's discoveries. For the grammarian so obsessed with accuracy that he wrote to Graves pointing out all the typographical errors in his most famous work of non-fiction, *The White Goddess*,[22] such a sloppy attitude may have proved intolerable.[23]

But then, though Graves was an immensely careful poet, he had little time for scholarly accuracy in the orthodox sense. His personal mythology meant that he privileged 'poetic' over 'scholarly' truth, even when his intuitively reached conclusions were unsupported by the facts. In poetic trance, he wrote, 'words come to life, and combine under the poet's supra-conscious guidance, into inevitably true rhythmic statements'.[24] They were the gift of the goddess, after all.

Take, for example, *The White Goddess*, published in 1948 and again in paperback in 1961. It was a work entirely of his poetic imagination, but his imperative style meant that the book was widely and eagerly accepted by an uncritical public as an accurate portrayal of ancient Druidic religion and bardic practice. It became one of the foundational texts of the revived Pagan religions of the twentieth century, many of which diligently worship Graves's invented triple goddess, and beseech her for poetic inspiration, with every passing moon.[25]

Wasson's notions seem to have been particularly potent stimulants for Graves's imagination, and he entered into feverish speculation about the supposed use of hallucinogenic mushrooms in the ancient world. He became increasingly convinced, on the basis of 'poetic' leaps of faith – dizzying even by Wasson's standards – that the fly-agaric had been used as a sacrament in the cult of Dionysus, in early Judaeo-Christian faiths, and by the Iron Age Druids. He regularly bombarded Wasson with his increasingly tenuous connections, once writing 'isn't it strange how I seem to act like a lightning-conductor for these things on your behalf?'[26] But, by the end, Wasson was less than thunderstruck by these flights of fancy that threatened to overshadow his own.

Nevertheless, back in the early, halcyon days of their friendship, Graves had played his part in the Wasson saga and the rediscovery of the Mexican mushrooms. His enthusiastic encouragement, for example, undoubtedly gave Wasson the conviction to pursue and develop the theory of the ancient mushroom taboo, while his unqualified support of the thesis counted towards its uptake on both sides of the Atlantic: Graves championed it at every opportunity.[27] But even more importantly, it was Graves who alerted Wasson to the existence of the Mexican mushroom rituals. In September 1952, he forwarded Wasson an article from a pharmaceutical journal, *Ciba Symposia*, detailing Schultes's identification of *teonanacatl*,[28] with the result that the delighted ethnomycologist immediately set about organising his first trip to Oaxaca.

Though he has received little recognition for his part in the story, Graves did at least get to try the mushrooms himself in January 1960, while a guest at Wasson's New York apartment (Wasson had previously visited Graves in Mallorca in October 1953, and received him in New York in 1957).[29] As we have seen, Wasson hosted a few of these mushroom soirées for invited friends during the years of his Mexican travels, at which he acted as hierophant, served up pills of dried mushrooms (in this instance, *Psilocybe hoogshagenii*) and directed the experience with slides of Sabina and recordings of her bemushroomed chanting. Solemnly, he began each session by advising his congregation to achieve a state of grace.[30]

Graves entered into the experience not only in a state of grace but fully in the expectation that he might be granted a vision of paradise. He had become convinced that the reason why there were so many 'similarities' between the various ancient descriptions of Elysium was that a common hallucinogenic experience lay behind them. It is not necessary to reiterate the reasons why this is unlikely to be true – but, armed with this expectation, Graves had exactly the paradisiacal encounter he craved. Writing to thank Wasson afterwards, he brightly exclaimed that it 'was not merely a red-letter day but a day marked with all the colours of a celestial rainbow'.[31]

Graves's account of his experience is not well known, buried as it is amongst his *Oxford Addresses on Poetry*, delivered during his time as Professor of Poetry; but it stands alongside those of De Quincey,

Huxley and, indeed, Wasson as one of the finest pieces of trip-lit in the English canon. His visions were vivid, for he found himself passing through a marble grotto replete with jewels, grimacing demons, and nakedly dancing caryatids. At one point, he grasped the knowledge of good and evil: 'my mind suddenly became so agile and unfettered that I felt capable of solving any problem in the world; it was as if I had immediate access to all knowledge everywhere. But the sensation of wisdom sufficed – why should I trouble to exploit it?'[32]

He was enraptured by the words of the *curandera* María Sabina who, unsurprisingly, became for him the embodiment of the Muse. 'Each song was followed by a pause,' he wrote of her chanting, 'and always I waited in a lover's agony for her to begin again, tears pricking at my eyelids.'[33] He watched, spellbound, as her words appeared to flow out as an intricate, golden-linked chain. 'Towards the end came a quick, breathless, cheerful song of creation and growth. The notes fell to earth but rose once more in green shoots which soared swiftly up, putting on branches, leaves, flowers – until it dominated the sky like the beanstalk in the fairy tale.'[34]

Graves's evident delight at finding paradise was, however, tempered by sentiments of guilt. He was terrified that the mushroom short cut might somehow jeopardise his ordinary day-to-day connection with the Muse. Poetic inspiration, though often more subtle and far less intense, was his *raison d'être* and, he concluded, was more lastingly worthwhile. He only ever took 'mushrooms', or rather pure psilocybin pills, one further time at a Wasson soirée, but the evening was unpleasant and anticlimactic. None of those present[35] enjoyed the experience and, perhaps tellingly, Graves watched as his wife Beryl's face became 'grotesquely distorted': she swore never to repeat the experience.[36] Blame was attributed to the chemistry of the drug, not that of the group, and the evening's horrors were sufficient to convince Graves that thereafter he should seek inspiration unassisted.

Nevertheless, he went on to argue, and in all seriousness, that the mushroom should be restored to its 'original (presumed) position in religion',[37] administered at puberty by way of initiation, at marriage to deepen the lovers' bond, and in old age to prepare the way for dying. 'Not that I should care to enrol myself in any such cult,' he added haughtily, 'which would imply ecclesiastical discipline and theological dogma, and force me into friendship with co-religionists not chosen by myself.'[38]

The co-religionists he did eventually elect to befriend, however, were nothing if not unusual, for during the 1960s Deià became a haven for hippies, freaks and other assorted members of the underground. Many were made welcome in the Graves household, young and attractive 'hippy-chicks' especially so, for they would be whisked hurriedly away to assist Graves with his musings.[39] The musicians Robert Wyatt, from free-jazz outfit Soft Machine, and Daevid Allen, later visionary leader of the über-hippy psychedelic band Gong, were two of the more famous visitors received at Deià.[40] The question that arises is whether Graves succeeded in inspiring anyone to go looking for magic mushrooms back home. That he was always happy to talk about magic mushrooms is clear, but whether anyone acted directly upon his words remains to be discovered.

Graves, however, was so enthused by magic mushrooms that he could not resist placing them into the revised, paperback editions of *The White Goddess* (1961) and *The Greek Myths* (1960). True to form, he put forward wildly inaccurate speculations as fact, stating, for example, that the fly-agaric, together with *Panaeolus papilionaceus* (at best, a capriciously active species), had been employed in classical times. Happily and erroneously, he declared that his own experiences with psilocybe mushrooms had mirrored the 'ancient toadstool mysteries' of the Celtic bards.[41]

Here, then, was all the information anyone needed to uncover the hallucinogenic properties of the Liberty Cap, bound up within a supposed tradition – invented, but delivered from an apparently reliable source – that proclaimed psilocybe mushrooms to have been safely used throughout antiquity. Both these books were widely read, and within the underground *The White Goddess* was regarded as something of a crucial text. It therefore seems probable that Graves *did* inspire someone to go foraging, either directly or through these printed works. One final accolade must be added, therefore, to those already heaped upon this brilliant but eccentric English poet: not only did he help Wasson discover the Mexican magic mushroom, but he led people to the British one as well. It seems he got his mushroom religion in the end.

It is not possible to say with any certainty whether the momentous mushroom article in *Oz* really does detail the very first intentional British mushroom trip – it is definitely the earliest on record – nor whether the experiences were the author's own. But written by a young hippy immersed in the underground and with a passionate interest in herbalism and the occult, it seems safe to presume that the article was autobiographical. Though it has not proved possible to trace him, documents show that the author, Lynn Darnton, spent much of 1967 living in North London with members of The Exploding Galaxy, 'a particularly Goonish dance-troupe'[42] run by kinetic artist David Medalla. Darnton was evicted from the house for transgressing the group's atypically stringent anti-drugs policy – he smoked dope on the premises[43] – but thereafter formed his own group called, appropriately enough, The Tribe of the Sacred Mushroom (most probably named after Andrija Puharich's book). The Tribe made news in 1968 by upping sticks from London's squats and bedsits, and setting up what *IT* called 'the first rural hippy commune' in Suffolk.[44] There they eked out a living making jewellery and psychedelic posters, and experimenting with macrobiotic diets.

The group were, in the spirit of the times, overtly spiritual. They adhered to their own particular blend of psychedelic mysticism, which mixed Druids, pyramids and Atlantis into its heady, patchouli-scented bricolage. Darnton was very much their high priest,[45] and during 1968 he led the group in a series of psychedelic rituals around the country, designed to answer the portentous appearance of the asteroid Icarus by 'raising vibrations'. In scenes worthy of Henry Livings, a ritual in London's famous club Middle Earth only succeeded in invoking the police.[46] When *The Times* caught up with the Tribe later that year to interview them for its series on the 'restless generation', it found them back in Notting Hill, waiting for a sign, a telepathic communication perhaps, that they should go and live in the more spiritual atmosphere afforded by the Peruvian Andes.[47] Presumably the call never came, their destiny lying at home, but thereafter the Tribe earned its title by devoting itself to fungal experimentation.

Somewhat anticlimactically, the main focus of Darnton's *Oz* article was not the Liberty Cap at all, but rather, as we saw in Chapter Eight, his fly-agaric-fuelled encounter with a parliament of gnomes. Psilocybin mushrooms appear almost as an afterthought, tacked onto the now familiar tales of bemushroomed Siberian shamans and Koryak tribesmen struggling to leap over cracks in the road. Even so, the mushrooms – presumably Liberty Caps – were picked in or around an unnamed English village, dried in front of a fire, and eaten with jam. Very few of their effects are described – Darnton is more concerned with his bemushroomed insights into the supposed kabbalistic origins of the village church – except the peculiar thought that flashed through his mind as the mushrooms took effect: 'my eyeballs have just been cleaned and see how new the world looks!'[48]

The effects of the article were similarly muted, for far from unleashing a mushrooming pandemic, the piece seems to have been met largely with indifference. This ambivalence was possibly due to Darnton's favouring the fly-agaric – a drug never destined to catch on – or to the drug snobbery of the day. For if cannabis served as a conspicuous marker of alterity, one – and only one – other drug conferred membership of the underground's inner sanctum: LSD.[49] To be experienced, in the Jimi Hendrix sense, was to have dropped Acid. In the underground, where status was measured in part by what drugs you took, mushrooms came a poor second to the real thing.

Darnton's article was eventually followed, four years later, by the first guidebook to the British psycho-mycoflora, Richard Cooper's *A Guide to British Psilocybin Mushrooms* (1974).[50] Just who Cooper was, and what his involvement with the underground might have been, remains unclear. The fact that he offered rudimentary descriptions of fungal biology and cultivation techniques, however, suggests that he had at least some mycological training. His somewhat patchily accurate publication, which is still in print, nevertheless very clearly identified the Liberty Cap, and suggested a dose, twenty mushrooms, as a safe starting point for psychedelic experimentation. The booklet was first advertised in the pages of *IT*, and over the years must have sold many thousands of copies, but its initial impact was underwhelming. A year after its publication, poet and author George Andrews bemoaned the fact that hardly anyone knew about or was making use of the psychedelic treasure growing freely in the hills.[51]

As drug pandemics go, then, the Liberty Cap made an unimpressive start. By the mid 1970s, its use was confined to small numbers of hippies, mainly in London and southern England, who were already turned on to Acid or part of the underground. Cooper's guide was helpful, but the news seems to have been passed on slowly as a form of folk knowledge, by word of mouth. All this was to change in the spring of 1976, however, for the magic mushroom suddenly made national news, and overnight it burst into popular consciousness.

*

Throughout the 1970s, there was an increasing sense of moral panic within mainstream culture about escalating drug use in Britain.[52] This growing sense of unease was optimistically accompanied by a feeling that the problem could be solved by simply tightening the prohibition noose. To this end Britain, already a signatory to the UN Single Convention on Narcotic Drugs of 1961, endorsed the UN Convention on Psychotropic Substances in 1971. Championed aggressively by the United States, both agreements defined drug use as a criminal, not a social or medical, problem. It was widely held that the combined efforts of police and customs officers would eradicate it if they were given sufficient time and resources.

In this tough climate, the British Drugs Squad made many high profile 'swoops' upon dealers and users alike that were hungrily reported by the media. Rock stars felt the brunt of this new hard-line policy: David Bowie was arrested for possessing cannabis in 1976; Keith Richards for heroin in 1977; Sid Vicious for amphetamines in 1978; and Paul McCartney for cannabis in 1980. But hardly a week went by without some horror story of heroin addiction amongst the upper classes, solvent abuse amongst the working classes, or LSD-triggered psychosis amongst the middle classes, splashed across the pages of the newspapers. Such was the climate that when a man and a woman appeared in court charged with possessing a new, hitherto unknown, drug – hallucinogenic mushrooms – the case was absolutely guaranteed to generate national media interest.

In April 1976, a young couple from Reading – Michael Garland, a civil servant, aged nineteen, and his landlady, Mrs Lois Wilkinson, twenty-one – became the first people ever to appear in a British court charged with possessing hallucinogenic fungi.[53] Their house had been raided by the Drugs Squad the previous October as part of a crackdown on cannabis use, for which Garland was successfully prosecuted and fined. However, the police had not expected to find magic mushrooms, and they were left scratching their heads over the legality of the dried fungi they discovered neatly wrapped in tissue and stashed away in jars. The defendants maintained that they had collected the mushrooms, along with leaves and nuts, as part of a nature study, and that they were ignorant of any hallucinogenic properties. That they had stored the dried mushrooms with their cannabis supply rather undermined this line of defence, and the trial eventually hinged on a point of law.

Both psilocybin and psilocin had been made illegal under the 1971 Misuse of Drugs Act in a move that, following the wording of the UN Convention on Psychotropic Substances, had outlawed all the synthetic psychedelics known at the time. The thorny question was whether the mushrooms themselves were prohibited for, following pressure from the Mexican government (who did not want to have to waste time and resources preventing indigenous mushroom use), the Convention had listed only the mushrooms' active ingredients, and not the species themselves. After two days of deliberation, the Judge, Peregrine Blomefield, directed the jury to find the pair not guilty. 'It may or may not be that you can get psilocin out of the mushroom,' he concluded. 'But psilocin is a chemical and mushrooms are not. You cannot find this man guilty of possessing psilocin in these circumstances.'[54] The decision opened the first chink of the legal loophole that would eventually allow magic mushrooms to be sold openly and legally in Britain.

Lurid headlines in the press – such as *The Times*'s 'Hallucinatory fungi not illegal, judge rules'[55] – would have been sufficient to ensure that the news about the mushroom was broadcast far and wide. But, almost as if to make sure, Britain's biggest selling popular science weekly, *New Scientist*, chose to repeat the story in an article published in September, just at the start of the mushroom season.[56] The piece included a detailed description and a black-and-white sketch of the Liberty Cap, and helpfully repeated Peter Mantle's portentous line about how many mushrooms one would have to consume for them to have an effect: twenty-five to thirty, it advised. The year 1976 was famously one of drought in Britain and beyond, so mushrooms of all kinds would have been scarce everywhere. But the following year reports of a new drugs craze began springing up in the north of England, Scotland, Norway, Finland, Germany and Holland, spreading out like a fan.[57] The magic mushroom had finally gone overground.

Whether or not this had been the magazine's intention is uncertain, but with a readership spread across Britain and Europe, the *New Scientist* must take its share of responsibility for popularising this new-found high. Two other events that took place in 1977, however, gave the mushroom additional momentum. The first was the publication of a new underground magazine, *Home Grown*, a British dope magazine modelled on the success of America's *High Times*. Appearing just at that transitional time when hippy 'love and peace' was being shattered and overturned by punk's 'anarchy in the UK', the magazine attempted to cater for both by appealing to their shared love of a good spliff. Articles by up-and-coming punk writers – like the one by Julie Burchill and Tony Parsons endorsing amphetamines – rubbed shoulders with more hippy-flavoured pieces about mysticism and LSD. In issue two, the poet George Andrews finally got his chance to tell the world about psilocybin mushrooms, which he did in a colourful article complete with taxonomic descriptions, dosages and trip reports.[58] The *Glasgow Herald*, bemoaning the arrival of the magic mushroom, wrote a piece deploring this article, to which it attributed at least some of the ensuing mushroom pandemic.[59]

The second, coincidental, event turned out to be one of the most famous drug busts of the decade: Operation Julie. In March 1977, no fewer than eight hundred police officers took part in simultaneous dawn raids on an Acid-manufacturing enterprise spread between London and Wales. The ensuing trial at Bristol Crown court captivated both the media and the chattering classes, for those arrested were neither gangsters nor junkies, but white, educated, middle-class professionals. The conspirators were motivated as much by a utopian belief in the transformative powers of LSD as by financial gain. One of the accused – a doctor, Christine Bott – told the court that LSD had had nothing but a beneficial influence upon the world, and that she wished everyone could take it.[60] The incredulous judge, handing out stiff sentences, thought otherwise, and so just as punk was savaging hippy idealism from below, the establishment delivered what it hoped would be its death-blow from above.[61]

The police, however, basking in the glory of a successful undercover operation, had rather overstated their case when they claimed to have removed Europe's biggest Acid manufacturer. Operation Julie had little impact upon supply because other illicit manufacturers simply stepped in to fill the gap created in the market. What the bust did succeed in doing was to add significantly to the growing sense of paranoia amongst drug users, punk and hippy alike. The fear of being stopped and searched, or of having one's door kicked down in the early hours, became a serious impediment to a good trip, and a major disincentive against psychedelic experimentation. Fears about 'bad Acid' – Acid supposedly cut with strychnine or other unpleasant adulterants – were rife. The news, therefore, that there was a powerful hallucinogenic mushroom that was plentiful, free, unadulterated and legal suddenly took on a much greater urgency than it had had seven years earlier. Condescending attitudes towards mushrooms as somehow lesser than LSD were blown away.

The autumn months of 1977 and 1978 apparently produced a bumper crop of Liberty Caps. Correspondingly, in large towns and cities where access to upland pastures was easy – as is the case in Manchester, Aberdeen, Glasgow, Cardiff and Dundee – hospital doctors noticed a strange new phenomenon: the admission of people having bad mushroom trips. Over the next five or six years, a flurry of papers in the medical and mycological literature traced this new curiosity.[62] Most of those admitted were young males, usually teenagers but occasionally school children. Most had eaten in the region of twenty to thirty mushrooms – some, hundreds – but all were having a terrible time. Some were coherent enough to admit themselves to hospital, others were brought in by police or concerned parents: one young man was apparently found wandering naked along a railway track.

Just as with their predecessors 150 years previously, medical staff were somewhat baffled by this phenomenon. All agreed that it was new, unheard of before 1977, and that what they were witnessing was the tip of a large iceberg, for they were only dealing with the casualties. But what to do with the afflicted, they were less certain. Sedating some, pumping the stomachs of others, it took a number of years before they realised that quiet reassurance was generally all that was required.[63] Troubling anecdotal reports suggest that 'gastric lavage' was sometimes used as a form of punishment for wasting precious hospital time.

Clearly, some of the patients' symptoms were highly distressing. One hapless individual managed to trigger a psychosis by taking mushrooms every day for a week, at the same time fasting and avoiding sleep: he was admitted in a clouded and distressed state, and took a further week to recover. Other cases were less grave. One man, having been put to bed, made a bid for freedom, but neglecting to put on his clothes ran off semi-naked into the night: he was never seen again. And others, the doctors wearily reported, eroded bedside patience with their sexual frivolity and 'fatuous comments'.[64]

From the late 1970s to the early 1980s, then, there was a substantial youth-centred magic mushroom craze in Britain, with knowledge passed on largely by word of mouth, and the mushrooms themselves passed around in pubs or sold on the black market. The number of mushroom-related enquiries to the National Poisons Information Service at London's Guy's Hospital increased steadily from 33 in 1978, to 47 in 1979, 96 in 1980, and 142 in 1981.[65] A troubling survey, undertaken in Scotland in 1981 at the peak of the craze, found that 66 per cent of school children in Tayside had at least heard about mushrooms, if not actually tried them.[66] There was a minor moral panic in the national press, fuelled by the unfortunate death of a sixteen-year-old London student, Byron Upton, in 1982. Under the influence of mushrooms, Upton had decided to walk home along the tunnels of the London Underground, where he was knocked down and killed by a train.[67] He was, thankfully, the only mushroom casualty.

It is no coincidence, I think, that this mushroom craze peaked during the early Thatcher years. Within a few short years of the Iron Lady's rise to power, the country was in recession, manufacturing industry was being dismantled, and the disgruntled unions were being tackled head-on. The result was that unemployment in the inner cities skyrocketed, while hope and good prospects became suddenly in short supply. For an increasingly disaffected youth, magic mushrooms were a convenient, illicit and exciting way of making life under Tory rule more tolerable, no better or worse than sniffing glue, the other great drugs craze of the era. But for another, possibly more middle-class section of society, the advent of the magic mushroom proved rather more promising. Contrary to what the punks were saying, hippiedom was not yet dead, and every summer it found expression at a number of occasionally uproarious free music festivals, where mushrooms were eagerly consumed.

The 'hippy festival' emerged as a cultural phenomenon in Britain during the 1970s, from its origins in the jazz and blues festivals of the previous decade.[68] Self-conscious attempts to emulate the success of Woodstock ('three days of love and peace') led to a calendar of commercial and free festivals in Britain, most famously the Isle of Wight, Windsor Free, Watchfield, Glastonbury Fayre, the Albion Fayres, and of course Stonehenge. Stonehenge was truly the mother of all festivals, running as it did for ten years from 1974 to 1984, before being famously and violently shut down by the Thatcher government. Lasting a month each year in a squatted field adjacent to the Neolithic monument, it attracted some thirty thousand people at its peak, and culminated with a mass invasion of the Stones, the symbolic heart of the festival, to mark the summer solstice.

Stonehenge was very much an experiment in alternative living. Bands and theatre groups performed on impromptu stages with Acid-fuelled sets lasting through the night. Druids and other assorted mystics, adherents of the same psychedelic spiritual bricolage as Darnton and the Tribe of the Sacred Mushroom, performed rituals, weddings and blessings. Strange new diets were catered for, while every type of alternative dwelling – from tepees and benders to converted live-in buses – was in evidence. Drugs of all kinds could be openly bought and sold, though heroin dealers were often forcibly removed, triggering some of the occasional outbreaks of violence that marred the superficial harmony. The authorities, meanwhile, could do little but stand on the sidelines and watch. A vision of utopia for some, a running sore for others, Stonehenge became the symbolic centrepiece of a whole free-festival movement, the hub around which the rainbow calendar of festivals revolved.[69] Many people made it their lifestyle to travel from one festival to another in converted buses and trucks, and thus was born the new-age traveller, a curious mixture of hippy idealism, spiritual yearning and, later, punk attitude.

As early as September 1977, the very first free magic mushroom festival was held, in Wales. The Psilly Fair, as it was called, took place in a secluded valley near the village of Pontrhydygroes, outside Aberystwyth.[70] Only about fifty people attended this first event, but by 1980 it was attracting eight hundred or so who brought with them all the trappings expected of a free festival: cafés, generators, a stage, bands, live-in buses and monumental quantities of hash cake. Its daily news sheet, the *Psilly Times*, kept people up to speed with the festival gossip: how the harvest was proceeding, levels of police activity, and where to find sympathetic locals.[71] The festival was organised (in as much as any free festival could be said to have been organised) by a group called The Tibetan Ukrainian Mountain Troupe, who were 'the creative, surreal, prankster circus of the travelling scene'.[72] Together with their offshoot band, The Wystic Mankers, they entertained punters with a not entirely serious mix of burlesque circus skills and improvised bemushroomed noise confusion.

Thereafter mushrooms, and mushroom-inspired bands, became inseparable from the free-festival scene. At Stonehenge in 1979, for example, one observer recorded that 'many folk were spaced out on mushrooms'.[73] Mushroom truffles joined hash cake as a festival staple, and festival-goers could munch them and groove to the mycelial space-rock of bands like The Ozric Tentacles, Boris and his Bolshy Balalaika, and the unambiguously named Magic Mushroom Band.

Of course, the government did everything it could to stop the free festivals and the emerging traveller lifestyle. Local opposition to the Psilly Fairs, for example, was intense, and succeeded in preventing the festival from happening in 1981 and again in 1982. The Stonehenge festival came to an abrupt halt in 1985 when the police, fresh from a year's clashes with striking miners, violently smashed the convoy of buses and trucks attempting to reach the Stones: the 'Battle of the Beanfield'. The law was tightened with the Public Order Act of 1986, and then the Criminal Justice Act of 1994, which effectively gave

Festival favourites: Delerium Record's compilation of psychedelic music, *Fun with Mushrooms* (1993), featuring a cover by cartoonist, Pete Loveday, and the track 'Toadstool Soup' by Boris and his Bolshy Balalaika; and the Magic Mushroom Band's subversively sleeved album, *Bomshankar* (1987).

police the power to prevent illicit gatherings from happening. Since that time, free festivals have been few and far between.

All this struggle, however, gave festival-goers a strength of purpose and identity – getting to the festival became a goal in itself – and they came to see themselves as some ancient pagan tribe, pitted against a brutal and oppressive fascistic regime. The magic mushroom slotted neatly into this romantic mythology, for hadn't the witches, the Druids, the stone-circle builders all used it, and hadn't they all been oppressed in turn by Roman invaders and then Christian missionaries? If the free festivals were the latest expression of this utopian, pagan impulse, then the magic mushroom was one of the principal tools by which the old pagan consciousness would be restored.[74] The key to fairyland, stated George Andrews, was now within anyone's reach,[75] and so no wonder the government wanted the door slammed shut.

The rediscovery of the mushroom in Britain, therefore, went hand in hand with a supposed history that placed its use in an unbroken countercultural tradition stretching back to the dawn of time. To any bemushroomed festival-goer watching the rays of the solstice sun slip past the lintels of Stonehenge, it was a history that made perfect sense, but of course it was a recent invention concocted by Wasson, made irresistible by Graves, and embroidered through endless retellings. It still appears in underground literature and even academic papers,[76] but gained widespread popularity when it was reiterated by the American science fiction writer Marion Zimmer Bradley (1930–1999) in her best-selling retelling of the Arthurian myths, *The Mists of Avalon* (1984).[77] In a move guaranteed to appeal to festival-goers, she recast the enduring story of King Arthur as a spiritual battle between an ancient matriarchal, Druidic, goddess-worshipping paganism, and a militant, newly arrived Christianity. Arthur's nemesis, Morgaine, is reinvented as the high priestess of this supposed pagan religion, an expert in herbal medicine, magic and Machiavellian political manipulation. Naturally wise to the local plant hallucinogens, she employs magic mushrooms as a means to gain 'the Sight' and so foretell the future.

Bradley derived her image of the past from talking to practitioners of modern, revived Paganism in America and the UK,[78] so what is portrayed in the novel is not the pagan past but rather a reasonably accurate record of how hippies and festival-goers, circa 1980, envisioned that past. Magic mushrooms appear in the book because she arrived in England to do her research just at the moment when the mushroom craze was at its peak. Without spotting that Bradley was holding up a mirror to their own beliefs, festival-goers read her book and accepted it as gospel.

Within the space of about twenty years, then, the magic mushroom went from obscurity to being an underground delicacy and then the drug of choice for a vocal, countercultural festival movement. By the 1980s, the Liberty Cap had become an iconic and conspicuous badge of alterity. Though it continued to be passed around in pubs as a cheap Friday-night high, it was more obviously conjoined with hippy, traveller identity, along with dreadlocks, piercings, outsized rainbow jumpers and army boots. Its distinctive goblin-capped shape appeared on festival flyers, T-shirts, postcards, album sleeves and the sides of buses. Defiantly, it proclaimed that those who knew and used the mushroom were an elect pagan few, pitted against the rising tide of consumerism and, like the mushrooms themselves, living on the margins. Its moment of fame did not last long, however, for during the late 1980s it was quickly eclipsed by a new, synthetic drug that ultimately caused far more excitement and moral outrage.

That drug was Ecstasy, or MDMA, millions of doses of which would soon be consumed every weekend across the globe. Ecstasy brought with it a whole new invigorating youth culture – 'Rave' or 'dance' culture – with its own forms of music and dancing, fashions and slang. Hardly any aspect of mainstream culture was left untouched by the Ecstasy revolution, and mushrooms were sidelined as a minority drug for dyed-in-the-wool hippies. It is all the more surprising, then, that it was on the back of Rave that the magic mushroom's most cogent advocate arrived. A geeky American armed with an idiosyncratic nasal drawl, a suitcase full of bizarre theories, and a Blarney-kissed ability to hold audiences spellbound, he did more than anyone since Gordon Wasson to pave the way for the current mushroom boom. His name was Terence McKenna.

14

The Elf-Clowns of Hyperspace

I assume that psychedelics somehow change our channel from the evolutionarily important channel giving traffic, weather, and stock market reports to the one playing classical music of an alien civilisation.

If the truth can be told so as to be understood, it *will* be believed.

Terence McKenna[1]

In the summer of 1989, when I was not quite nineteen, I went to the Treworgey Tree Fayre, a one-off hippy music festival in Cornwall.[2] The first such festival to be held in the south-west for a number of years, it was eagerly anticipated as a commercial and hence legal event that would nevertheless maintain the unique atmosphere of its free predecessors. And so it did. Bands cranked out the expected festival melange of space-rock, punk, dub-reggae and folk from wonky and improvised stages. The usual array of drugs were on sale – hash, Acid, Speed, and Special Brew – with stalls openly offering chillums and 'hot knives', and grungy entrepreneurs happy to deliver hash fudge to your tent door, morning, noon and night. The ever-present collection of battered trucks and buses were parked up in a potato field along with a smattering of horse-drawn wagons for authenticity's sake. And, as usual, the organisation was chaotic, the toilets were execrable, the police stood by helplessly, and outraged local council officials bristled with indignation – and writs – over the festival's many petty infractions.

Within a day, the fences had been torn down, the festival was declared free, and the landowner was nowhere to be seen, having absconded until the resulting furore abated. All in all it was an absolutely typical festival of the 1980s, and it lived up admirably to everybody's expectations.

What made the festival unusual, however, was the unexpected appearance of a sound system. It was set up unofficially on the back of a lorry, from which DJs, not bands, pumped out a seamless mix of unfamiliarly repetitive dance music, and in front of which a handful of

Festival programme from the Treworgey Tree Fayre, Liskeard, Cornwall, 1989. Though a commercial event, the fences were very rapidly torn down and, much to the annoyance of the local authorities, the festival was declared free. Many mushroom-inspired psychedelic bands played, but it was also one of the first 'hippy' festivals to feature an unofficial sound system with Acid House DJs.

people moved jerkily and self-consciously. The arrival and incursion of Acid House into the festival scene was not, as I recall, met with much enthusiasm. The staunch hippies amongst us thought House too much the product of the metropolitan south-east, of affluent yuppies, of VW, not 2CV, drivers. 'It'll never catch on,' we said dismissively, 'it's too synthetic, too repetitive, too cool.' We could not have been more wrong.

Within three years, House music, in all its derivative genres and forms, had become the music of the underground, with most festivals transformed into raves, and DJs eclipsing bands in popularity. Dodging the police to dance illicitly in a field all night became *the* way to spend the English summer, and a whole new cross-section of people from the cities and suburbs found their way into the festival scene, reviving it, but also transforming it in the process. The epitome of this crossover took place in 1992 when thirty thousand people, travellers and clubbers alike, turned up for the Castlemorton Free Festival which erupted near Malvern, much to the consternation of the then Conservative government who brought in the somewhat draconian Criminal Justice Act (1994) to outlaw free festivals, raves and what they stuffily called 'repetitive beats'.

If Rave brought together fashion-conscious urban clubbers and metrophobic be-dreadlocked travellers, then the catalyst for this unlikely alliance was a new drug: Ecstasy.[3] From its invention in 1912 as an appetite suppressant, MDMA, or 3,4-methylenedioxy-N-methylamphetamine to give it its full name, went largely ignored until the 1960s when its empathic and euphoric properties were noticed by the maverick Californian chemist Alexander Shulgin. Shulgin – who has made the discovery of new synthetic psychoactives his life's work – popularised MDMA as a tool for psychotherapy, but the drug eventually found its way onto the streets and into the clubs of Chicago and Detroit, where it collided explosively with a new style of music.

Acid House, as it came to be known, reached London clubs, by way of Ibiza, in the late 1980s, triggering a second summer of love. The underground writer Fraser Clark, clearly taken by the euphoria of those early days, enthusiastically declared it 'a fuckin' movement that's bringing together . . . the best in new age, the best in psychedelic and the best in green ecology. It's an evolutionary rollercoaster that's going to go on growing through the '90s and into a new age!'[4] A rollercoaster it may have been, for it left virtually no corner of popular culture untouched, but a new age . . . well, probably not.

Nevertheless, what the Ecstasy revolution succeeded in doing was to make drug-taking a much more mainstream pursuit. Of course, there was still a small vanguard of psychedelic enthusiasts, or 'psychonauts' as they now called themselves, diligently experimenting with new and obscure designer drugs that were appearing in the marketplace (mostly from the Shulgin stable[5]) and, at the other extreme, there was a vigorous anti-drugs campaign given urgency by occasional Ecstasy-related deaths. But for most people in the middle – and as Noel Gallagher famously observed – taking drugs became like having a cup of tea. By the mid 1990s, millions of doses of Ecstasy were being consumed every weekend, with the result that interest in psychedelics of all kinds was significantly revived.

Ecstasy and Rave culture, then, paved the way for the second, current magic mushroom boom with its loved-up social attitudes. It also created a platform for one of the most original psychedelic thinkers to have emerged since Gordon Wasson. For Terence McKenna, mushrooms were not just a 'drug' but a portal to a shamanic realm where gnomic spirits strained to impart information of quite possibly earth-shattering importance. He built his career by spreading the news that the elf-clowns of hyperspace were really out there, just a few mushrooms away.

Mushroom advocate Terence McKenna, during one of his improvised bardic 'raps' at the 'Secret McKenna Workshop', London, 1993.

Of all the psychedelic thinkers, writers, gurus, conspiracy theorists and proselytisers to have emerged, then, Terence McKenna (1946–2000) was perhaps the most idiosyncratic. Possessed of an unusually acute mind, a sparkling wit, and that rarest of qualities for an American, self-irony, his message was genuinely different. Moreover, he was a brilliant speaker with the power to transfix audiences with his free-flow improvised 'raps', as he called them, delivered in a characteristic nasal drawl that was the envy of every media-hungry scientist the world over.

He was born in 1946 into a Catholic family in a small town in Western Colorado. An archetypally geeky child, with thick bifocal spectacles and a passion for geology and science fiction, he was something of a loner who discovered an enduring love of nature while hunting for fossils and collecting butterflies. Bookish and intellectually precocious, he had by the age of fourteen rejected Christianity in favour of Sartre, Camus and Nietzsche. Had events turned out differently, he would almost certainly have become an academic, perhaps a philosopher, but most probably a geologist.

It was his terrible eyesight that led him to read Aldous Huxley's book about the Bates method for correcting vision, *The Art of Seeing*. Won over by the author's writing style, he worked through the rest of Huxley's novels and essays, eventually alighting upon *The Doors of Perception*. Coincidentally, at exactly this time a minor moral panic erupted about a new drug craze: psychedelic Morning Glory seeds.[6] McKenna immediately procured himself a packet from the local hardware store, ground them up, swallowed them with a glass of water and had his first psychedelic epiphany.

He found that his natural affinity for studying things – rocks, fossils, insects and beetles – was enhanced a hundredfold, and he delighted in the fact that a multitude of faces smiled out at him from the weave of his mother's favourite curtains. Even more remarkably, on closing his eyes he was suddenly immersed in extraordinary visions that welled up and flooded across his mind's eye. Amongst other vistas, he saw 'ruined cities covered with creeping jewelled lichen . . . inhabited by shining-eyed creatures'.[7] Through a set of happy coincidences, he had stumbled upon the mystery that would preoccupy him for the rest of his life.

He left home at sixteen to go to San Francisco and the Experimental University, arriving in plenty of time for the summer of love: he was there when Leary addressed the crowd at the Human Be-In in San Francisco's Golden Gate Park in 1967. Presented with the classic sixties choice between political or spiritual agitation, he chose Leary and the latter. Discovering both cannabis and LSD, and the ability to consume them in what he would later call 'heroic doses', he began his psychedelic career in earnest. Eventually tiring of Acid, which he found 'abrasively psychoanalytical',[8] he tried smoking DMT, a supply of which had just been 'liberated' from the US Army. It would be his second great epiphany.

In terms of sheer intensity, DMT is in a league of its own. N,N-dimethyltryptamine, to name it properly, is the most powerful hallucinogen so far discovered. Paradoxically, the drug occurs naturally in the brain; its ordinary presence there is little understood but, like psilocybin, it is structurally similar to serotonin. It forms the active ingredient of dozens of the psychoactive snuffs and potions employed throughout the Amazon. On its own, it is not orally active, for it is destroyed by enzymes in the stomach, but the synthetically produced extract can be smoked through a glass pipe. Taken thus, its effects, which are characterised by overwhelmingly intense three-dimensional hallucinations, come on rapidly within the space of a minute or so and diminish almost as quickly. The entire trip is over in about ten or fifteen minutes. This was how McKenna took it in early 1966, but not even his hundred and fifty Acid trips could have prepared him for what happened.

With eyes closed, he saw what he later termed 'the chrysanthemum', a revolving yellow-orange floral mandala of light, texture and form, towards which he was hurtled at great speed. Then, with a sound that he described as like the tearing of a plastic membrane, he felt himself puncturing through to another world entirely. He appeared to be in a domed circus-like space, lit from the side and apparently filled with bouncing, babbling creatures.

> I had the impression of bursting into a space inhabited by merry elfin, self-transforming machine creatures. Dozens of these friendly fractal entities, looking like self-dribbling Fabergé eggs on the rebound, had surrounded me and tried to teach me the lost language of true poetry. They seemed to be babbling in a visible and five-dimensional form of Ecstatic Nostratic, to judge from the emotional impact of this gnomish prattle. Mirror-surfaced tum-

> bling rivers of melted meaning flowed gurgling around me . . . Under the influence of DMT, language was transmuted from a thing heard to a thing seen.[9]

The elf-clowns of hyperspace were calling to him, and their message was this: copy us and do what we do, transform your language and sing out in solid, resonant, ectoplasmic, semantic stuff!

Pulled back almost as quickly as he had arrived, he re-entered the humdrum world unscathed, except that his otherworldly crash course in syntax and poetry had left him dumbstruck. Most of his subsequent work, his unorthodox application of science and philosophy, can be said to be an attempt to make sense of this mind-boggling encounter. Throughout his life, he took DMT about a further thirty times, until he was, in his own words, no longer able to summon the raw courage, but on every occasion he found the elves were waiting for him.

In the early 1970s he began a short-lived attempt at smuggling cannabis, which ended abruptly when a consignment shipped from Jerusalem to Aspen fell into the hands of customs officials. He entered a three-year period of self-imposed exile, sojourning in India, Nepal and Indonesia, ostensibly collecting butterflies and studying Buddhism, but actually using the time to take psychedelics and to ponder the mystery of those confounded machine elves. It was while still something of a fugitive that he made the trip to Colombia, which was to prove the final turning point of his life: for it was here that he alighted upon both the magic mushroom, and the rest of his outlandish ideas.[10]

He travelled south with a small party that included his brother Dennis, four years his junior, but just as bright and equally as fascinated by psychedelics, DMT and shamanism. Dennis, the quieter of the two, later went on to become a prominent neuropharmacologist in his own right, conducting much of the known research into the neurological effects of the Amazonian, DMT-containing, ayahuasca brew. But at the time, neither of the brothers, though well educated, had any formal scientific training. They wanted to pursue the hunch – baked from a mix of science fiction, natural history, alchemy and magic, and fired in a cannabis-fuelled crucible – that perhaps, in some mysterious way, the tryptamines worked their peculiar and idiomatic effects by temporarily bonding with DNA. They went to the Amazon in the hope of shedding light on this rather unorthodox hypothesis, for if they found cultures where the use of psychedelics had been institutionalised, who knew what insights they could gain from those indigenous technicians of the soul? In particular, they hoped to find people willing to share the secrets of a DMT preparation called *oo-koo-hé*, rumoured to produce encounters with diminutive spirits: perhaps these last were the very same DMT creatures the brothers had already confronted? So it was that in the early months of 1971 they arrived in the remote jungle outpost of La Chorrera, already struck by the presentiment that something of great moment was to happen.

In fact they never found the fabled *oo-koo-hé*, nor the diminutive spirits, which remained the secret of the taciturn locals. What they discovered instead was a bountiful supply of magic mushrooms, *Psilocybe* (then, *Stropharia*) *cubensis*, sprouting from the cowpats spread abundantly across the pastures, unnoticed and ignored by the indigenes. Upon eating his first mushroom, Terence knew that he had found the drug he had been looking for: less abrasive than Acid and as profound as DMT, but not quite so hard on one's ontology.[11]

Nightly ingestion of mushrooms, together with much joint-smoking, led to some fevered speculation by the brothers McKenna. They concocted a bizarre hypothesis: perhaps, while under the influence of psilocybin, if it were augmented by a secondary tryptamine, harmaline, it might be possible permanently to bond psilocybin into one's neural DNA, and so effect an enduring expansion of consciousness. By singing at the exact resonant frequency of the relevant molecules, they hoped to set up a standing wave that would somehow slip the psilocybin into the DNA, rather as we might plug in an additional piece of software to augment an existing computer program.

With an idea sounding more like science fiction than science fact, the brothers were nevertheless on a slightly manic creative roll, and conducted their 'experiment' on 4 March. They ate mushrooms, drank a harmaline-containing brew (which they made from the bark of the vine *Banisteriopsis caapi*), and sang at the exact frequency of the strange buzzing noise that subsequently rang through their ears.

Something odd definitely happened. Dennis was catapulted into a lingering state of expanded consciousness in which, amongst other anomalies, he could telepathically communicate with anyone in the world. Terence lost all need to sleep, was buzzed by a flying saucer, and was led to pursue startling notions about the nature of time by an alien presence, the mushroom spirit, which spoke to him and egged him on to make yet further arcane discoveries. For the McKennas, this was all concrete proof that their experiment had worked. By most other reckonings the brothers had gone temporarily mad.

This preternatural state of excitement, whether graciously rapturous or dangerously psychotic, did not last more than a few months, but the brothers then poured their energies into a book that attempted to rationalise their discoveries and experiences – *The Invisible Landscape*, first published in 1975. 'Dense. Technical. Fascinating. Infuriating. Marvellously weird' is how the writer Jay Stevens described it.[12] Like some alchemical text – and by Dennis McKenna's own confession, a mix of science, pseudo-science, and pure hokum – the book covers an extraordinary wealth of subjects. The details of their experiment are outlined, Terence's fractal theory of time presented, along with its mathematical 'proof', and the nature of shamanism explored. Most significantly, the book defends the brothers' experiences against the cultural charge of schizophrenia: they had not gone mad in La Chorrera, they had simply entered into shamanistic domains of noesis and experience, for which Western culture lacks the necessary explanatory frameworks. Thereafter, convinced that something of great ontological importance had indeed occurred in Colombia, Terence became a passionate consumer of magic mushrooms, taking them as often, and in as high doses, as he felt was constitutionally prudent, which worked out at about twice a month in colossal doses of five dried grams.

It was only later when pressured by the need to support a family – he eventually married artist, ethnobotanist and fellow psychedelic explorer Kat Harrison, with whom he had two children – that he discovered his innate ability for communicating his passions and ideas through public speaking. As he put it, in the poetic and purple prose that was to be his hallmark: 'I realized that my innate Irish ability to rave had been turbo-charged by years of psilocybin mushroom use . . . I had apparently evolved into a sort of mouthpiece for the incarnate Logos. I could talk to small groups of people with what appeared to be electrifying effect about . . . peculiarly transcendental matters.'[13]

He wrote a further three books, and co-authored one more with his brother, but it is in the realm of the spoken word that his true talent shone. During the late 1980s he became a regular fixture of the American psychedelic circuit, speaking at camps and conferences, seminars and soirées. There are legions of cassettes, CDs and MP3s available to the McKenna fan, officially recorded or simply bootlegged from the stalls, that capture the brilliance of his long, improvised and entertaining bardic raps about mushrooms, DMT, the elf-clowns of hyperspace and the nature of time.[14] Blarney-kissed (or, less romantically, fortified by the odd spliff or two), he could speak for hours at a time, and his inspired rhetoric, way with words and pithy sound bites cried out to be sampled. Catchphrases were digitally picked up by DJs and musicians looking for choice morsels to spice up their dance tracks and to inject them with the requisite amount of psychedelic credibility. McKenna's unmistakably nasal, sing-song voice became de rigueur, cropping up regularly in Rave anthems, and so his unorthodox message, along with his celebrity, was broadcast far and wide.[15]

The most famous was surely 'Re-evolution', released in 1993 by the British psychedelic dance band The Shamen, who began life as a credible rock act before their transmutation into tacky chart-toppers and subsequent demise into obscurity. This seven-minute mix of sinister swirling analogue synth loops features an extended McKenna rap on the nature of the apocalypse, and begins with the ominous, if platitudinous, words 'if the truth can be told so as to be understood, it *will* be believed'.[16] Released at the height of the band's success, and with McKenna in full rhetorical flow, it was this record that did more than any other to propel the man and his ideas into popular consciousness.

Tragically, McKenna's life was cut short at the relatively young age of fifty-four by the intercession of a massive brain tumour, a diagnosis that he realised would give his detractors powerful ammunition. While he knew of no evidence to link mushrooms or DMT to cancer, he worried, rightly, that people would inevitably arrive at that conclusion. His brother Dennis remains keen to point out that the illness, a particularly rare form of tumour, could have been caused by all manner of factors, not least Terence's lifelong affliction with migraines.[17] Whatever the cause, the Western world was suddenly and surely deprived of one of its most charismatic eccentrics.

I was lucky enough to meet McKenna in February 1993 during one of his occasional visits to London. By luck rather than judgement, I managed to get myself invited to 'The Secret McKenna Workshop', a talk held, appropriately enough, in a squatted vegetarian café called Fungus Mungus on Battersea Park Road. The café, now defunct, was a deliberate flashback to the 1960s, complete with swirling paint, retro furniture and nylon curtains to assail the senses.

The event was organised by Fraser Clark, self-styled 'zippie', editor of the almost insufferably upbeat magazine *Encyclopaedia Psychedelica International*, and founder of the legendary clubs Megatripolis and The Parallel Youniversity, at which McKenna spoke to rapturous audiences.

My diary records that the invited audience (as the flyer said, 'by recommendation only!') comprised the great and the good from London's revived psychedelic scene: a founder of the *International Times*; an ex-hash-smuggler; several Acid House DJs; a German baron; a lord (who had hot-footed it over from a debate in the Upper House); and several alternative therapists and practitioners. With his protruding eyes and bushy beard, McKenna looked for all the world like a character from Gilbert Shelton's comic strip *The Fabulous Furry Freak Brothers*. But he spoke eloquently for an hour or more, never losing his train of thought once as he bamboozled us with his outlandish trip tales and Byzantine theories about the extraterrestrial origins of the magic mushroom. 'Far out,' I wrote afterwards, somewhat self-consciously. 'McKenna just blew my mind.'

I freely admit that in my early twenties I was a fan, won over, if not entirely by his ideas, then by the man who very quickly became a figurehead for psychedelic enthusiasts across the world. But what are we to make of those ideas now? Do they merit serious consideration, or should they be consigned to the lunatic fringe where they can be safely ignored as the ravings of a brilliant mind hopelessly unhinged by drugs?

Unlike Gordon Wasson, McKenna did not suffer from scholar-envy and *knew* that he was an intellectual, well read as he was in philosophy and literature (for example, one of his great regrets was that he never had enough time to get to grips with Heidegger[18]). Unlikely as it sounds, he regarded himself as wedded to science, rationalism and a sceptical outlook. 'I think the proper way to contact the Other,' he once said, 'is with hard-headed rationalism exercised under weird conditions.'[19] He regularly lambasted the New Age movement for its rejection of rationalism, and ridiculed the American UFO-hunting community as 'a magnet for screwballs.'[20]

He was, however, self-aware enough to know that his own ideas were, at best, off the edge, and would in all likelihood be similarly categorised as the worst the New Age had to offer. Candidly, he wrote: 'I am in the absurd position of being either an unsung Newton or completely nuts. There is very little room to manoeuvre between these two positions.'[21] His hope was that by setting his ideas in the concrete language of the Western intellectual tradition, within the interstices of science and philosophy, he might at least get a fair hearing. As detractor Will Self put it: 'Certainly Terence McKenna is a silly ass. But his heart is so clearly in the right place, and so much of what he says is a fresh synthesis of a collection of sixties ideas . . . that his book[s] . . . deserve some careful attention.'[22] It would be churlish not to grant McKenna that much.

There is an immediate problem when trying to distil McKenna's core message. His preferred mode of delivery – the improvised rap – though rarely waffling, and only occasionally repetitive, necessarily lacks the concision of the written word. His prodigious recorded output means that there are literally hundreds of hours of material to compare and contrast, and given that such a vast range of subjects interested him, and fell under his musings, it is hard to know exactly where to start. He was as likely to talk about the psychedelic experience as he was about prehistoric religion, crop circles, the mechanisms of evolution, and the reasons for the fall of the Trebizond empire – often in the same breath. He did at least set out his principal ideas in his written works, and so, with a little sifting, it is possible to arrange them into three broad categories: speculation about the nature of the psychedelic experience and arguments for the merits of psilocybin and DMT; speculation about the role of psilocybin mushrooms in human evolution; and speculation about the metaphysics of space, time, cosmology and eschatology.

Firstly then, drugs. The interesting thing about McKenna is that while he was a staunch advocate for the magic mushroom (above all other drugs), he was not a psychedelic proselytiser in the same way as Leary. He never demanded that everyone 'turn on', because he thought psychedelic shamanism, as he labelled it, was a calling. 'If you like watching your world shredded before your eyes and made into nonsense, if that makes you feel liberated and secure, then you can sign up for that carnival,' he once jokily advised. 'If that alarms you then you can stick to the tried and true.'[23] He only really preached to the converted, working to raise the status of the magic mushroom from Acid's poor relation to the workaday choice for the would-be shaman.

Before McKenna, and taking LSD as their cultural model, aficionados consumed mushrooms for Dionysian ends: to heighten their appreciation of art, music, nature, dancing and sex; to find religious epiphanies in some Learyesque setting; or to proclaim their alterity at free festivals. But for McKenna, the point of taking mushrooms was solely and simply a shamanic quest for visions. His prescription was five dried grams (of *Psilocybe cubensis*, or an equivalent dose of another species), taken alone on an empty stomach, in silent darkness, with eyes closed. No music, no art, no sex. With the mushroom taken in such a manner, and at such high doses, he suggested two things would happen. First, the swirling phantasmagoria of ever-changing coloured lights, repeated figures and juddering visuals ordinarily seen behind closed eyelids would condense and deepen into more profound *visions*. That is, as at the climax of *2001: A Space Odyssey*, one would be propelled into other realities, into strange, alien landscapes, futuristic scenarios or ancient cities; or perhaps through a process of death and rebirth so real that it left one drenched in sweat, gasping for air and overwrought with ecstasy. Second, clear as day and in broad English, one could expect the mushroom to speak.

In his account of the trip to La Chorrera, *True Hallucinations* (1993), McKenna included a passage, written many years earlier, which he claimed had been dictated to him directly by 'the mushroom' and copied down verbatim. In an extraordinary speech, the mushroom claimed to be of extraterrestrial origin, its spores drifting through space and colonising new planets. 'The mushroom which you see,' it said, 'is the part of my body given over to sex thrills and sunbathing . . . [The] occurrence of psilocybin and psilocin in the biosynthetic pathways of my living body,' it went on, 'opens for me and my symbiots the vision screens to many worlds,' and it enabled faster-than-light communication with all other mycelial bodies across the cosmos.[24]

McKenna was as surprised by this as anyone, but hedged on the rather important issue of the provenance of this voice. Sometimes he seemed convinced that it really was the mushroom speaking. At others, he thought it came from some hidden aspect of his mind, or from the mind of nature, or from Gaia, or from an alien civilisation. He was not obfuscating, he simply did not know, and besides, provenance mattered less than the message itself.

Strange as it seems, there is some clinical evidence that a statistically significant number of people do 'hear' voices under the influence of high doses of psilocybin.[25] The parsimonious and culturally sanctioned view is that the drug interferes with cognitive processing in some unknown fashion, giving the aural 'illusion' of a discarnate voice. But that is not how it feels, said McKenna. It feels as if one is in the presence of some autonomous, external, overbearing, alien entity with which it is possible to have an I/Thou relationship. This was the voice that spoke to him and egged him on to continue his theorising. This was the Logos that he claimed lay behind his improvised raps.

For McKenna, then, what made the tryptamines (and mushrooms and DMT in particular[26]) superior to all other drugs was not only the depth of the visions they occasioned, but also the overwhelming force with which they brought about this encounter with the Other.[27] Even to think of them as drugs was wrong. They were rather independent agentic souls, plant-allies or spirit-teachers, with whom a symbiotic relationship would yield ecstasy and gnosis. The shamanic world view, in other words, was right all along.

This concern with the 'natural' leads to the second area of speculation that preoccupied him: the possible role of mushrooms in human history and evolution. These ideas are spelled out in his *Food of the Gods: The Search for the Original Tree of Knowledge*, published in 1992. The book – dotted with his usual stories about the mind-boggling nature of DMT and mushrooms – is a curious, disjointed and ultimately disappointing mix of well-researched twentieth-century drug history (including a passionate argument against prohibition) and a 'conspiracy-theory'-type invention of a presumed psilocybin mushrooming tradition stretching back to the Palaeolithic. Its principal argument is that human evolution was accelerated by our ancestors' use of magic mushrooms.

When human hominid ancestors left the forests for the plains of Africa, argued McKenna, they encountered and experimented with new and unusual foods, which included the psilocybe mushrooms growing abundantly in cattle dung. At low doses the mushrooms, which have been shown to sharpen visual acuity, conferred an evolutionary advantage by improving hunting ability. At medium doses, he continued, the mushroom's supposed aphrodisiac qualities conferred an evolutionary advantage by upping the reproductive success of those who ate them. But, most dramatically, at high doses the encounter with the Logos occasioned by the fungi stimulated our ancestors' latent ability to speak, and so drove forward the evolution of language. For McKenna, the magic mushroom, eaten by our ancestors in the dim and distant past, brought out in us the best and noblest qualities that we now possess: they are what made us human.

He went on to suggest that early prehistoric religions in Africa, the Mediterranean and the Middle East continued to use the mushroom, and lived in a goddess-worshipping, matriarchal, 'partnership' society, such as that supposed by some archaeologists to have existed in Minoan Crete: an Eden where war and conflict were unknown, and where we lived in a harmonious ecological relationship with the earth. A climatic change led to a gradual warming and drying of the prevailing weather, with the result that mushrooms became scarce. Patriarchal, mead-quaffing, 'dominator' cultures invaded from the north, and with that we fell from Eden into the clutches of linear time, history and ultimately the ecological mess of the modern industrial world. McKenna therefore called for an 'archaic revival' whereby the reintroduction of a magic-mushroom-based shamanism would erode the 'calcareous tumour of ego'[28] and restore society once more to a partnership paradise.

By now it should be clear, from our consideration of Wasson's supposed prehistoric mushroom cult and the various mushroom conspiracy theories, that there is scant evidence for any of this speculative prehistory. McKenna was no archaeologist and, like the conspiracy theorists, he took scraps of evidence, denuded them of their context, and shoehorned them into his preferred justificatory narrative, with little thought for alternative interpretations. His twist on the Wasson theory was to substitute psilocybin mushrooms for the fly-agaric, and to imagine their use in ancient matriarchal religions – the latter having been proposed by American feminists during the gender-riven years of the 1970s. Thus, he borrowed the terms 'partnership' and 'dominator' societies from the writer Riane Eisler, and the idea of a Mediterranean matriarchy overturned by male-dominated Indo-European marauders from Lithuanian archaeologist Maria Gimbutas.

Putting to one side the likelihood of this feminist-inflected view of the past (it remains contentious within both feminist thought and archaeology[29]), McKenna's thesis here rests on the notion that psychedelics produce societal harmony. By dissolving boundaries, he claimed, they make us less ego-centred, less selfish and more empathetically in touch with the needs of the other denizens of the natural world. In his rather disparaging review of *Food of the Gods*, the writer Will Self (no stranger to drugs himself) considered this a curious form of Marxist dialectical materialism: the type of drugs you take determines the type of society you get: mushrooms good, alcohol bad.[30]

The problem with McKenna's materialism is that indigenous societies with institutionalised psychedelic drug use are very often far from harmonious. Amazonian tribes take ayahuasca as often to curse as to cure,[31] mushrooms failed to prevent the petty Mazatec vendettas, and even McKenna noted ruefully that, for all his bemushroomed revelations, he had been unable to save his marriage.[32] As we saw when considering Timothy Leary, the idea of a psychedelic golden age, and of psychedelics as a panacea or magic pill, must remain a gloriously beguiling chimera; but without it, McKenna's thesis starts to look very shaky.

As for his notion about mushrooms propelling human development by conferring evolutionary advantages and instilling in us our most elevated qualities, this is the same category of reasoning as is employed by evolutionary psychologists. This popular school of thought, very much on the ascendant, argues that selective pressures acting upon human ancestors in some imagined African past determine the way we behave today. Countless scenarios are conjured up to 'explain' why people commit adultery, or why homosexuality exists, or why men are supposedly better at parking cars and women at multi-tasking, and so on and so forth.

The difficulty is that, however plausible they seem, these are untestable hypotheses, which places them in the realm of storytelling, not science. In McKenna's case, it rather stretches credulity to believe that a bemushroomed hunter could actually pick up a spear, let alone walk or hunt. One could equally well imagine a scenario in which mushroom use was selected *against*, the bemushroomed being so incapacitated by fits of the giggles as to provide easy pickings for predators. Yes, mushrooms might have been used by our ancestors, with unexpectedly beneficial effects on our development, but then again, they might very well not have been. At present, there is absolutely no way of knowing.

The final, and most important, area of McKenna's thought concerns metaphysics, the nature of time, and the end of the universe. McKenna – who you will recall was, as a child, fascinated by geology and science fiction – was in adult life profoundly interested in cosmology and the evolving complexity of the universe. Considering this, he thought increases in the universe's structural complexity had occurred suddenly and dramatically at discrete, identifiable intervals: the formation of hydrogen atoms from the initial plasma of the Big Bang; the condensation of stars; the creation of the other elements in stellar furnaces; the formation of planets; the emergence of DNA and complex life forms; the evolution of consciousness; the evolution of language, then culture and technology. Borrowing terminology from the process philosophy of the British mathematician Alfred North Whitehead (1861–1947), he called these events 'ingressions of novelty', or 'revelatory intervals'.[33] After each event, the universe existed in a more complex state of organisation than it had before; and each new level enabled the next, unexpected level to emerge.

At the same time, a recurring motif in his bemushroomed visions was of some kind of metaphysical object, a transcendental vortex perhaps, into which we are being irresistibly drawn. Borrowing from chaos theory, he called it the 'attractor' or 'dwell point'; from the process theology of the French Jesuit Pierre Teilhard de Chardin (1881–1955), the 'omega point'; but most commonly, in his own poetic way, 'the glittering object in hyperspace' or 'the glittering object at the end of time'. Considering this object, he became convinced that time and history were not being inched forwards by causality, as is ordinarily supposed in classical physics, but were being drawn ineluctably towards this transcendental object lying ahead in the future, like water down a plug hole. Our whole notion of time, in other words, was topsy-turvy.

He married these two ideas by suggesting that the omega point represented the final, ultimate ingression into novelty, the apocalyptic point after which nothing in the universe would ever be the same again. Furthermore, he believed that these ingressive events occurred at mathematically predictable intervals, which meant that it might be possible to determine when what he wryly called the 'Big Surprise' would happen.

So far (his mushroom visions aside), he was on reasonably stable, if unorthodox, ground, drawing upon ideas of progression and endings that have infused the Western intellectual tradition, in some form or another, from antiquity onwards: in Aristotelian metaphysics, Christian eschatology, Hegelian idealism, Marxism and Darwinism. What tipped him over the edge of epistemological orthodoxy was his belief that the mathematical formula lay encoded in the ancient

Chinese oracle of the *I Ching*, and that the end of the world would occur on 22 December 2012.

The exact way in which he arrived at this date is explained fully in *The Invisible Landscape*, but the baroque and complex details of his methodology need not concern us here. Suffice it to say that the end result of his calculations, based upon the perceived internal structure of the *I Ching* divination system, was a complicated fractal wave, the 'timewave', which, he maintained, accurately mapped the ingression of novelty within its ebb and flow. According to his model, time was wrinkled and crenulated, winding itself in ever tighter spirals as it juddered towards its abrupt end in 2012. 'The twentieth century is the shudder that announces the approaching cataracts of time over which our species and the destiny of this planet is about to be swept,' he ominously predicted.[34]

What would happen at this ultimate denouement was less clear: like many futurists and prophets, he was somewhat vague on the details. He proposed various scenarios throughout his many works: the laws of physics would be utterly transformed; humanity would leave the planet; we would leave our bodies for a higher dimension; we would discover time travel, at which point time would collapse and have no meaning; language would be transformed into something not heard but seen and beheld (like the language of the machine elves); the archaic revival would be complete, and we would live in harmony in accordance with the will of the planetary 'Gaian' mind.

Conveniently, perhaps, he argued that multiple hypotheses were unavoidable because it was impossible to predict how the universe would look after an ingression of novelty had occurred. Could one, say, have predicted Reality TV by watching the stars forming? The big concrescence was coming, but as it lay beyond the event horizon, it remained in 'an abyss of ambiguity'.[35] The best anyone could do was try to catch glimpses and intimations. McKenna was like a small child straining to peep over the garden wall: he leapt and leapt again, but never knew if he had seen aright.

One image he had was of the 'object' as some celestial glitter-ball sending off distorted reflections of itself backwards through time. In his own unique take on Aldous Huxley's 'perennial philosophy' – the idea that all religions are just different but commensurate paths up the same spiritual mountain – he suggested that mystics, saints, visionaries and prophets were simply people caught by the sudden light of those reflections. The Buddha, Mohammed, Moses and Jesus were just unusually sensitive people 'in a relationship of resonance with the transcendental object, such that they, in a sense, embodied it'.[36] Those less naturally tuned in could peer at hyperspace through the telescopic lens of the mushroom.

What, then, are we to make of this extraordinary scenario, one that has gripped the hardcore psychedelic community on both sides of the Atlantic, and has elevated McKenna to the status of sainthood in psychedelic eyes? Objections fall into two categories: mathematical and metaphysical. Because of the fringe nature of McKenna's ideas, and the opaque text of *The Invisible Landscape*, the timewave went for twenty years without being properly scrutinised. It was a brilliant young mathematician, Matthew Watkins, now at Exeter University, who made the first proper analysis in 1994. Then a PhD student and McKenna fan (later a mendicant mystic, travelling through Ireland with little but a donkey, a cart and a Turkish saz), his critique is known on the Internet as 'the Watkins Objection'.[37]

Again sparing the mathematical details, Watkins found three problems with the timewave. The first, weak objection was that it was not fractal, as McKenna had claimed, but was a 'complex piecewise linear progression' no less. Though this took some of the shine off, it did not completely undermine the timewave. More problematically, in order to produce the wave, McKenna had performed a 'half-twist' on the data: the mathematical equivalent of the mountebank switching the cups while the audience isn't looking. Watkins could find no justification for this piece of arithmetical legerdemain, other than that without it a different, trivial and uninteresting wave was produced. And finally, Watkins found that the end point of 22 December 2012 was not calculated but *assigned* arbitrarily, because it apparently gave the 'best possible fit' to 'known' ingressions of novelty. In fact, McKenna chose this date because it already has significance within the psychedelic movement as the supposedly portentous date on which the ancient Mayan cyclical calendar ends.

When moving the wave to and fro to find the best fit, McKenna was not content to restrict his definition of novelty to changes in the physical organisation of the universe. Rather, he stretched 'novelty' to include historical events, cultural trends, and even fads and fashions. Thus, he defined 'the sixties' as a partial ingression of novelty, and likewise the Acid House days of the 1980s and 1990s. His timewave had become a grand unifying theory of everything, of physics, cosmology, culture, religion and consciousness. But it is not hard to spot the cultural bias and Americo-centricism inherent in his construction of significant dates. Nor to see that the considerable margins of error in, say, trying to pinpoint the exact moment when stars began to form, would make it impossible to place the timewave with any degree of accuracy. In fact his timewave is so crenulated, and the vicissitudes of history so convoluted, that such margins of error are irrelevant: one could randomly drop the wave onto the actual timeline, and always find *something* significant happening on the predicted dates.

McKenna was certainly mining a rich and valid vein of thought by considering the physical evolution of the universe. He was correct to point out that the cosmos has become increasingly complex through time, and could, just conceivably, be right in his assertion that novelty has advanced in discreet, mathematically predictable intervals: after all, he would not be the first to suggest that the universe runs like clockwork. However, the process by which he made his prediction of 2012 as the date of the apocalypse turns out to be nothing more than an elaborate piece of circular numerological chicanery. The date may yet prove significant, but we should certainly call time on the timewave.

Of course, sceptics might wonder why a theory constructed on numerological speculation about an ancient and opaque fortune-telling system, the *I Ching*, should merit serious consideration at all. McKenna would surely reply that if, as he supposed, the metaphysical object at the end of time actually existed, then there was no reason why some essence of its nature should not be written into the divinatory system of a calendrically obsessed culture like the ancient Chinese. We have reached the nub of all McKenna's arguments, for whether he was talking about the machine elves or the discarnate mushroom Other, the evolution of humanity or the end of time, all his ideas rested upon the same assumption that runs throughout this history: that there was a metaphysical truth that could be accessed through the use of hallucinogens.

For McKenna, the truth resided in the metaphysical object at the end of time, and a new facet of it was encountered with every mushroom trip. 'When you take psilocybin,' he once cried, 'it takes you. You are participating in all the trips that it ever induced in anyone.'[38] The insights gained on magic mushrooms were therefore genuine and true, and the fact that the Other often chose to broker this information through the mouths of elves and gnomes was neither here nor there (though it did reveal the universe to have an idiosyncratic sense of humour).

But this assumption highlights the philosophical dilemma that lies at the heart of the debate as to whether the use of magic mushrooms and other hallucinogens should be tolerated in contemporary society. Enthusiasts believe passionately that mushrooms enable them to puncture the envelope of ordinary consciousness, to transcend the self, and to acquire knowledge that was not known before. Just as one downloads new songs to an iPod, mushrooms hook up an Ethernet connection to the metaphysical realms. Mainstream society, wedded to a scientific materialist view of consciousness, denies this, and maintains that all a drug such as psilocybin can do is put the iPod of consciousness on shuffle, so that its existing contents are mixed up and juxtaposed in novel and surprising ways. At best one might acquire pleasing poetic insight, at worst a psychotic reaction, but in no sense can one gain new empirical knowledge.

The two positions are hard to reconcile. The matter would surely be resolved if the mushrooms or the machine elves imparted information that was testable. At first sight, the 2012 prediction would seem to satisfy this requirement, but McKenna's vagueness about what exactly will happen, coupled with the rather elastic boundaries surrounding his definition of novelty, means that it will be extremely difficult to recognise if anything significant has actually occurred. Indeed, it is not entirely clear that he acquired the date while under the influence of mushrooms. He never emphatically explained how he alighted on it (probably because it was so arbitrarily assigned), sometimes saying that it was given him by an elf-troop, sometimes that he chose it because it was the end of the Mayan calendar, and sometimes that it fell out of his timewave calculations. In truth, while most enthusiasts claim to have gained significant and important insights while tripping on mushrooms, few, as far as I know, claim to have been given testable empirical data. As McKenna himself admitted, the 'character of the mushroom experience is almost entirely that of understanding',[39] which is not the same thing as knowledge.

The uncertainty as to whether mushrooms really can bestow transcendence is magnified by the fact that so many of McKenna's ideas are historically and culturally contingent. That is, for someone so concerned with metaphysics, his ideas are remarkably of their time. We have already seen how his feminist construction of the past, and his nod to new-manism in his vision of the archaic revival, were shaped by contemporary late-twentieth-century American Goddess spirituality. But he also couched his ideas in the language of chaos theory (of fractals and strange attractors) and of ecology (especially the Gaia hypothesis), which happened to be enthusing popular culture at the time he was writing. His vision of shamanism was very much the standard Western, Eliadean, twentieth-century construction, and his concern with nature and transcendence places him squarely in the American Romantic tradition of Emerson and Thoreau. His notion that the future was mathematically predictable was entirely a modernist assumption, while his apocalyptic concern with the end of the world could easily be seen as a product of a Catholic upbringing, and the background pre-millennial tension that suffused the 1990s.

Furthermore, it is hard not to hear McKenna's own voice when reading the famous monologue he attributed to the mushroom. Would an indigenous *curandeso* have described the mushroom as 'the part of my body given over to sex thrills and sunbathing'[40]? Certainly the prudish María Sabina, for whom sex-thrills and sunbathing were unheard of, never said such a thing. Could it be that the space-age vision of intergalactic citizenship and hyper-light travel were not the product of an alien mind, but of the man who, as a geeky child with bad eyesight, discovered an enduring love of geology, astronomy and science fiction? Could it be that the machine elves' obsession with language was actually McKenna's own?

On the other hand, might we not expect a superior intelligence to tailor its message to our historically bound cultural expectations, to tell the truth in such a way that we would understand it, and believe? And if the elves are real, with genuine free will and agency, might they not decide to withhold from us empirical data? According to McKenna, the 'mushroom' desired to share the knowledge of hyper-light travel and to grant us the keys to intergalactic citizenship, but we had not reached the required level of cultural development to merit such a dispensation.

The frustrating thing is that experience, psychedelic or otherwise, is a private affair, and so there is no basis upon which to judge the truth of McKenna's claims about what he had beheld. There is an inseparable ontological gulf between the believers who have had the experience, and the doubters and onlookers who have not. The latter see it as all in the mind; the former, as something so powerful, moving, and overwhelmingly other, that it cannot but originate outside. If that leaves them with a certain humility, or a set of praxes by which they can better lead their lives, then perhaps the provenance of the machine elves and the glittering object at the end of time is ultimately irrelevant. As McKenna said: 'The elves and the gnomes are there to remind us that, in the matter of understanding the self, we have yet to leave the playpen in the nursery of ontology.'[41] Anything that dents human pride a little must surely be a good thing.

One final philosophical objection to McKenna's reasoning concerns free will. If, as he maintained, the next catastrophic ingression of novelty is to occur, come what may, on 22 December 2012, then why do we need the archaic revival of psychedelic shamanism that he called for? How can our actions make any difference to this dreadful event, whatever it is, the nature of which is set in metaphysical stone? If, on the other hand, an archaic revival would mollify or ameliorate the event in some way, perhaps by making us better prepared for it, then the idea of time being ineluctably pulled forward outwith causality cannot be true. Either the future is already determined, or it isn't: he can't have it both ways.

Such dilemmas, quandaries and philosophical wranglings will doubtless do little to unseat McKenna from the position he now occupies within the psychedelic pantheon. He is seen as a visionary and a prophet, and I have even heard him called the 'Jesus of the Mushroom'. The late American psychonaut and comedian Bill Hicks (1961–1994) built an entire routine around McKenna's ideas, and I am sure they will be with us for as long as people continue to use mushrooms.

It is not hard to see why his message has struck such a chord with enthusiasts. He reinforced the idea that salvation is to be found, not through prayer or devotion or political agitation or social change, but through higher doses, taken more often. By insisting that genuine shamanism is based on psychedelic ingestion, he elevated psychedelic enthusiasts to the status of shamans. Shroomers were now visionaries, users of the very plant that had shaped human destiny, prophets who would help humanity ride the rapids of the eschaton. And, essential for any apocalyptic prophecy, his timewave theory held within it the ultimate get-out clause: for if the revolution failed to materialise, then that was because there was another, bigger and better one waiting round the corner. These are potent myths of identity for a section of society usually lambasted as flakes, cranks, malingerers and wasters.

How, then, are we to judge Terence McKenna? He was certainly a rare and brilliant thinker, but sadly, few of his ideas stand up well under academic scrutiny. His version of the past, of the role of mushrooms in human evolution and in creating a golden age of early civilisation, is based on scant evidence and is untestable. The arcane numerology of the timewave, meanwhile, is best forgotten. As for the machine elves, we cannot rule out the possibility of their independent existence, but the burden of proof must surely lie with the believers, and not the gainsayers who have the weight of the culturally sanctioned scientific-materialistic discourse behind them.

Finally, 2012. I agree with Watkins that the demolition of the timewave does not preclude something interesting, unusual, or even of great magnitude from happening as predicted, but this has to be set in the wider context of the doom-mongers' atrocious track record in these matters. Then again, with the alarming predictions about the speed of global warming, and the West's apparently gleeful intention to resurrect medieval vendettas as it presses for global democracy – and oil – 2012 seems as good a date for the apocalypse as any.

Where does that leave McKenna? He was a visionary, certainly . . . but a scientist? a philosopher? Though he might protest, and though he was well versed in both areas, I do not think we should see him as either: these drab disciplines could never have contained or fulfilled his restless and colourful imagination. Rather, I think it best to see him as a *storyteller*. He was an Irish-American seanachie, blessed with inspiration and gab-gifted in the best tradition of the bards, who deftly spun a golden mythology from the straw of our rationalist culture, and who held us transfixed with the beguiling power of his words. He believed wholeheartedly in the stories he told, for like Ossian or Thomas the Rhymer, he had been to the otherworld and seen it with his own eyes. But the masterful way he told them, placing them just on the edge of plausibility and setting them in the dominant theological language of our time – fairy tale explained as science – meant that they could be both understood *and* believed. Who else but a storyteller would have had the brilliant idea of naming the machine creatures 'elves'?[42]

For those struggling to find a way through the existentially barren and disenchanted secular world, he painted a remarkable vision of hope and redemption, science and magic, accessible to all through five dried grams. He once wrote that 'we have the peculiar good fortune of fulfilling the wish conveyed in the Irish toast "may you be alive at the end of the world"'.[43] The great tragedy is that he won't be: he so dearly wanted to know if he was right.

But if the jury is still out on the question of whether McKenna really was an unsung Newton, or was just plain nuts, there is one of his achievements, made with his brother, that we have yet to mention and that most definitely can be said to have changed the world. For, tinkering around in the shed at the bottom of his garden, the brothers McKenna hit on their equivalent of alchemical gold: they cracked the problem of how to grow unlimited supplies of magic mushrooms. Within twenty years of their discovery, cultivated hallucinogenic mushrooms were being openly sold in Holland and the UK, and to the horror of politicians and, one hopes, Terence's eternal satisfaction, the second magic mushroom boom was underway.

15

Muck and Brass

It's sort of an odd fact about mushrooms that the qualities you need to take them are precisely the qualities that you will inculcate into yourself if you learn how to grow them: punctuality, cleanliness, attention to detail, so forth and so on.

Terence McKenna[1]

We grow our mushrooms on a mixture of shit and straw . . .
Our turnover in 2004 was a million pounds.

British magic mushroom grower

In July of 1971, while he was still in the preternatural state of excitement caused by the 'experiment' at La Chorrera, Terence McKenna returned to the remote village hidden in the depths of the Colombian jungle. The original trip had been rather dramatically cut short when the rest of McKenna's party, unnerved by the brothers' apparently deteriorating mental health, arranged to have them all airlifted to safety. But this left unfinished business. Whatever bizarre psychic process had been initiated by the experiment, it had yet to be resolved, hence Terence's pressing desire to return.

He sojourned there with his then girlfriend, Ev, for five months, during which time his excitement eventually bore fruit in the form of the timewave theory, apparently revealed under the aegis of the cosmic mushroom overseer. The actual, physical mushrooms were, however, in much shorter supply than they had been six months previously, a fact that possibly contributed to his general psychic cooling down, but definitely prompted him to collect spore samples. Back home, the news about the American *Psilocybe* species had yet to become public knowledge, and synthetic psilocybin was unavailable (at least, to those less well-connected than Gordon Wasson), so finding a sustainable supply of his treasured *cubensis* mushrooms was a top priority. On his return, he hoped to discover a way to cultivate them.

The timewave, and McKenna's reversion to a state of relative normality, complete, the couple returned from their Edenic existence in the jungle. The pressures of modern life, however, disrupted their horticultural plans, and the carefully collected spores sat neglected in a freezer for a number of years. Eventually, together with his brother Dennis, Terence began tinkering in the greenhouse at the bottom of his Berkeley garden, but with disappointing results. They succeeded in growing a few mushrooms, but yields fell far short of the amounts deemed sufficient for their psychedelic needs. Then, by chance, Terence stumbled across a new technique for growing commercial field mushrooms on sterilised rye grains, of all things,[2] and, stumped for any other ideas, adapted the technique for his *cubensis* spores.

Halfway through his experiments, however, he split up with Ev, and their messy break-up left him depressed, downcast and beset with migraines. All thought of mushrooms was put on hold – that is, until the day he decided to go and clear out the greenhouse, which he thought would be overrun with detritus and mould. Imagine his surprise when he pulled open the door to discover row upon row of *cubensis* mushrooms sprouting merrily from the abandoned trays. His troubles were very quickly cast aside. 'I was neck deep in alchemical gold!' he wrote. 'The elf legions of hyperspace had ridden to my rescue again. I was saved! As I knelt to examine specimen after perfect specimen, tears of joy rolled down my face. Then I knew that the compact was still unbroken, the greatest adventure still lay ahead.'[3] He had cracked the secret of how to grow magic mushrooms.

At first it might seem curious that it took anyone so long to work out how to cultivate hallucinogenic fungi, but mushrooms of all kinds are surprisingly difficult to grow. Agriculture has been with us for many thousands of years, mycoculture for just a few hundred, beginning in the West only during the late seventeenth century. It was then that French gardeners worked out a rudimentary method of growing button mushrooms (*Agaricus bisporus*) in the cool, damp quarries and tunnels that percolate the foundations of Paris. Spawn was taken from existing mushroom colonies and mixed with a well-watered compost of straw and rotted manure, after which the gardeners sat back and hoped for the best. Similar techniques were used in Japan up to the early twentieth century to grow shiitake mushrooms on rotted wood, and are still used in South East Asia where cow dung is collected and watered to encourage *cubensis* mushrooms to sprout, for the lucrative tourist trade. Such hit-and-miss methods were the only ones available until the early twentieth century when the mysterious lifecycles of the various fungi were finally unravelled, and sterile techniques for culturing them in the lab perfected.

Several difficulties lie in the way of the would-be mushroom cultivator. First, and most importantly, spores have to be collected, germinated and cultured without contamination by any of the other microscopic organisms – viruses, moulds and bacteria – that routinely fill the air, and that will hungrily colonise a freshly prepared dish of agar, or a jar full of rye grains, intended for the fungus. A sterile laboratory is required, ideally replete with antiseptic wipes and bleaches, special air-conditioning filters, and some means of autoclaving the necessary jars, dishes, scalpels and inoculating loops of the trade. Second, it must be possible to fool the species into producing mushrooms by imitating the correct, usually stressful, environmental cues that would ordinarily trigger reproduction. This is far from given, and so while some species may be easily fooled by, say, their mycelia being refrigerated to simulate frost, others refuse to cooperate: no one has found a way to persuade the fly-agaric, *Amanita muscaria*, to yield mushrooms in the lab. And finally, for this to be more than a labour of love, the process must generate a high yield of mushrooms.

As luck would have it, the species that Terence McKenna stumbled upon growing so abundantly in La Chorrera was *Psilocybe cubensis*, the one that has proved the most compliant with human needs. It is the easiest magic mushroom to grow, and produces the most bountiful harvests. Whether this discovery was fluky, serendipitous or portentous depends entirely upon your point of view.

The first person to cultivate magic mushrooms in the lab was Roger Heim, who grew carpophores of *Psilocybe mexicana* from specimens collected with Gordon Wasson in Mexico:[4] thirty-two of these, you will recall, caused Albert Hofmann to see his physician transformed into an Aztec priest. Laboratory cultivation is, of course, the mycologist's stock in trade – *Psilocybe mexicana* mushrooms were also grown by Rolf Singer's rival team in America[5] – but both teams' yields were low, and their methods necessarily complicated. An underground handbook appeared in the United States in 1968 summarising Heim's procedures and ostensibly making them publicly available: *The Psychedelic Guide to Preparation of the Eucharist.*[6] It is all very well having a recipe book, but not much use if you need all the accoutrements of a scientific laboratory, and a university training, to use it. The book went pretty much ignored.

What was so revolutionary about the McKennas' method was that it could be easily followed by the non-specialist at home and, unlike the lab techniques, it was pretty well guaranteed to produce a high yield of mushrooms. It was, the brothers claimed, only marginally more complicated than making jam, the hardest part being obtaining a supply of spores to begin with.

Their method had three stages. First, spores were collected from a mushroom specimen, and then germinated in Petri dishes of sterilised agar jelly. When, after a few weeks the jelly was suitably infused with healthy and uncontaminated mycelium, pieces were transplanted to jars filled with sterilised rye grains. (Though it may seem like quite a lateral step to grow a dung-loving species on rye, many fungi species from diverse habitats will thrive quite happily on grain in the laboratory. For example, spores of the Liberty Cap, naturally a grassland species, can be germinated on cool, damp corrugated cardboard and then transferred successfully to a substrate of wood chips.) Finally – and this was the masterstroke – the grain was 'cased' with a layer of sterilised soil. Casing is one of those techniques used to fool a fungus into producing mushrooms. By depriving the mycelium of light and oxygen while, at the same time keeping it moist, the McKennas quadrupled the yield of mushrooms from a single jar of spawn. No wonder Terence was delighted.

The brothers published their method – together with illustrations by Terence's then wife, Kat Harrison, and detailed photographs by Jeremy Bigwood – as *Psilocybin: Magic Mushroom Grower's Guide*, in 1976, under the pseudonyms O. T. Oss and O. N. Oeric. Since then it has been reprinted eight times, and by 1981 had sold over a hundred thousand copies. The brothers estimated that there must be five thousand people worldwide following the method and growing magic mushrooms at home. It is hard to be accurate with these sorts of figures, but certainly interest in home cultivation became sufficiently large for the American corporate reflex to spring into action. By the autumn of 1976, spores and growing kits were being advertised by several companies in the dope mag *High Times*.[7]

The McKenna book was followed by others, including one on a similar method put forward by Stephen Pollock in 1977, and an exhaustive guide to growing mushrooms of all kinds commercially,

The Mushroom Cultivator, by the impresarios Paul Stamets and Jeff Chilton in 1983.[8] This, I am reliably informed, remains the bible for today's magic mushroom farmer. Needless to say, while spores could be traded legally in America, growing magic mushrooms, particularly with intent to supply, remained a felony. Like the secret stills supplying moonshine in the days of prohibition, magic mushroom cultivation became celebrated as an illicit folk art. In that spirit, anyone who defied the law to perfect growing techniques and develop best-quality moonshine mushrooms stood to become an underground hero. That is exactly what happened to Robert McPherson.

Robert McPherson (1947–) is better known on the Internet as PF, or Psylocybe Fanaticus, famed for his so called PF-Tek growing technique.[9] A jazz-blues guitarist, he was a hippy living in the Haight-Ashbury during the height of the psychedelic sixties, and discovered magic mushrooms sometime during the 1970s. He obtained Pollock's book and succeeded in growing a handful of mushrooms, more by luck than judgement. Realising that the problem with the cased-rye method was the likelihood of contamination when the rye was inoculated, he wondered whether it might not be possible to germinate spores *directly* onto the grain. He found that it was, provided that you carefully injected a solution of spores down into the medium, and then covered the rye with a layer of dry vermiculite to protect it from contamination. Vermiculite is an inert substance, most commonly used as cat litter. It possesses an extraordinary ability to absorb water and does not shrink on drying, and so makes an ideal admixture for mushroom compost. It keeps the substrate moist and aerated, and it more faithfully mimics the natural environment by forcing the mycelium to stretch across its inert particles in search of more nutriment.

The upshot was that the PF-Tek proved a great success. Growing times were shortened, every jar produced a high yield, and the method was so easy that McPherson claimed that the mushrooms would grow virtually by themselves. It was also financially successful. McPherson was a canny businessman, for while he copyrighted the method, he distributed it freely, ensuring that there was a huge demand for the spore syringes produced by his company. At one point he was earning an alleged $30,000 a month. The bubble could not last and, following a tip-off from concerned parents, the Seattle-based company was busted by a large police operation in 1993. Though the spores were technically legal, he was charged with, and pleaded guilty to, distribution and manufacture of psilocybin, an offence with a maximum of twenty years in prison. In the end, he received six months' home detention and three years' probation, since which time he has melted away from the public gaze.

His infamy and the PF-Tek live on, however. The method is widely available in book form and on the Internet, feted amongst a dedicated if anorak-bound community of (probably) teenage, male, mushroom hobbyists, for whom PF is a folk hero. Putting a figure on the numbers of Tekies is hard, but Internet chat rooms and message boards groan under the weight of discussion about the merits of this or that mushroom strain, autoclave efficiency and the other finer points of Teking. Any concerted attempt to crack down on hobbyists would be a Herculean task.

It is fair to say that home cultivation has so far been more popular in America than in Europe, where proximity to naturally occurring species makes it much easier to meet demand. That was until the 1990s, when enterprising enthusiasts in Holland spotted a loophole in the law, and hit upon the novel idea of growing magic mushrooms commercially. It proved successful beyond their wildest dreams.

The Dutch are famous for their pragmatism, and this is nowhere more evident than in their approach to the matter of illicit drugs. Recognising that a drug-free society, though perhaps desirable, was an unobtainable ideal, the Dutch government acted in the 1970s to separate 'drugs with acceptable risks' from 'drugs with unacceptable risks', both in law and in the marketplace.[10] Putting cannabis in the former category led most famously to the opening of the Dutch 'coffee-shops', cafés where small amounts of cannabis could be bought and sold openly without fear of prosecution. Their hands tied by the UN Single Convention on Narcotic Drugs (1961), the Dutch authorities were unable to legalise cannabis, but the half-measure of decriminalisation freed them to maximise their efforts in preventing 'unacceptable' drug use: of heroin, cocaine, LSD, and later Ecstasy.

It was within this liberal climate that a young entrepreneur, Hans van den Huerk, opened Holland's very first 'smart shop', Conscious Dreams, in Amsterdam in 1993.[11] A disillusioned IT worker, volunteer drugs charity worker, and Rave enthusiast, van den Huerk envisioned a shop that would sell a mixture of vitamins and supplements together with the legal synthetic 'designer' drugs that were creeping onto the market (created as underground chemists worked their way through Alexander Shulgin's two enormous recipe books[12]).

For during the early 1990s, at just the time when McKenna was arguing for a return to the natural, there was a concomitant Ecstasy-inspired rush of excitement about the possibilities of synthetic, or 'smart', drugs. This sci-fi, cyber-cultural vision of the future saw, not a world of archaic revivalists, but one of neuromancers armed with both laptops and the pharmacological power to adjust brain chemistry in any manner of their choosing, just as a computer programmer might tinker with a piece of software. There would be drugs to help us sleep, dream, wake up again and concentrate, to enhance and prolong sex, to improve our memory and intelligence. Drugs would enable us to fulfil our limitless potential by giving us mastery and absolute control over that most unruly of subjects, the self.[13]

Thus, the Dutch smart shops were born, selling everything from vitamin pills, natural plant stimulants such as guarana and ephedra and plant hallucinogens like *Salvia divinorum* and the San Pedro cactus, to synthetic drugs with arcane names such as 2C-B and 2C-T-2. A game of cat and mouse ensued, in which new drugs would come onto the market only to be criminalised a few months later by the ever-watchful authorities: the hunt was always on to find legal loopholes and drugs that might slip through them. Not long after opening, van den Huerk was approached by a home-growing magic mushroom enthusiast with an offer to supply magic mushrooms. After legal consultation, van den Huerk concluded that the law was sufficiently grey for him to put mushrooms on the market (following the 1971 UN Convention on Psychotropic Substances, the Dutch Opium Act only proscribes psilocybin and psilocin, and not the mushrooms themselves). They very quickly became a best-seller, at one point accounting for 60 per cent of the shop's sales.

As the market grew, Conscious Dreams opened two further shops in Amsterdam and expanded into the wholesale mushroom trade. It was quickly followed by other competitors, such as De Sjamaan (based in Arnhem). Most mushrooms sold were strains of *Psilocybe cubensis*, occasionally with the harder-to-grow but very strong Hawaiian species *Copelandia cyanescens*, and also 'truffles', that is sclerotia, of *Psilocybe mexicana* and *Psilocybe tampanensis*. Dried Liberty Caps, *Psilocybe semilanceata*, which are rare in Holland and problematic to grow, were imported in huge quantities from the UK.

At the same time, there was a crash in the price of ordinary, humdrum supermarket mushrooms. Dutch mushroom farmers could only get the equivalent of about €1 per kilo for button mushrooms, while cultivated magic mushrooms were selling for fifty or sixty times that amount. With such a price differential, it is not hard to see why some made the switch to growing hallucinogenic mushrooms: with the expertise and the infrastructure already in place, they could grow them on an industrial scale. The biggest growers now produce hundreds of kilos a week from large environmentally controlled sheds, with supply tailored to demand through computer forecasting. In Holland, the magic mushroom is big business.

The Dutch authorities have so far tolerated this, although van den Huerk was arrested and prosecuted in 1997 for supplying dried mushrooms, which, the authorities argued, constituted an illegal preparation. A protracted legal battle ensued, which van den Huerk eventually lost: in Holland, fresh mushrooms can be bought and sold with impunity, but dried ones remain illegal.

There is, of course, an economic pressure for smaller growers, who do not have the luxury of computer forecasting, to dry their mushrooms. Mushrooms, however well kept and refrigerated, have a shelf-life of only a few days, after which time they must be discarded – or dried. When I visited Amsterdam in the spring of 2005, dried mushrooms were being sold along with other illicit synthetics under the counter at various smart shops (not Conscious Dreams, I hasten to add). The game of cat and mouse continues.

In Britain, it took budding entrepreneurs a little longer to catch on to the idea of retailing mushrooms, though the law was no less ambiguous. You will recall that at the conclusion of the very first court case in 1976, the judge, Peregrine Blomefield, ruled that possession of mushrooms and the prohibited substance psilocybin were not the same thing. This ruling was reinforced by an analogous case involving cannabis, the so-called *Regina* v. *Goodchild* case of 1978.[14] Goodchild had been arrested for possession of immature cannabis plants that lacked the buds and flowering tops proscribed by law. The police, eager to see the case come to court, charged Goodchild with possessing THC, the active chemical ingredient of the plant. Possession of pure THC was then a Class A offence, whereas cannabis was Class B, so somewhat unfairly Goodchild stood to have a stiffer sentence than if his plants had been in bud and usable. On appeal, the law lords eventually ruled that the police were in error, and that possessing the plant was not the same as possessing THC (the law on cannabis was subsequently changed to prohibit all parts of the plant, though more recently cannabis has been downgraded to a Class C drug by the Labour government: it has effectively been decriminalised).

The upshot of the Goodchild case was that, by analogy, possession of a magic mushroom and of its proscribed chemical ingredients could not be regarded as the same thing. A test case in 1983 settled the matter. Kelvin Curtis, aged thirty-one, was acquitted after being tried for growing *Psilocybe cubensis* at home.[15] Thereafter, fresh magic mushrooms were technically legal, with only the act of preparation rendering them otherwise.

The Camden Mushroom Company, formed in 2002, were the first to exploit this loophole by selling mushrooms cultivated at home from Dutch grow-kits (grow-kits are ready-prepared trays of mycelia and growth medium, requiring little but warmth and water for mushrooms to sprout: one kit may produce a few kilos of mushrooms).[16] They sold their wares to head shops (shops selling cannabis-related paraphernalia), and then openly themselves from barrow stalls at Camden Market and Portobello: a move that generated a certain amount of press interest, especially amongst the more streetwise broadsheets.[17] The company eventually realised that mushrooms could be bought wholesale from the Dutch who, with a keen eye on the economic ball, had been looking to expand the market beyond their borders. Other companies sprung up, and virtually overnight, it seems, mushrooms were being sold in around four hundred shops and market stalls across Britain, with kilo-loads of mushrooms arriving

Flyer advertising commercially grown magic mushrooms, London 2005, prior to the change in the law which made such sales illegal.

from Holland every week. Demand was such that some British companies began growing mushrooms themselves. By 2004, the Camden Mushroom Company alone were selling on average 100 kg a week, with an estimated five times that amount being sold nationwide. As an average street dose of *Psilocybe cubensis* is 20 grams, this works out at roughly 25,000 mushroom trips per week. The second great British magic mushroom boom was underway, and inevitably it caught the attention of police and politicians alike.

One difference between the Dutch and British approaches to the magic mushroom trade is that in Holland far more negotiation has taken place between the authorities and the smart-shop owners. This is not to say that arrests never take place, or that the smart shops never get their knuckles publicly rapped. But Dutch pragmatism still concludes that it is better for mushrooms to be sold openly and transparently, so that sales can be monitored and levels of harm more easily assessed.

In Britain, negotiation has been less in evidence, partly because the mushroom sellers organised themselves less well than did their Dutch counterparts – the Dutch smart-shop owners very quickly established a trade organisation to give them more lobbying power – and partly because of a very different and entrenched prohibitionist stance on the behalf of the British government.

It was apparent from the beginnings of mushroom-trading in Britain, however, that the legal situation was one of disarray. The Camden Mushroom Company received a letter from the Home Office to the effect that they would not be prosecuted for selling *fresh* magic mushrooms, a move that effectively gave them the go-ahead to begin trading. Then in August 2004, a Customs and Excise ruling decreed that magic mushrooms were not a food but a drug, and therefore subject to VAT sales tax.[18] Officials estimated that they would be looking to collect arrears in the region of £1 million from traders. At almost the same time, the Home Office performed a volte-face and sought to deter trading by warning sellers that they were breaking the law. Thus, the government foolishly appeared to be demanding tax on a product that it deemed illegal, the left hand unaware of quite what the right was doing.

In the same year, a small number of shops were raided by the police and charged with illegally selling hallucinogenic fungi, a Class A drug. The first of these cases – concerning Dennis Mardle and Colin Evans, who ran the head-shop Collector's Choice in Gloucester – came to trial in December, when it was promptly thrown out by the Crown court recorder, Ms Claire Miskin.[19] The law was so ambiguous, she ruled, that it would be an abuse of process for the two men to stand trial. In a carefully aimed rebuke, she chided the government for expecting the judiciary to sort out a legal mess of Parliament's making. Suitably admonished, the Home Office moved quickly to add a clause about mushrooms to their Drugs Bill (now Act) 2005, though not without some lively debate in the House of Commons. In a speech that may yet come back to haunt him, the Labour MP for Newport West, Paul Flynn, attacked the bill as illogical and poorly thought through. 'We cannot make nature illegal,' he told the House. 'Magic mushrooms are part of the natural world. Some might describe them as a gift from God.'[20]

Flynn's remonstrations had little impact. Specifically, the Act has now made possession or sale of 'all fungi containing psilocybin' a criminal offence, so the loophole has been firmly closed. The speed with which the bill achieved royal assent, however, has left many disgruntled parties – not least the police, unhappy at having to enforce a law regarded as an unnecessary burden on overstretched resources – who feel, correctly, that the matter was never properly debated in the upper and lower chambers. The imminence of the May 2005 general election meant that the bill was hurried through Parliament lest New Labour appeared to be weak on crime, weak on the causes of crime: being seen to tolerate brazen mushroom-dealing on the streets would have been an electoral liability too far.

The new Act came into force in July 2005, with the effect that open mushroom-trading in Britain has ceased. As I write, the mushroom season has not yet begun, so the extent to which police will try to stop people harvesting naturally occurring Liberty Caps is unclear. It is also too early to tell whether dried mushrooms will appear on the black market, as one Dutch wholesaler predicted. Certainly, the Dutch were loath to see the market closed, and smart shops are already hunting for other legal plant hallucinogens that might be substituted for mushrooms. At the same time, they have been looking to open new markets in other European countries where the law is sufficiently vague – Germany perhaps, or Spain. Spores may still be traded legally in Britain (they contain no psilocybin), and special legal, idiot-proof grow-kits, containing

Flyer advertising mushroom grow-kits sold by the Psychedeli – an offshoot of the Camden Mushroom Company, London 2005.

spores and substrate, are being developed for export from Holland. The Camden Mushroom Company intends to sell spores and the rest of the accoutrements needed for home growing, but concedes that the market will be small. Whatever scams and schemes people come up with to try to get past the law, however, by the time you are reading this, the British magic mushroom boom will almost certainly be over.

To what extent were the government's actions justified? Statistics from Holland suggest that, however visible magic mushrooms are in the marketplace, the numbers of people actually using them remain comparatively small. In Amsterdam (where drug use of all kinds is atypically high compared to the rest of the Netherlands), the percentage of the city's population that had ever tried mushrooms stood at 7.6 per cent in 2001, a rise from 6.6 per cent in 1997.[21] (This compares with 1.3 per cent for heroin, 8.7 per cent for Ecstasy, 10 per cent for cocaine, and 38.1 per cent for cannabis: 2001 figures.) Most users were in their twenties. The percentage of people who had taken mushrooms during the previous thirty days, however, stood at a mere 0.3 per cent. This implies that though increasing numbers of people are trying mushrooms, obtained from the city's smart shops, very few take them more than once or twice in a lifetime. Of those 7.6 per cent of lifetime users, only 8.4 per cent had taken mushrooms more than twenty-five times and so counted as 'experienced users'. Terence McKenna was right: psychedelic shamanism was a calling, and most people found that it was not for them.

There is not yet any equivalent British data for comparison, but there is no reason to think that the situation would be different. Had mushrooms remained legal, numbers of one-time users would have steadily crept up, but the proportion of regular users would almost certainly have remained small. Twenty-five thousand trips a week sounds rather alarming, but it is proportionally insignificant and would, in any case, have levelled off. Without underplaying the possibility of bad trips, of lingering psychiatric complications resulting from mushroom use, or of injury sustained under their influence, the current boom has not produced the corresponding increase in hospital admissions that was seen during the early 1980s. Though this may be a time-lag effect, medics having yet to publish their data, it rather suggests that people have learnt how to take mushrooms safely. The average street dose is a tolerably low fifteen grams of *Psilocybe cubensis*, less than a quarter of that preferred by McKenna-type psychonauts.

The risks, in Dutch terminology, would seem to be acceptable, and in that light the British response appears heavy-handed, motivated more by political concerns than by any sensible assessment of the evidence.

Indeed, prohibition may prove to be a retrograde step in terms of harm reduction. The likelihood of novices picking the wrong mushrooms, or of spurious mushrooms finding their way onto the black market, will undoubtedly be increased, while a whole new cross-section of society will have to face up to the legal consequences of their actions, consequences that far outstrip any health repercussions. Given that, by any measure, prohibition has been an abject failure in preventing the escalation of drug use – the illegal drugs trade in Britain alone is worth $4 billion a year[22] – perhaps the time has come for a more sober and rational debate modelled on the pragmatic approach of the Dutch. In the current climate, where any call for decriminalisation is met with a barrage of invective from the tabloid press, and an unseemly political tussle to occupy the moral high ground, such a move would still seem a very long way off.

This, then, pretty well concludes the story of the magic mushroom, of how it went from being an obscure poison to being the drug of choice, hawked on street corners in Britain and Holland, and grown in cellars the length and breadth of America. A long correspondence between an English poet and an American banker led the latter, obsessed as he was with rewriting religious history, to go to Mexico. There, he found the magic mushroom, which had been in use since the time of the Aztecs, and the *curandera* María Sabina, and he broadcast his discovery to the world. He located the best minds in the field to identify the mushrooms and reveal their chemical secrets. The social and cultural revolutions promulgated by LSD in the 1960s, and Ecstasy in the 1980s and 1990s, turned on successive generations to drugs in general, and psychedelics in particular, and opened up a radical new way of understanding the hitherto shunned effects of the mushroom. In an effort to meet this new-found demand, techniques for cultivating them developed, and industrial-scale production was perfected. In a comparatively short period of time, mushrooms were remoulded and transformed by the cold hand of capitalism from a revered sacrament to that most pliable of artefacts, a commodity.

This is how it happened, but the question that remains is 'Why?' Why have we found value in those peculiar bemushroomed visions, when all those in the West that came before us discarded them as worthless? What is so distinctive about contemporary Western culture that it has embraced the magic mushroom? Why us, and why now?

Epilogue: Love on a Puffball

Though some poems, melodies, works of art, love-affairs and fever dreams may give glimpses of a lost magical reality, their spell is short lasting . . . The hard, dirty, loveless synthetic world reasserts itself as the sole factual truth.

Robert Graves, 'The Universal Paradise'[1]

And as the mushroom told its tale she screamed 'My God!
My body is made of sunlight
There can be no other way
My body is made of love!'

Circulus, 'My Body is Made of Sunlight'[2]

Towards the end of his acutely observed radio play *Under Milk Wood* (1954), the poet Dylan Thomas briefly introduces the character Mae Rose-Cottage, 'seventeen and never been sweet in the grass'. Lying back in a field, and dreaming of forbidden pleasures to come, 'she blows love on a puffball'.[3] In this arresting image, which skilfully conjures memories of awakening desire, Thomas is drawing upon a rich cultural fund of earthy and enchanted fungal associations that stretches back through Shakespeare and beyond. But the darker, opposing side of fungi has never been far away, and in the uncertain post-war days in which Thomas briefly flourished it was another, altogether less comfortable mushroomic image that was uppermost in most people's minds.

On 16 July 1945, at Los Alamos in New Mexico, the world's first nuclear bomb was successfully exploded. As the deadly symmetry of the first mushroom cloud rose above the desert – to be followed a month later by two more over the Japanese cities of Hiroshima and Nagasaki – a new symbol for the dark and terrifying power of human scientific ingenuity was thrust upon the world. At the same time, it was as if the baleful side of mushrooms, so long known about and feared, had finally risen triumphant, a conquest written large upon the sky.

The oft-recycled image of the mushroom cloud was a constant reminder, throughout the Cold War, of what was at stake in the ongoing political games and manoeuvres played out between East and West. That particular danger has passed – the Cold War is an increasingly distant memory, seemingly as incomprehensible as a bad dream – but the twenty-first century has brought with it a new raft of anxieties. Terrorism and climate change, not the threat of all-out nuclear war, are what cause sleepless nights now. But interestingly, just as the nightmarish fear of the mushroom cloud has receded, so, it seems, has the magic mushroom made an unstoppable rise. It has brought with it a steady drift back towards the other cultural view of fungi as carnivalesque eruptions, enchanted sentinels of the weird, benign and gnomic things upon which to blow love.

Mushroom enthusiasts today typically portray humanity as being at a crossroads, our choice of direction framed in exactly these terms: modernity, the mushroom cloud and annihilation, or an archaic revival, the magic mushroom and salvation. Daniel Pinchbeck, New York journalist and latest pretender to the throne of psychonaut-in-chief, puts it like this: 'The destruction of the World Trade Center may turn out to be a prelude or birth pang, first in a series of convulsions, before modern civilisation is expunged from the planet and forgotten forever, unless there is a quick and unlikely reversal of trends.' His prescription for change? A rebirth of psychedelic shamanism that will 'suck the spiritual poison from our social body'.[4] Though I do not share Pinchbeck's confident belief in the power of the magic mushroom – or any other psychedelic, for that matter – to act as magic eco-pill or societal panacea, I do nevertheless think that this stark dialectic contains within it a clue to the magic mushroom's unexpected popularity, and why the West has taken to it with such enthusiasm.

The reasons that people take the drugs they do are complex, and not easily broached in a short concluding chapter such as this. As we have seen, attitudes towards psychoactive substances and their effects are not fixed or stable, but shift according to the interplay of power and knowledge. Changing constructions of what constitutes pleasure, desire, intoxication, addiction and transgression work to legitimate different substances – and the sanctioned ways of using them – at different times. An appetite suppressant like MDMA can suddenly become a party drug. An addictive stimulant like cocaine can, at once, be the scourge of the inner cities and yet the fashionable choice of celebrities. A harmful and addictive drug like alcohol can be construed as an essential cornerstone of adult, civilised life. A poisonous mushroom can be reinvented as a portal to other worlds.

As we have seen, statistics from Holland suggest that, with regard to contemporary mushroom use, a large number of one-time or occasional users surround a small but dedicated hub of enthusiasts or habitués. This pattern – a European trend that is not mirrored in the US, where stricter laws have kept mushroom use firmly underground – has arisen very much on the tail end of Ecstasy and Rave, a movement that, for better or worse, made illicit drug use an acceptable part of life for the under forties. If caffeine is the drug that pumps the working week, then mushrooms have joined Ecstasy, Acid, amphetamines, cannabis, cocaine, ketamine and alcohol as the means of venting pressure at the weekend. Tolerably weird, the effects of mushrooms (at least, at moderate street doses) are interesting enough to last the length of an all-night party, but not so overwhelming as to prevent one from returning to work on Monday morning. (By contrast, the effects of, say, the legal plant hallucinogen *Salvia divinorum*, or of DMT, appear to be too disturbing and bizarre for these drugs ever to become co-opted by the mainstream.) Nevertheless, the open sale of mushrooms in Europe has given them an unintended legitimacy and apparent societal seal of approval, such that people who might not otherwise buy illicit drugs, or who have no leanings towards psychedelia or Rave, have been tempted to try them. The mushroom craze has been driven by sheer availability as much as by socio-cultural factors or accidents of pharmacology.

Ease of acquisition, visibility and changed social attitudes go some way towards explaining the recent boom, but not really why people on both sides of the Atlantic took to the mushroom so enthusiastically in the 1970s, when mushrooms still had to be actively sought out. Acid and psychedelia provided the favourable context here, while the latest advances in fungal taxonomy and pharmacology swept away any residual fears that these mushrooms might be poisonous. Mushrooms also arrived at a time when interest in, appreciation of and concern for the environment were all increasing significantly. This was the decade of Greenpeace, the Whole Earth Catalogue, *The Good Life* and self-sufficiency. Mushrooms came to be cast as a 'natural', 'organic' and hence better, healthier and more authentic alternative to synthetic LSD – a perception that remains in place today and accounts for at least some of their popularity.

But it is here, in this last point, that a revealing contradiction becomes apparent. On the one hand, by overturning the historically predominant poisoning discourse, science delivered up the magic mushroom to the West. Advances in microscopy, mycology, chemistry, pharmacology and myco-culture led directly to its recognition, reinvention and rise. On the other, the very people who took to the mushroom most ardently in the 1970s, and whose interest ensured its eventual popularity, were those who were challenging or even rejecting science's claims to epistemological exclusivity, or who saw science's destructive power – as revealed by weapons of mass destruction and ecological degradation – as something in dire need of restraint.

Rather like the impact of matter and antimatter in a particle accelerator, science and anti-science have collided throughout the history of the magic mushroom. Just as the ricochets and trajectories of subatomic particles reveal something of how the physical universe is constructed, so, I think, do the reverberations caused by the magic mushroom expose something fundamental about our cultural universe, about the attitudes and sensibilities that shape our time. That we in the West, in the twentieth and twenty-first centuries, have found the mushroom's litany of peculiar effects desirable is, I would suggest, symptomatic of a broader underlying craving for meaning – more specifically, for enchantment – that sits somewhat awkwardly within our supposedly rationalist, scientific and technological culture.

In his essay 'Science as a Vocation', written during the First World War, the pioneering German sociologist Max Weber (1864–1920) noted that 'The fate of our times is characterised by rationalisation and intellectualisation and, above all, by the "disenchantment of the world".' By this he meant that there was no longer any need to invoke mysterious forces to explain the phenomena of the world because 'one can, in principle, master all things by calculation'.[5] The rising scientific tide had steadily scoured away belief in fairies, demons, spirits, djinns, God even, and with it any notion that our place in the world was ordained or had meaning. Though Weber died some twenty-odd years before its advent, the nuclear bomb threw this process into grim relief, revealing in dramatic form the godlike power of science and, by contrast, the impotence and obsolescence of other explanatory frameworks.

If Weber seems implicitly to have accepted the supremacy of science, then he nevertheless gave people dispensation to make the 'intellectual sacrifice' and retreat back into the welcoming arms of religion should the 'fate of the times' prove unendurable. In doing so, he showed more forbearance than later theorists, particularly those of the 1960s, who thought the tide of science and secularisation both desirable and unstoppable. Such a conclusion now appears hasty, for culture seems to be increasingly parenthesised by two irreconcilable, absolutist positions: fundamentalist religion on the one hand, and fundamentalist science on the other.

The former is evidenced by the rise of the Christian Right in America and of militant Islamism elsewhere, the latter by the presence of vocal champions for science, such as writer and evolutionary biologist Richard Dawkins. A militant atheist, Dawkins regards religion as an abrogation of rational thought, an intellectual sacrifice too far. The cold message of Darwinian natural selection, he admits, is that human existence, and indeed the fact of life, is meaningless, the product of random molecular collisions and selective pressures acting over vast timescales. Nevertheless, comfort may be found in knowing how light splits into the colours of the rainbow, how stars form, how the fig tree reproduces or how the mammalian eye evolved; that is, in science's unbridled capacity to explain. For Dawkins, science reveals the world to be so full of wonder that there simply is no need to invent mythical forces, deities or beings.[6]

Weber thought the 'tension between the value-spheres of "science" and the sphere of the "holy" . . . unbridgeable',[7] but in truth people have managed to adopt an array of 'softer' positions between the extreme poles of religious and scientific fundamentalism. Even so, there remains a significant residuum for whom processes of disenchantment have left an unaccountable feeling of loss. Possessed of romantic longings, and an appetite for mystery that science can never fulfil – its *raison d'être* is to banish mystery, after all – they are, nevertheless, unable to return to traditional religion. Caught between science on the one hand and religion on the other, they are continually forced to find re-enchantment elsewhere.

The prevailing cultural discourses, however, do not make this easy. Anyone who, say, stands up and proclaims the existence of hives of self-transforming machine elves in a parallel dimension to ours is likely to meet with a cynical and derisive response. Culture can only tolerate elves and faerie denizens if they are bound firmly into film, the pages of fiction – children's, sci-fi, fantasy or the knowing conceit of magic realism – or the symbolic language of the unconscious, where the ridiculous possibility of their actual physical existence need never be raised. As Graves noted, the dirty synthetic world has asserted itself as the sole factual truth.

And yet, there is a palpable thirst for these oases of enchantment. Children's literature and fantasy fiction, for example, have never been so eagerly lapped up by adults. Witness the huge popularity of Philip Pullman's *His Dark Materials* trilogy, J. K. Rowling's *Harry Potter* romps and Peter Jackson's epic screen version of Tolkien's *Lord of the Rings*. According to some critics, we are all 'kidults' now, stuck in a permanent 'middle youth'. Think also of the burgeoning smorgasbord of alternative spiritualities – New Age, Pagan and neo-Shamanic – elective and eclectic assemblages that sit somewhere outside science and traditional religion, and yet proclaim a special place for us in the greater scheme of things. For all his tolerance of religion, Weber could not bear the New Age precursors of his time: he took a sideswipe at those who 'play at decorating a sort of domestic chapel with small sacred images from all over the world, or . . . produce [religious] surrogates through all sorts of psychic experiences to which they ascribe the dignity of mystical holiness'.[8] Still lambasted by the mainstream, these alternatives are, nonetheless, genuine attempts to answer the dilemma of our times.

And this is where mushrooms fit in. For whether they sweep back the veil to reveal the world as it really is, as enthusiasts maintain, or push one dangerously close to the edge of madness, as society ripostes, they indisputably occasion experiences that nowadays only happen in movies or stories, and with an immediacy that makes them seem, to all intents, real. One small cup of mushroom tea can assuage the most ardent craving for enchantment.

Take the following example, recounted to me by a woman who was, at the time of the story, living in Glastonbury. This small town in rural Somerset is famous not only for its eponymous music and arts festival (actually held in the neighbouring village of Pilton) but also for its Arthurian associations, its hippies, Pagans and New Agers, and its mammiform Tor topped with a ruined church tower. It is the ancient, mystical Isle of Avalon, and as such remains an island refuge of enchantment in all its guises. Hopeful of gaining an insight into the Buddhist concept of 'the void' or 'emptiness', this plucky explorer went out alone and ate forty-five fresh Liberty Caps. Sitting on the lower slopes of the Tor, Chalice Hill to be exact, she watched as the stars dutifully winked out one by one. At just the moment when nothingness, horror and panic threatened to overwhelm her, she saw a lozenge shape glowing reassuringly in the grass at her feet. Looking up, she saw that it was, in fact, a scale, one of thousands on the back of a gigantic beast that looped away and over the Tor. To her surprise she realised that the Tor really was, as some myths say, a sleeping dragon. The enchantment continued for some time, during which two moons rose into a flashing purple sky, and a small piece of woodland became magically transformed into a forest, through which she stumbled, lost and pook-ledden, for what seemed like hours. It was, she said, one of the most terrifying, and yet beautiful, experiences of her life.

Though the mushroom trip is never an easy ride – as capable of inducing horror as delight – enthusiasts are adamant that it contains a message: the world is not as sterile as science maintains, meaning resides within all things, and the essential quality of the universe, its quiddity, is one of enchantment. Or as Daniel Pinchbeck puts it: 'The nature of reality is spiritual, not physical. Everything we see around us is animated by sentient essences, dainty sub-Planck-length flimmers of cosmic wit.'[9] For those in need, the magic mushroom seems to answer existential angst like the key to a lock. No wonder it has proved so enticing.

In the summer of 2005, to round off my research for this book, I went to America specifically to attend the Telluride Mushroom Festival. This mycological celebration takes place every summer in, as the name

Identifying the morning's pickings. A dawn fungus foray at the Telluride Mushroom Festival, 2005.

suggests, the small ski town of Telluride, high up in the Rocky Mountains of Colorado, and has done so now for twenty-five years. On the surface it is very much like any other gathering of amateur mycologists. Every day there are forays, a chance to wander off into the forests and collect mushroom specimens, with an expert on hand to help with identification. The day's pickings are diligently wrapped in wax paper and carried back to the town centre, where they are formally identified, labelled and laid out in a marquee for all to see. Some of the top American mycologists lecture on all aspects of fungal taxonomy and ecology, or conduct workshops on cultivation: the latest hot technique is apparently growing mushrooms on newspaper and straw. There are cookery classes, and a trophy awarded to the local restaurant that crafts the best wild mushroom dish. And, to cap it all, there is a procession through the town, where festival-goers put on fancy dress as their favourite mushroom – usually, but not exclusively, the fly-agaric – and wave cheerily at the somewhat bemused locals.

But look a little closer, and you will find that the festival is not quite what it seems. Though Telluride is a smart ski resort, its eco-café, organic supermarket and anti-war graffiti betray the fact that it is also a bit of a hippy town, and back in the early 1980s the hippies rather successfully infiltrated and subverted the mushroom festival. Now, not everyone who attends the workshops intends to grow gourmet or edible fungi; not everyone who studies identification wants to know how to pick wood blewitts and morels.

Uniquely amongst American mycological gatherings, Telluride blows love across puffballs, chanterelles and shaggy manes – Death Caps, even, and fly-agarics – but, most importantly of all, across Liberty Caps, *cubensis* and all the other psilocybe species. Telluride celebrates mushroom-kind in all its diversity, and in its understated way has become one stop on a much larger and thriving psychedelic circuit. It is certainly the place to network and discover what's hip and happening in the mushroom underworld. But while the lectures and workshops are all fascinating and well-attended, it is clear that, since the death of Terence McKenna in 2000, the psychedelic movement has had no obvious figurehead: the magic mushroom currently lacks a cogent and vocal champion.

There is, of course, no shortage of contenders, with a new generation of enthusiasts ready to take on the McKenna mantle. Daniel Pinchbeck – the journalist who began writing about shamanism, but ended up deciding he wanted to be one – is perhaps the brightest star, but Jeremy Narby – the latest anthropologist to have taken ayahuasca and made a sudden departure from epistemological orthodoxy – is also in the ascendant.[10] Erik Davies adds a Californian perspective, with his intellectually more rigorous consideration of psychedelics, technology and magic.[11] The Wassonian torch is being carried forward in Italy by Giorgio Samorini, and in Germany by Christian Ratsch – both prodigious writers – while in Britain, Simon Powell has become a cogent proselytiser for mushrooms as our ecological redeemers.

During my visit to Telluride, however, the star performer was undoubtedly ethnobotanist and illustrator Kat Harrison, who has, until recently, been weighed down with the burden of being 'Terence McKenna's ex-wife'. She is, however, a brilliant speaker and thinker in her own right, and is beginning to be recognised as such. At Telluride she brought a refreshing womanly perspective to what has been hitherto a male-dominated field, and in a gentle way laid out the case for the rightness of indigenous, animistic approaches to the use of psychedelics in general, and mushrooms in particular.

Whoever it is that eventually assumes the throne (I rather suspect that Harrison does not want it), they can be certain that they will be joining an illustrious line of poets, scholars, artists, travellers and dreamers – romantics, one and all – whose dalliances with the mushroom have, if nothing else, injected a bit of sparkle into a loveless, synthetic world. The magic mushroom has been with us in the West for just fifty years, but in that time it has caused upheavals far greater than its diminutive size should rightfully allow. As long as there is a yearning for magic and enchantment, its place in culture seems assured, no matter what prohibitions are introduced. Who knows what fabulous theories it will inspire next? At the very least we shall, I think, be blowing love on puffballs for some time to come.

Appendix

Chemical structures of some psychedelic compounds contained within magic mushrooms, and their close relatives.

The mushrooms most commonly consumed are those which have psilocybin and psilocin as their principle active ingredients. These indole alkaloids are structurally similar to the most famous psychedelic drug, LSD; all three are similar to the endogenous brain neurotransmitter, serotonin, whose action they mimic. The fly-agaric, on the other hand, contains the psychoactive compounds ibotenic acid and muscimol, which act on another part of the brain entirely. See chapter 2.

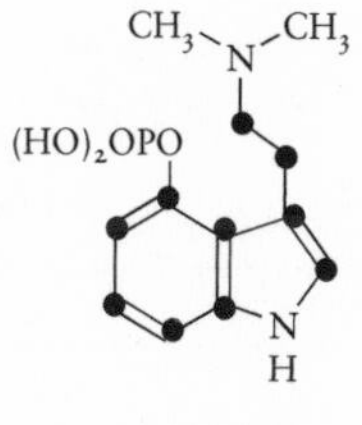

Psilocybin

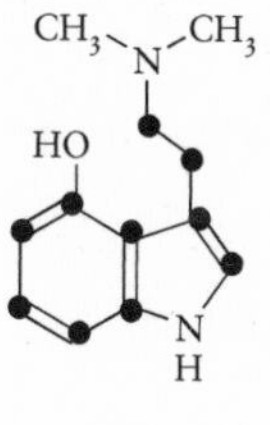

Psilocin

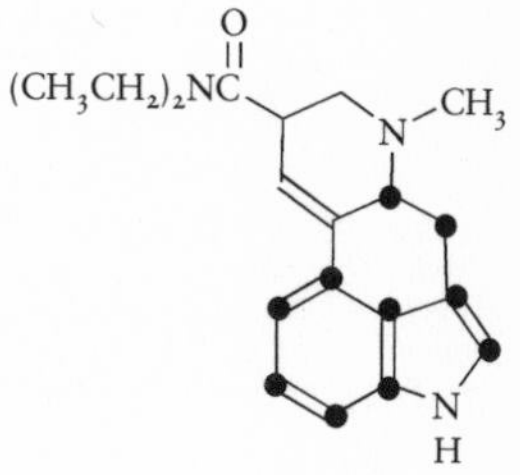

LSD

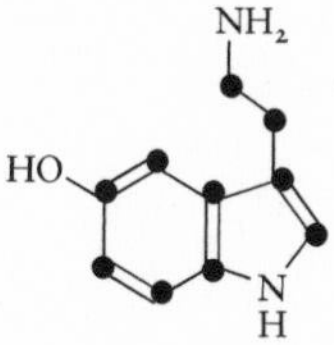

Serotonin

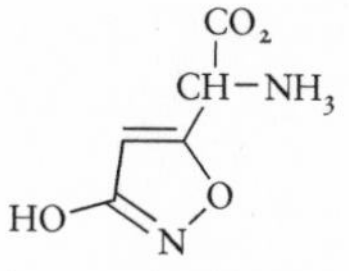

Ibotenic acid

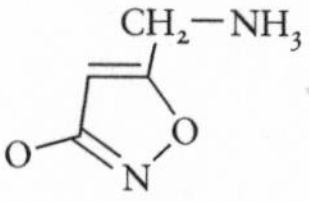

Muscimol

Notes

Material from the R. Gordon Wasson Archives, Harvard Botanical Libraries, Harvard University, Cambridge, Massachusetts is referenced here as Wasson Archive, with the name of the archive file given in brackets.

PROLOGUE

1 Douglass 1917.
2 Douglass may have been mistaken, for when he had the mushrooms checked they were identified by Dr Murrill as *Panaeolus semiglobatus*. The psychoactivity of *Panaeolus* species is, however, notoriously capricious and unpredictable.

ONE

1 Dent 2004.
2 See, for example, Cooper 1977; McKenna 1992; Powell 2003.
3 Alexopoulos et al. 1996.
4 Hawksworth 2001.
5 Smith et al. 1992.
6 Johnson 2005.
7 Chilton 1975a; Hammelev 1986.
8 Findlay 1982.
9 Prokhorov 2004.
10 I am indebted to Clare Brant for drawing this to my attention.
11 Briggs 1993.

TWO

1 Plant 1999, p. 4.
2 All statistics are from Guzmán et al. 1998.
3 Technically, the primary active ingredient should be spelt, as originally named, psilocybine. The final 'e' seems to have been lost, so that the accepted, if inaccurate, spelling is psilocybin. See Ott 1996.
4 Guzmán et al. 1998.
5 Keay and Brown 1990.
6 These figures are from Stamets 1996, who collates data from a variety of published scientific assays. See also Hofmann et al. 1958; Mantle and Waight 1969.
7 A list of common names may be accessed at www.mushroomjohn.org
8 Stamets 1996.
9 Bigwood and Beug 1982.
10 Powell 2003.

11 For a critique of these terms, see Letcher 2004.
12 Passie et al. 2002.
13 Ibid., p. 363.
14 Passie et al. 2002.
15 Stamets 1996.
16 Badham 1984.
17 Riper and de Kort 1999.
18 Curry and Rose 1985.
19 Hasler et al. 2004. See also Borowiak et al. 1998.
20 Passie et al. 2002.
21 Smart et al. 1982, p. 419; but see also Benjamin 1979.
22 Riper and de Kort 1999.
23 Vollenweider et al. 1997.
24 See, for example, Sink et al. 1983.
25 Spitzer et al. 1996.
26 Honigsbaum 2005.
27 Moreno et al. 2003.
28 For a far-sighted essay on the relationship between psychedelics and mental illness, see Lipton 1970.
29 Mount et al. 2002.
30 One of the best and most thoughtful guides to the identification of psilocybin mushrooms is Stamets 1996.

THREE

1 Hancock 2005.
2 Huxley 1994, pp. 42–3.
3 La Barre 1970; Weil 1972; Siegel 1989.
4 Devereux 1997.
5 Cooper 1994; Cohen 2001.
6 For a more thorough discussion of this, see Letcher 2004.
7 Simmons 2001.
8 Ramsbottom 1953; Schultes and Hofmann 1992.
9 Reay 1960.
10 Heim and Wasson 1958, 1965; Thomas 2002.
11 Thomas 2002.
12 See Schechner 1993.
13 La Barre 1970; Schultes and Hofmann 1992.
14 Laura Rivel, pers. comm.
15 Halstead 1967.
16 Wasson 1971, 1973b.
17 Li 1978.
18 Fowler 2002.
19 Wasson 1971, pp. 238–9.
20 Watling 1975.
21 Blanchette et al. 1992.
22 Sherratt 1991, 1995; Harding and Healy 2003; Merlin 2003.
23 Ibid.

24 Moffat 1993. Recently this finding has been challenged. See Long et al. 2000.
25 Dronfield 1995b; Creighton 2000; Wallis 2003.
26 Green 2001.
27 Sherrat 1991, 1995.
28 Hutton 2001.
29 McKenna 1994.
30 See Wallis 2003.
31 Gartz 1996; Stamets 1996.
32 Lasko 1972.
33 Henrietta Leyser, pers. comm.
34 Kaplan 1975.
35 Goskar et al. 2003.
36 Ibid.
37 Wallis 2003.
38 McKenna 1992; McKenna and McKenna 1993.
39 Lajoux 1963.
40 Harrison, pers. comm.
41 Lajoux 1963.
42 For example, see the images in Southworth 1998, especially those on pages 55 and 82.
43 See Helvenston and Bahn 2003, for example.
44 Lewis-Williams and Dowson 1988.
45 Oster 1970.
46 For example, Reichel-Dolmatoff 1978.
47 Wasson 1981; Hofmann 1983.
48 Bradley 1989; Whitley 1992; Dronfield 1995a,b, 1996; Creighton 2000.
49 Lewin 1991; Syal 1996.
50 For example, see Devereux 1997.
51 See Helvenston and Bahn 2003. One archaeologist I spoke to happily described the three-stage model as 'bullshit'.
52 See Wasson et al. 1978.
53 Graves 1972.
54 Wasson et al. 1978.
55 Samorini 1998; McKenna 1992.
56 Jay 1999.
57 Graves 1969, p. 4.
58 Karana 1968; Wallis 2003.
59 Hutton 1993.
60 See Hutton 1993, 1996b; Green 1997.
61 Hutton 1996b.
62 Wasson 1957.
63 Hutton 2003.
64 Hutton 1993, 1999, 2003.
65 See Clifton 2004.
66 Levack 1995, p. 48.

67 Schultes and Hofmann 1992, p. 90.
68 For example, by Edward Tylor. Tylor 1871 Vol. 2, p. 379.
69 Clark in Murray 1921, pp. 279–80.
70 Barnett 1965.
71 Harner 1973.
72 See Harvey 1997.
73 Harner was not quite the first to make the suggestion about broomsticks being used as dildos. He seems to have picked up the idea from Michael Harrison's 1973 book *The Roots of Witchcraft*. See Clifton 2004.
74 Ginzburg 1992; Mann 1994.
75 Wasson 1957; Cohen 2001.
76 Harner 2003 [1980]; see also Wallis 2003.
77 Schultes and Hofmann 1992.
78 Levack 1995; Sharpe 1996; Briggs 2002.
79 Sharpe 1996; Clifton 2004.
80 For example, see Rudgley 1998; Jay 1999.

FOUR

1 Gerard, cited in Findlay 1982, p. 15.
2 Marsh 1973; Piearce 1981; McKenzie 1997.
3 Cited in Wasson and Wasson 1957, p. 238.
4 Ibid., p. 239.
5 Pickering 1755, p. 586.
6 Ibid.
7 Ibid., p. 586.
8 Anon. 1757.
9 Heberden 1772, p. 380.
10 See Ainsworth 1976.
11 Stephen Harris, pers. comm.
12 Brande 1800, p. 41.
13 Sowerby 1803.
14 Though Singer's identification has become generally accepted, it is worth noting that Roger Heim, the French Professor of Mycology who was responsible for the identification of the Mexican psilocybin mushrooms, thought *Stropharia semiglobata*, the Dung Roundhead, the more likely candidate. Heim had previously found the presence of psilocybin in samples of this species, but all attempts to replicate his findings had failed, so the capricious presence of psilocybin would fit perfectly with the father's contention that he had eaten the mushrooms many times with no ill effect. Of course, the father may well have been mistaken, but as Heim's paper has never been translated into English, his doubts have made little impression upon scholarly and popular opinion. Consequently, it is generally accepted that this remains the first clear incident of *Psilocybe semilanceata* consumption. See Heim 1971.
15 Sowerby 1803, p. 2.
16 Sowerby 1832.
17 Ibid.
18 This incident was rediscovered by Wasson and Wasson 1957.

19 Glen 1816, pp. 451–2.
20 Salisbury 1815.
21 See Heim 1963; Heim et al. 1966; Pollock 1976.
22 Anon. 1815, pp. 397–8.
23 Cooke 1894, p. 89.
24 Edwards 1836, p. 512.
25 Ramsbottom 1953, p. 13.
26 Cited in Ramsbottom 1953, p. 13.
27 Findlay 1982, p. 4.
28 Ramsbottom 1953.
29 Ramsbottom 1953, p. 13.
30 Findlay 1982, p. 11.
31 Ramsbottom 1953, p. 17; Findlay 1982, p. 24.
32 Ramsbottom 1953; Findlay 1982.
33 Badham 1863, p. 43; *The Times* 10 August 1841, p. 7; *The Lancet* 13 November 1852, p. 434; 8 September 1906, p. 663.
34 Cited by Lightfoot 1777, p. 1016.
35 Watson 1744, p. 56.
36 Hayter 1968.
37 Hayter 1968; Jay 2000.
38 Anon. 1757.
39 *The Times* 13 October 1851, p. 6.
40 *The Times* 13 October 1859, p. 7.
41 Badham 1863, p. 38.
42 Brande 1800, p. 41.
43 Hay 1887, p. 15.
44 Badham 1863, p. 39.
45 Smith 1891.
46 See Ainsworth 1976; English 1987.
47 Cooke 1997 [1860].
48 *Guardian Review* 30 November 2002.
49 Cooke 1997 [1860], p. viii.
50 English 1987.
51 Cooke 1997 [1860], p. 259.
52 English 1987.
53 Cooke 1894, p. 25.
54 Ramsbottom 1923, p. 78, 1945, 1953.
55 Smith 1891, p. 163; 1910.
56 Cooke 1904.
57 *The Lancet* 6 January 1934, p. 44.
58 Ford 1909.
59 Indeed, this provides one of the clearest lines of evidence that magic mushrooms were not intentionally used until the 1970s. *The Collins Guide* was only revised to reclassify the Liberty Cap as late as 1993, in its sixth edition, while the first guide to list the Liberty Cap as hallucinogenic was Geoffrey Kibby's *Colour Encyclopaedia of Mushrooms and Toadstools*, published in

1977, one year after the famous court case that propelled the mushroom into popular consciousness. See Chapter Twelve.

FIVE

1 Sabina, in Estrada 2003, p. 38.
2 Account in Verrill 1914, p. 409.
3 Pollock 1976; Stamets 1996, pp. 79–80.
4 McIlvaine 1900; Douglass 1917; Imai 1932; Southcott 1974.
5 McIlvaine 1900, p. 386.
6 Ibid., p. xiii.
7 Verrill 1914.
8 Verrill 1914, p. 408.
9 On the Aztec pharmacopoeia, see Elferink 1988.
10 For a complete discussion of the chronicles, see Wasson and Wasson 1957; Wasson 1980.
11 Translation from Wasson 1980, p. 200.
12 Ibid., p. 201.
13 Cited in Wasson and Wasson 1957, p. 218.
14 Ibid., p. 219.
15 See Wasson 1980; Heim et al. 1967; Lowy 1971.
16 Wasson 1980, p. 112.
17 De Borhegyi 1961, 1963; Lowy 1971.
18 Wasson 1980.
19 Cited in Wasson and Wasson 1957, p. 226.
20 The history of the rediscovery of Mexican psychoactive mushrooms has been told many times. See Schultes 1940; Wasson and Wasson 1957; Singer 1958; Guzmán 1983.
21 Safford 1915.
22 Ibid., p. 293.
23 Ibid., p. 294.
24 Schultes 1940, p. 440.
25 Schultes 1940.
26 Ibid., p. 441.
27 Cited in Schultes 1940, p. 441.
28 Weitlaner-Johnson 1990.
29 Schultes 1940.
30 Singer 1958.
31 The party consisted of Johnson, his fiancée and Weitlaner's daughter Irmgard, Louise Lacaud and Bernard Bevan.
32 Johnson 1939.
33 Weitlaner-Johnson 1990, p. 135.
34 Brown 1990.
35 See, for example, Wasson 1986, p. 17. Wasson retold the story in almost all his books.
36 For example, see Pinchbeck 2002.
37 Britten 1990.
38 Wasson and Wasson 1957.

39 The first letter was from Dr Hans Mardersteig of Verona regarding a mushroom stone held in Zurich, the second was from Robert Graves. Wasson and Wasson 1957, p. 275.
40 Pike and Cowan 1959.
41 Wasson 1981.
42 María Sabina could neither read nor write, but interview transcripts were compiled by Álvaro Estrada. Estrada 2003a, p. 13.
43 Munn 1973; Rothenberg 2003.
44 Halifax 1979, p. 131.
45 Sabina, cited in Halifax 1979, p. 131.
46 Translated in Halifax 1979.
47 From Halifax 1979, pp. 133–4.
48 Wasson 1957, p. 3.
49 Wasson 1957, p. 3; Wasson and Wasson 1957, p. 289.
50 Wasson 1957, p. 3.
51 Sabina ate thirteen pairs.
52 Wasson and Wasson 1957, p. 293.
53 Ibid., p. 295.
54 Wasson 1957.
55 Wasson et al. 1974.
56 Wasson had eaten specimens of the locally preferred 'landslide' variety, *Psilocybe caerulescens* var. *maxatecorum*. See Heim and Wasson 1958; Heim et al. 1967.
57 Hofmann et al. 1958. Hofmann accompanied Wasson to Mexico in 1962, and persuaded Sabina to hold a *velada* using only his prepared psilocybin pills. He was gratified when, after some initial doubts, Sabina concluded that the pills contained just the same power as the mushrooms. Hofmann 1990, p. 123.
58 Wasson 1963.
59 Hofmann 1990.

SIX

1 Wasson Archive: letter from Gordon Wasson to Robert Graves, 13 May 1950 (Graves).
2 Cited as a warning by Wasson 1971, p. 8.
3 Wasson Archive: correspondence between Albert Hofmann and Gordon Wasson, passim (Hofmann).
4 Richardson 1990.
5 In fact the term 'mycophile' was first used by a French mycologist, Mangin, in 1907 to refer to members of the Société Mycologique de France. He distinguished 'mycophages', those whose interest in mushrooms was solely gastronomic, 'mycologues', amateurs with the status of professional mycologists, and 'mycophiles', members who were interested in larger fungi for their own sake. See Ainsworth 1976, p. 284.
6 Wasson 1971, p. 172.
7 See Bowman 1995.
8 Boyes 1993; Hutton 1996a.

9 To this day morris dancing is widely believed to be an ancient fertility rite, rather than the Tudor court dance it once was (see Hutton 1996a). Quite why any Mother Earth goddess would choose a rag-tag of middle-aged, pot-bellied, bearded lovers of real ale as her champions of fecundity remains a thealogical mystery.
10 Hutton 1999.
11 Wasson 1980, p. 53; 191–2. For Graves's notion of taboo, see Graves 1959.
12 Barfield 1997 (see entries for Boas, Frazer, and evolution).
13 Hutton 1999.
14 Lincoff and Mitchel 1997.
15 J. Ramsbottom, letters to *The Times*: 28 August 1942, p. 5; 3 August 1943, p. 5.
16 Though Leary later claimed Wasson as one of his formative influences, Wasson actually wrote a memorandum condemning Leary's profligate use of psilocybin and urging that he be reined in by his Harvard superiors. Wasson Archive: memorandum for Mr Thomas Lamont, 13 September 1961 (Leary).
17 For accounts of the pilgrimage, see Swain 1963; de Solier 1965; Bresser 1969; Finkelstein 1969; Nada 1969; Sandford 1972; Big D. Unlimited 1976; Weil 1980; Krippner and Winkelman 1983.
18 Finkelstein 1969, p. 55.
19 Ibid.
20 Sabina in Estrada 2003a, p. 71.
21 Estrada 2003, pp. 57, 64–7.
22 Halifax 1979, p. 129.
23 Wasson 1980, p. xvi.
24 Wasson 1970.
25 Wasson 1980, p. 48.
26 Wasson 1981, p. 20.
27 Wasson 1980, p. xvi.
28 Richardson 1990.
29 Wasson Archive: letter to Robert Graves, 26 August 1955 (Graves).
30 Wasson Archive: memorandum for Mr Thomas Lamont, 13 September 1961 (Leary).
31 Wasson Archive: letter from Philip Wootton to Pantheon Books Inc., 16 January 1957 (Life).
32 Wasson Archive: letter to Philip Wootton, 27 February 1956 (Life).
33 Wasson 1980, p. xvi.
34 Wasson and Wasson 1957, p. 254.
35 Metzner 1971, p. 104.
36 V. Wasson 2000, pp. 158–9.
37 Wasson Archive: letters to Albert Hofmann, 27 April 1975; 3 June 1975 (Hofmann).
38 For example, Wasson failed to recognise that, in Mazatec culture, vomiting during a *velada* was regarded as an essential and positive sign of getting well, rather than a sign of being unwell, as it is in the West (Wasson 1980,

p. 25; Weil 1980; Estrada 2003a, p. 28). He also claimed that Sabina's daughter was following in her mother's footsteps by becoming a *curandera*. She was not, and Sabina made it clear that hers was not a vocation that could be inherited (Wasson and Wasson 1957, p. 133; Estrada 2003a, p. 61).

39 Harrison 1999.
40 Estrada 2003b, p. 130; Halifax 1979, p. 129. See also Swain 1963, p. 222; Finkelstein 1969, p. 56; Krippner and Winkelman 1983, p. 226.
41 Estrada 1981, p. 202 n. 3; Estrada 2003a, p. 48; Wasson and Wasson 1957, pp. 311, 314.
42 Wasson 1980, p. 28.
43 Wasson 1980, p. xxiii.
44 Wasson 1980, p. xviii.
45 Sabina in Estrada 2003a, p. 63.
46 Ibid., p. 25.
47 Rothenberg 2003.
48 Yépez n.d.
49 Yépez n.d.; *The Times* 29 September 1967, p. 7.
50 Yépez n.d.
51 Ongaro in Halifax 1979.
52 Wasson Archive: letter from Eunice Pike, 11 October 1956 (Pike).
53 Palmer and Horowitz 2000, p. 239.
54 The opera 'María Sabina' by Leonardo Balada, with a libretto by José Cela, was premiered at Carnegie Hall on 17 April 1970, performed by the New York University Choruses and the Symphony Orchestra from the Manhattan School of Music. It starred María Soledad Romero in the lead role.
55 Yépez n.d.
56 Harrison 1999; White 2004.
57 Having waded through a tiny percentage of his archived correspondence, I can confirm that this is true.
58 Riedlinger 1990a.
59 Halifax 1990, p. 113; Ott 1990, p. 191. See also contributors to Riedlinger 1990a; Smith 2003, p. 53.
60 Wasson 1980, p. xxiv.
61 Wasson Archive: letter to Robert Graves, 31 May 1956 (Graves).
62 In fact he did receive a bursary from the Geschickter Fund for Medical Research to support his trip to Mexico in 1956, but this body was actually a front for the CIA – see Chapter Nine.

SEVEN

1 Recorded by Heiki Silvet in 1988, and cited in Saar 1991.
2 From Blyth 1973, p. 103.
3 Some fine examples may be found in Taylor-Hawksworth 2001.
4 Michelot and Melendez-Howell 2003; Imazeki 1973.
5 See Ott 1996.
6 For example, see Arthur 2000; Heinrich 2002.
7 Ogloblin, cited in Wasson 1971, p. 234.
8 See Hutton 2001.

9 Flaherty 1992.
10 Flaherty 1992.
11 Cited in Wasson 1971, p. 235.
12 Goldsmith 1762, p. 55.
13 Ibid.
14 Cited in Wasson 1971, p. 237.
15 Cited in Wasson 1971, p. 244.
16 Cited in Wasson 1971, p. 236.
17 Cited in Wasson 1971, p. 249.
18 Lindley 1853, p. 38; James 1855 Vol. 2, p. 170.
19 Cooke 1997 [1860], p. 257.
20 Cooke 1894, p. 83.
21 The writer Michael Carmichael has suggested that Carroll learnt about the fly-agaric directly from reading Cooke's *Seven Sisters*. Carroll made his only recorded visit to Oxford's Bodleian library in June 1862, where a copy of *Seven Sisters* had been deposited. To date, the only pages that have been cut are those of the final chapter on the fly-agaric, and Carmichael thinks that this is the book Carroll went to the library to read. In truth, there are so many other places from which Carroll could have learnt about the mushroom that Carmichael's inference can only be speculative. Carmichael 1997.
22 Holmes's cocaine use is revealed in *The Sign of Four*, published in 1889.
23 Jay 2000.
24 Kingsley 1866.
25 Martineau 1997.
26 Carter 1994. I am indebted to Matilda Leyser for drawing this to my attention.
27 Carter 1994, p. 238.
28 Garner 2004, p. 45.
29 Said 1995.
30 Ronald Hutton has written the authoritative work on the matter. See Hutton 2001.
31 Ibid.
32 Johnson, cited in Hutton 2001, p. 30.
33 Hutton 2001.
34 Ibid.
35 See Hutton 2001.
36 See Wallis 2003; Blain 2002.
37 Hutton 2001, p. 25.
38 Hutton 2001.
39 Saar 1991.
40 Saar 1991; Dunn 1973.
41 Saar 1991.
42 Ibid.
43 Ibid.
44 Salzman et al. 1996
45 Gary Lincoff, pers. comm.

46 Ibid.
47 Graves 1972, p. 96.
48 Ott, 1976a, p. 97.
49 Rogan Taylor, 'Who is Santa Claus?', *Sunday Times Magazine* 21 December 1980, pp. 13–17.
50 Morgan 1986, p. 45.
51 Hutton 1996a, pp. 118–19.
52 Hutton 2001.
53 Hutton 2001; Eliade 1989.

EIGHT

1 Translated by Doniger O'Flaherty 1981, p. 131.
2 Rig Veda 8.48, verses 3 & 6, translated by Doniger O'Flaherty 1981, pp. 134–5.
3 Doniger O'Flaherty 1971.
4 Huston-Smith 1972.
5 Riedlinger 1993.
6 Wasson 1971, p. 15.
7 Wasson 1971, plates viii a & b.
8 Wasson Archive: various correspondence between Wasson and Richardson (Richardson).
9 Rig Veda 9.74, verse 4, cited in Wasson 1971, p. 29.
10 Wasson 1971, p. 176.
11 In fact many psychoactive drugs may be excreted unchanged in urine, not least psilocin. It seems that no cultures outside Siberia made use of this surprising fact. See Sticht and Käferstein 2000.
12 Lévi-Strauss [1970] 1994; Needham 1974, p. 122.
13 La Barre 1970.
14 La Barre 1972.
15 Dash and Pahdy 1997.
16 Olivelle 2005, Chapter 4, verse 222; Chapter 5, verse 19.
17 Dallapiccola 2002.
18 Brough 1971.
19 Brough 1971, p. 341.
20 Wasson 1972a.
21 Brough 1971.
22 Wasson 1971, p. 174.
23 Wasson 1979; Keewaydinoquay 1978.
24 See Navet 1988.
25 Wasson first wrote to Keewaydinoquay in July 1975 on the recommendation of John Nichols from the University of Wisconsin: Kee had just published a paper on the ethnobotanical and ethnomycological knowledge of the Ojibway (more properly, the Ahnishinaubeg people). There ensued a lively correspondence between the two, culminating in Wasson visiting Kee at her summer retreat – a remote hut on Garden Island, Lake Michigan – in the summer of 1976. Though he was a widower in his late seventies (she was in her sixties), the two appear to have become lovers.

Piecing together the details of their affair is made difficult by the fact that Wasson – normally so assiduous in keeping copies of his letters – removed from his archive a substantial number of his letters to Kee. As many of Kee's to Wasson are, however, anguished love letters, and as she sent him several audio tapes that are considered too personal for non-family members to hear, it is reasonable to assume that the relationship did not run altogether smoothly. The uncomfortable conclusion one reaches on reading what correspondence remains is that though Wasson was clearly fond of Kee, he used her affections as a lever with which to extract otherwise secret information about Ojibway beliefs and practices, including those to do with *miskwedo*, the fly-agaric. Kee, for example, prepared Wasson a birch bark scroll – a sacred object amongst the Ojibway – illustrating a *miskwedo* story. In an undated letter from 1977, she made it clear that the passing on of such information amounted to a betrayal of her people, but that she was happy to do it for love's sake.

Wasson arranged to have a narrative account of Ojibway ethnomycology, written by Kee and titled *Puhpohwee*, published by the Harvard Botanical Museum (Keewaydinoquay 1978), but he continued to press her for information so that he could author one final book on the Native American use of fly-agaric. Exceptionally, even by Wasson's standards, only one copy of this book – *Miskwedo* – was ever printed, and mysteriously, though it was published either in 1981 or 1982, no one may look at it until a period of twenty-five years has elapsed; it remains hidden in one of Harvard's many libraries. It is no secret what the book contains, however, for various drafts are to be found in Wasson's archive: the book publishes supposedly authentic *miskwedo* stories and anecdotes provided by Kee; a reproduction of the birch bark scroll, a description of an actual *miskwedo* ceremony conducted by Kee and attended by Reid Kaplan, and a critical commentary by Wasson. That the book is locked away suggests that it contains information so sensitive that, were Kee still alive on its release, she would be ostracised by her community for this act of betrayal. In the light of Wasson's treatment of María Sabina, this episode would appear to mark one final stain upon his already questionable reputation, the comparison to Cortés and Malinche, for once apposite.

The situation is, however, a little more complicated. Kee was an intelligent woman and was well aware that Wasson's affections for her were mixed. She knew that once she had provided him with all the information he needed, he would break off the affair (indeed, their correspondence stops abruptly in 1982, after the publication of *Miskwedo*). My belief is that, as a consequence, she drew out the process of providing him with the anecdotes and stories he so wanted, feeding him in dribs and drabs over a five-year period. Presumably, she sought to prolong the relationship in the hope that her affections might eventually be reciprocated. It is entirely possible, however, given the lack of any other evidence to indicate fly-agaric use amongst the Ojibway, that she fabricated much of this material and tailored it to meet her reluctant lover's eager expectations. If true, the twenty-five-year moratorium on the book, imposed at her insistence, might be there to prevent her exposure as a fraudster. Wasson's draft commentary suggests that, for once, he was at least a little suspicious of Kee's testimony.

Though there was clearly an imbalance in their relationship, and though she comes across in her letters as needy, and at times desperate, Kee should not be seen as an entirely helpless victim. Indeed, before she met Wasson she was unhappy and lonely, but by the end of their relationship she had, with Wasson's assistance, completed a PhD in Anthropology at the University of Michigan, and had taken up a teaching position that she relished. She discovered a new lease of life through this newfound career, and gained a reputation as an indigenous healer in her own right. In many ways her circumstances changed for the better as a direct result of her encounter with Wasson.

He, on the other hand, does not come out of the affair so well, for he seems to have been happy to exploit that power imbalance for his own ends. Though we shall never know the full story until such time as his letters become available, his treatment of Kee was, at best, exploitative and betrays an ingrained chauvinistic attitude that must necessarily colour any reading of his encounter with Sabina. That he was more than prepared to behave in ethically unacceptable ways for the sake of promoting his ideas – and his fame – necessarily undermines the credence we might give them. Sadly, and in the absence of any corroborating evidence, Kee's evidence must be considered highly suspicious. See Wasson Archive, passim (Keewaydinoquay).

26 A surprisingly good summary of these debates can be found at http://en.wikipedia.org/wiki/Rig_Veda#More_Recent_Indian_Views
27 Wasson 1971, p. 68.
28 The psychoactive effects of Syrian rue have been thoroughly investigated by Mike Jay 1999.
29 Greene 1992.
30 Greene overlooked the fact that the active ingredients of cannabis are also soluble in fat, and that the process could equally describe the making of bhang.
31 McKenna usually referred to this species by its older, outdated name of *Stropharia cubensis*. McKenna 1992.
32 Wasson 1982.
33 Jay 1999.
34 Cited in McDonald 1978, p. 217.
35 See Huston-Smith 1972.

NINE

1 Heinrich 2002, p. 6.
2 Arthur 2000, p. 9.
3 Ott 1996.
4 Wasson Archive: letter to Andrija Puharich, 3 November 1958 (Puharich).
5 Wasson agreed to try to 'contact' Puharich telepathically during his anticipated Mexican mushroom sessions in 1955. He wrote to Puharich with details of a 'vision' of a New England barn (Wasson Archive: letter to

Andrija Puharich, 11 July 1955), which he 'received' during one of his mushroom trips with María Sabina. He seems to have distanced himself from Puharich after the publication of *The Sacred Mushroom*, regarding the parapsychologist as a pseudo-scientist.

6 Hermans n.d.
7 Ruck 1981, 1982, 1983, 1986.
8 Hajicek-Dobberstein 1995.
9 *Sunday Mirror* 19 April 1970, pp. 34–5. The other issues were: 15 February, p. 1; 12 April, pp. 10–11; 26 April, pp. 28–9.
10 *The Times* 19 May 1970, p. 2d.
11 Ibid.
12 Cited in King 1970, p. 11.
13 Letter to *The Times* 26 May 1970, p. 9c.
14 King 1970, p. 190.
15 Ibid., p. 147.
16 Ott 1996.
17 *Sunday Times* 15 February 1970.
18 See King 1970, p. 22.
19 Allegro 1970b, p. 12.
20 Ibid., p. 7.
21 Ibid., p. 11.
22 Wasson 1971.
23 Arthur 2000, pp. 6–7.
24 Heinrich 2002, p. 121.
25 For an early version of this objection, see Stabell-Kulø 1980.
26 Wasson 1971, p. 210.
27 Bowden and Mogey 1958, p. 145.
28 Ott 1996.
29 Chemists Peter Waser and Scott Chilton dosed themselves with ibotenic acid and muscimol respectively, to test whether they were the responsible agents. Waser 1967; Chilton 1975.
30 Hall and Hall 1994.
31 Tsunoda et al. 1993b.
32 Chilton 1994.
33 Festi and Bianchi 1991; Saar 1991; Tsunoda et al. 1993a.
34 Tsunoda et al. 1993a.
35 Michelot and Melendez-Howell 2003.
36 Allen 1999.
37 See Samorini 1997; Festi and Bianchi 1991.
38 McIlvaine 1900.
39 Curtis 1777–98.
40 I am indebted to Toni Melechi for drawing this episode to my attention.
41 Samorini 1997.
42 Horne and McCluskie 1963.
43 See Hutton 2001.
44 Letter to Gordon Wasson, cited in Wasson 1972.

45 Karana 1968, *International Times* 35, 12–25 July, p. 5.
46 Moorcraft 1970.
47 Darnton 1970.
48 From 'Magick Mushrooms' by Heathcote Williams, *International Times* Vol. 3(1) June 1975, p. 5.
49 I am indebted to Anthony Henman for drawing the film to my attention.
50 Weil 1977; Pollock 1975a,b; Ott 1976b.
51 Ott 1996.
52 Hawk and Venus 2003; see also www.somashamans.com
53 Heinrich 2002.
54 Wasson Archive: letter from Keewaydinoquay to Wasson, 16 March 1977 (Keewaydinoquay).
55 Michelot and Melendez-Howell 2003.

TEN

1 Gelpke 1981 [1962].
2 Stevens 1989.
3 Melechi 1997.
4 Osmond 1957, pp. 428–9.
5 Holmstedt 1967.
6 Hofmann 1983.
7 Stevens 1989; Lee and Shlain 1992.
8 Hofmann 1958, p. 279.
9 Hofmann 1983, pp. 112–13.
10 An identical problem beset Arthur Heffter (1860–1925) when he attempted to isolate the active ingredients of peyote. He was forced to test all the extracts upon himself in a very similar process of 'Prussian roulette'. Using this method, though, he successfully identified mescaline.
11 Heim 1957, p. 603.
12 Ibid.
13 See the various accounts in Heim and Wasson 1958.
14 Cailleux 1958, p. 285.
15 Delay et al. 1958; Delay et al. 1959a,b,c,d.
16 Delay et al 1958, p. 293.
17 Delay et al 1959c.
18 Ibid.
19 Plant 1999.
20 Horowitz and Palmer 1999, p. 176.
21 Berge 1999, 2002.
22 Ibid.
23 Berge 1999, p. 264.
24 Heim and Thévenard 1967.
25 Ibid., p. 210.
26 Heim and Thévenard 1967, p. 204.
27 Ibid.
28 Berge 1999.
29 Stevens 1989.

30 Riedlinger 1990b.
31 Marks 1979.
32 Richardson 1990, p. 199.
33 Singer 1958; Ott 1996, pp. 301–3.
34 Ott 1978; Singer 1982; Wasson 1982b.
35 Stein 1958.
36 One is reminded of the editor of *Life* magazine, Henry Luce, who said that thanks to LSD he found God on the golf course. See Green 1999, p. 108.
37 Guzmán 1983; Guzmán et al. 1998.

ELEVEN

1 Leary 1995, p. 335.
2 Stevens 1989.
3 Inglis 1975; Hofmann 1983; Stevens 1989; Lee and Shlain 1992; Melechi 1997; Green 1999; Plant 1999; Black 2001; Davenport-Hines 2001; Farber 2002; *Q Magazine* 2005.
4 Leary 1995, p. 34.
5 Ibid., p. 41.
6 Leary 1983, p. 37.
7 Leary et al. 1963.
8 Leary 1983, p. 48.
9 Leary worked at the Center for Personality Research at the Morgan Prince building, 5 Divinity Avenue. The original building has been moved to 6 Prescott Street.
10 Leary 1995, pp. 338, 328.
11 Leary 1995.
12 Gable 1993, p. 45.
13 See Leary 1970, 1989; Wilson 1990.
14 Leary 1995, p. 333.
15 Leary 1970, p. 117.
16 Ibid., 149.
17 Stevens 1989.
18 Leary 1995.
19 Anon. 1963, p. 242.
20 Doblin 1999.
21 Ibid., unpaged.
22 Doblin 1991.
23 Smith 2003, p. 104.
24 Smith 2003, p. 101.
25 Cited in Doblin 1991, unpaged.
26 Green 1999, p. 213.

TWELVE

1 Robbins 1973, p. 211.
2 Timothy Egan, 'Down on the Farm with the Slugs, Cows and Magic Mushrooms'. *University of Washington Daily*, 29 October 1973; accessed at www.mushroomjohn.org

3 Thomas Kuhn's model of 'scientific revolutions' is a case in point. Kuhn 1996.
4 Gladwell 2000.
5 Pollock 1977/78.
6 See, for example, Cohen 1973; Hetherington 2000.
7 'Going native' was, until very recently, regarded as the ultimate anthropological heresy. See Young and Goulet 1994; Blain et al. 2004.
8 As I have already argued at length, the application of the term 'shamanism' to cultural practices ostensibly similar to those found in Siberia is problematic. For simplicity's sake, and to avoid unnecessarily overburdening the reader, I have adhered to convention and referred to Central and South American spiritual practices as shamanism throughout. Nevertheless, the problems remain.
9 See Harner 2003.
10 Harner 1980.
11 For a balanced appraisal, see Wallis 2003.
12 De Mille 1976; see Wallis's review of the evidence, p. 2003.
13 Ott 1996, p. 285.
14 Wasson 1969, 1972, 1973, 1974.
15 Wasson 1972b.
16 Harner 1973.
17 See, for example, Ott and Bigwood 1978; Chilton, Bigwood and Jensen 1979; Bigwood and Beug 1982; Beug and Bigwood 1982.
18 Ott 1990.
19 Pollock 1978.
20 See the obituary accessed at www.mushroomjohn.com
21 Pollock 1975, 1976, 1977/8; Ott and Bigwood 1978; Ott 1996; Weil 1975, 1977.
22 Pollock 1975, p. 77; Pollock 1977/8.
23 Stamets 1996, p. 3.
24 Ott and Bigwood 1978.
25 Pollock 1978, p. 25.
26 Pollock 1976.
27 Jacobs 1975, p. 35.
28 Egan 1973, unpaged.
29 Robbins 1973.
30 Weil 1977, p. 137.
31 Anon. 1976.
32 Pollock 1977–8, p. 29.
33 Guzmán and Ott 1976.
34 Sarchin 1973.
35 Pollock 1977/8, p. 28.
36 Kendrick 1976, unpaged.
37 Ott 1976, p. 14.
38 Stamets 1996.
39 Kendrick 1976, unpaged.

40 Anon. 1976.
41 Stamets 1996, p. 3.
42 Ibid., p. 4.
43 Pollock 1975, 1976, 1977/8; Lowy 1977; Ott 1996; Marcano et al. 1994.
44 Pollock 1975, 1977/8.
45 Anon. 1961.
46 Allen 1993.
47 Cited in Southcott 1974, p. 470.
48 Allen 1999.
49 Ibid.
50 Cox 1981; Allen and Merlin 1992.
51 Cox 1981.
52 Allen and Merlin 1992.

THIRTEEN

1 Bradley 1984, p. 682.
2 Andrews 1977, p. 18.
3 Darnton 1970. With two exceptions, British magic mushrooms are notable by their absence from the obvious places where you would expect to find them – namely, the publications of the underground press. These include the magazines *Oz* (1967–73), *International Times* (1966–75), *Gandalf's Garden* (1968–9), *Albion* (1968) and *Friends/Frendz* (1969–72); all of Nicholas Saunders's guides to *Alternative London* (Saunders 1970); and *The IT Book of Drugs* (Green 1974).
4 *Oz* famously scandalised the establishment with its pornographic 'school kids' issue, and after a trial at the Old Bailey, the editors were briefly imprisoned. *IT* was no less radical, and regularly had its offices raided and ransacked by the police.
5 Koestler, cited in Leary 1995, p. 151.
6 The thirty-minute programme was broadcast at 9.00 p.m. on Wednesday 19 April 1961. Sadly, no copy of the documentary survives, but screen shots may be found in the Wasson Archive at Harvard University, in the BBC file.
7 On the Powick research, see Melechi 1997.
8 Taylor 1961.
9 'Land of Magic Mushrooms', *The Times* Friday 29 September 1967, p. 7a.
10 Martin 1967.
11 Sandford 1972.
12 Livings 1965.
13 I am indebted to Dave Todd, both for alerting me to the film and for finding me a copy.
14 Hofmann et al. 1963.
15 Ibid., p. 303.
16 Mantle, pers. comm.
17 O'Prey 1984, p. 215.
18 Ibid., p. 85.
19 Wasson Archive: letter from Wasson to Graves, 5 June 1962 (Graves).
20 Seymour-Smith 1982, p. 464.

21 Wasson Archive: letter from Wasson to Ricardo Sicre, 9 September 1955; letter from Robert Graves to Gordon Wasson, *c.* 10 October 1958 (Graves).
22 Wasson Archive: letter from Wasson to Graves, 6 June 1950 (Graves).
23 See Seymour-Smith 1982; O'Prey 1984; Graves 1996.
24 Graves 1972, p. 92.
25 See Hutton 1993, 1999.
26 O'Prey 1984, p. 213.
27 Graves 1959; Graves 1961a; Graves 1972, passim.
28 Heizer 1944.
29 Also present were a Miss Jess McNab and Jerome Robbins. It took place on 31 January 1960.
30 Graves 1961, p. 122.
31 O'Prey 1984, p. 189.
32 Graves 1961a, p. 124.
33 Ibid., p. 125.
34 Ibid.
35 This soirée, on 6 May 1960, was attended by Wasson, Robert and Beryl Graves, and Esteban Frances. Ann Heathers, a friend of Esteban, watched over the proceedings without taking the drug.
36 Wasson Archive: Experiment with Synthetic Psilocybin, 8 May 1960 (Graves).
37 O'Prey 1984, p. 190.
38 Graves 1972, p. 91.
39 Theo Sloot, pers. comm. Sloot was a school-friend of Graves's son, Tomas, and spent several summers at Deià during the 1960s.
40 Coleman 1971.
41 Graves 1961, p. 45.
42 Nuttall 1970, p. 200.
43 Darnton 1968a.
44 *International Times* 25, 2–15 February 1968, p. 3.
45 Firsoff 1980, pers. comm.
46 Death's Jester 1968; Firsoff 1980.
47 More 1968.
48 Darnton 1970, p. 11. It is quite possible that, erring on the side of caution, Darnton took only a low dose of mushrooms, which may explain why the experience was underwhelming.
49 See Green 1999, p. 173; Snow 1999.
50 The only other widely circulated booklet on magic mushroom identification was published in 1979 by the drugs charity Release. This was several years after the mushroom had gone overground, and so it was not responsible for bringing the mushroom to widespread attention. See Release 1979.
51 Andrews 1975, p. 249 n. 5. His observation is supported by a medical survey that showed no recorded incidents of magic mushroom poisoning in the five years before 1977 (Mills et al. 1979), and by the fact that field guides to Britain's mushrooms and toadstools only listed the Liberty Cap as hallucinogenic from 1977 onwards (see Kibby 1977, 1979; Phillips 1981).
52 See Davenport-Hines 2002.

53 *The Times* 13 April 1976, p. 4; 14 April 1976, p. 4.
54 Cited in *The Times* 14 April 1976, p. 4.
55 *The Times* 14 April 1976, p. 4.
56 Carter 1976.
57 Christiansen et al. 1981; Stijve and Kuyper 1985; Ohenoja et al. 1987.
58 Andrews 1977.
59 Freeman 1978.
60 *Guardian* 14 February 1978.
61 *Guardian* 9 March 1978.
62 Hyde et al. 1978; Benjamin 1979; Mills et al. 1979; Cooles 1980; Harries and Evans 1981; Lawson 1981; Peden et al. 1981; Young et al. 1982; Peden et al. 1982; Bennell and Watling 1983; Francis and Murray 1983; Murray 1984.
63 Francis and Murray 1983; White and White 1995.
64 Smart et al. 1982, p. 419; Bennell and Watling 1983.
65 Francis and Murray 1983.
66 Peden et al. 1981, p. 544.
67 *Sunday Times* 28 November 1982, p. 14; *The Times* 9 December 1982, p. 3.
68 McKay 1996.
69 On Stonehenge, see McKay 1996; Hetherington 2000; Chippindale 2001.
70 See festival archive at http://festival-zone.ocatch.com/PsilocybinFestival.html
71 Ibid.
72 Ibid.
73 Mayes 1980, p. 10.
74 Andrews 1977; Cooper 1977.
75 Andrews 1977, p. 18.
76 See, for example, Cooper 1977; Cohen 2001.
77 Bradley 1984.
78 In keeping with academic convention, I use lower case when referring to ancient, pre-Christian paganism, and upper case when referring to modern, revived Paganism.

FOURTEEN

1 Sheldrake et al. 2001, p. 17; The Shamen with Terence McKenna 1993.
2 The festival ran from 28–30 July, at Old Treworgey Farm, two miles outside Liskeard.
3 See Collin 1997; Metcalfe 1997; Saunders 1997; Malbon 1999.
4 Fraser Clark, cited in Rosenberger 1990, p. 13.
5 His discoveries are listed in two large volumes, *Pihkal* and *Tihkal*. Shulgin and Shulgin 1997, 2000.
6 Those wanting to experiment with Morning Glory seeds should know that they are usually sprayed with an emetic before being sold.
7 McKenna n.d.
8 McKenna 1992b, unpaged.
9 McKenna 1993, p. 7.
10 The story is recounted in full in McKenna 1993.
11 McKenna n.d.

12 In McKenna and McKenna 1993, p. xi.
13 McKenna 1993, p. xi.
14 Many can be found at http://mckenna.otterly.com; http://deoxy.org/mckenna.htm; http://mckenna.drugtext.org; and http://mckenna.psychedelic-library.org
15 An extensive, though probably not exhaustive, list of tracks including McKenna samples can be accessed at http://www.cmays.net/tmbib.shtml
16 The Shamen with Terence McKenna 1993.
17 Dennis McKenna on the Art Bell Show. Mp3 available at http://mckenna.otterly.com
18 Interview with Terence McKenna by Jon Hanna and Sylvia Thyssen, at the 1999 AllChemical Arts Conference, Kona Coast, Hawaii; accessed at http://www.maps.org/news-letters/v10n2/10206mck.html
19 McKenna n.d.
20 Ibid.
21 McKenna 1993, p. 225.
22 Self 1994.
23 McKenna 1992b, unpaged.
24 McKenna 1993, pp. 210–14.
25 Beach 1996/7.
26 McKenna was also a great advocate of cannabis, even though it comes from a completely different chemical and pharmacological family. See McKenna 1992b.
27 He claimed that at extremely high doses – eight dried grams of *Psilocybe cubensis* – one entered the same elf-nest as was encountered on DMT. In other words, both took one to the same ontological space. McKenna n.d.
28 McKenna 1992b, unpaged.
29 Hutton 1999.
30 Self 1996.
31 Johnson 1995; Laura Rivel, pers. comm.
32 McKenna 1993, p. 226.
33 McKenna and McKenna 1993, p. 189.
34 The Shamen with Terence McKenna 1993.
35 McKenna and McKenna 1993, p. 193.
36 McKenna n.d.
37 Watkins n.d.
38 McKenna 1992b, unpaged.
39 Ibid.
40 McKenna 1993, p. 210.
41 Sheldrake et al. 2001, p. 108.
42 Robert Hunter, for example, named them, with far less aplomb, 'the klaxton men'. McKenna and Hunter 1996, p. 3.
43 Sheldrake et al. 2001, p. 157.

FIFTEEN

1 McKenna 1992b, unpaged.
2 San Antonio 1971.

3 McKenna 1993, p. 207.
4 Heim and Wasson 1958.
5 Singer 1958.
6 Brown 1968.
7 For example see the full page advert in *High Times* September 1976, Issue 13, p. 18.
8 Pollock 1977; Stamets and Chilton 1983.
9 Information on Robert McPherson, his arrest and trial, and the PF-Tek itself can be found at http://fanaticus.fungifun.com/notice.php?ref=/index.htm
10 Abraham 1999; Riper and de Kort 1999.
11 Hans van den Huerk, pers. comm., 20 April 2005.
12 Shulgin and Shulgin 1997, 2000.
13 On cyberculture, see Rushkoff 1994; Davis 2004.
14 *Guardian* 9 August 1978.
15 *The Times* 31 March 1983, p. 3.
16 I interviewed a representative of the Camden Mushroom Company on 27 May 2005. The representative wished to remain anonymous.
17 For example, see Honigsbaum 2003.
18 Verkaik 2004.
19 Honigsbaum 2004.
20 Hansard 18 January 2005, column 712.
21 All statistics from Abraham et al. 2003.
22 Travis 2005. See also figures in Davenport-Hines 2001.

EPILOGUE

1 Graves 1972, pp. 78–9.
2 Circulus 2005.
3 Thomas 2000, p. 54.
4 Pinchbeck 2002, pp. 291, 296.
5 Gerth and Wright Mills 1991, pp. 155, 139.
6 See, for example, Dawkins 1998.
7 Gerth and Wright Mills 1991, p. 154.
8 Ibid.
9 Pinchbeck 2002, p. 291.
10 Pinchbeck 2002; Narby 1999.
11 See Davies 2004.

References

Abraham, Manja D. 1999. *Places of Drug Purchase in the Netherlands.* Paper given at the 10th Annual Conference on Drug Use and Drug Policy, Vienna, September 1999. Accessed at www.cedro-uva.org/lib/abraham.places.pdf

Abraham, Manja D., Hendrien, L. Kaal, and Cohen, Peter D. A. 2003. *Licit and Illicit Drug Use in Amsterdam, 1987 to 2001. Development of Drug Use in Amsterdam as Measured in Five Population Surveys Between 1987 and 2001.* Amsterdam: CEDRO/Mets & Schilt. Accessed at www.cedro-uva.org/lib/abraham.licit.pdf

Ainsworth, Geoffrey Clough. 1976. *Introduction to the History of Mycology.* Cambridge: Cambridge University Press.

Albert-Puleo, Michael. 1978. Mythobotany. Pharmacology and Chemistry of Thujone-Containing Plants and Derivatives. *Economic Botany* 32: 65–74.

Alexopoulos, Constantine John, Mims, C. W., and Blackwell, M. 1996. *Introductory Mycology.* New York: John Wiley & Sons Inc.

Allegro, John. 1970a. *The Sacred Mushroom and the Cross.* London: Hodder & Stoughton.

— 1970b. *The End of a Road.* London: MacGibbon & Kee.

Allen, John W. 1993. Copelandia and Other Psychoactive Fungi in Hawaii. *Psychedelic Illuminations* 4: 61. Accessed at www.mushroomjohn.org

— 1999. *Magic Mushrooms of Australia and New Zealand.* Accessed at www.erowid.org/library/books_online/magic_mushrooms_aunz/magic_mushrooms_aunz.shtml

Allen, John W., and Merlin, Mark D. 1992. Psychoactive Mushroom Use in Koh Samui and Koh Pha-Ngan, Thailand. *Journal of Ethnopharmacology* 35: 205–28.

Andrews, George, ed. 1975. *Drugs and Magic.* St Albans: Panther Books Ltd.

Andrews, George. 1977. Psilocybin. *Home Grown* 1(2): 18–19.

Anon. 1757. *Gentleman's Magazine* XXVII: 385.

— 1757. *Gentleman's Magazine* XXVII: 431.

— 1815. Letter. *Gentleman's Magazine* LXXXV (ii): 397–8.

— 1864. Mushrooms – Information about. *The Times* 17 September: 11.

— 1961. 'Sacred Mushrooms' are Sought on Kaui. *Honolulu Star-Bulletin,* Wednesday 11 March: 11a. Accessed at www.mushroomjohn.org

— 1963. Four Psilocybin Experiences. *Psychedelic Review* 1: 219–43.

— 1976. Mushroom Madness Packs Pickers in Psychedelic Pastures. *The Oregonian,* 27 November 1976: 15b. Accessed at www.mushroomjohn.org

Arthur, James. 2000. *Mushrooms and Mankind. The Impact of Mushrooms on Human Consciousness and Religion.* Escondido, CA: The Book Tree.

Badham, Charles D. 1863. *A Treatise on the Esculent Funguses of England.* London: Lovell Reeve & Co.
Badham, Edmond R. 1984. Ethnobotany of Psilocybin Mushrooms, Especially *Psilocybe cubensis. Journal of Ethnopharmacology* 10: 249–54.
Barfield, Thomas, ed. 1997. *The Dictionary of Anthropology*. Oxford: Blackwell Publishing Ltd.
Barnett, Bernard. 1965. Drugs of the Devil. *New Scientist* (22 July) 27: 222–5.
Beach, Horace. 1996 –7. Listening for the Logos: A Study of Reports of Audible Voices at High Doses of Psilocybin. Newsletter of the Multidisciplinary Association for Psychedelic Studies MAPS 7(1): 12–17. Accessed at http://www.maps.org/news-letters/v07n1/07112bea.html
Benjamin, C. 1979. Persistent Psychiatric Symptoms After Eating Psilocybin Mushrooms. *British Medical Journal* 19 May: 1319 –20.
Bennell, Alan P., and Watling, Roy. 1983. Mushroom Poisonings in Scotland. *Bulletin of the British Mycological Society* 17(2): 104–5.
Berge, Jos Ten. 1999. Breakdown or Breakthrough? A History of European Research into Drugs and Creativity. *Journal of Creative Behavior* 33(4): 257–76.
— 2002. Jekyll and Hyde Revisited: Paradoxes in the Appreciation of Drug Experiences and Their Effects on Creativity. *Journal of Psychedelic Drugs* 34(3): 249– 62.
Berkeley, M. J. 1860. *Outlines of British Fungology. Containing Characters of Above a Thousand Species of Fungi, and a Complete List of All That Have Been Described as Natives of the British Isles.* London: Lovell Reeve.
Beug, Michael W., and Bigwood, Jeremy. 1982. Psilocybin and Psilocin Levels in Twenty Species from Seven Genera of Wild Mushrooms in the Pacific Northwest, U.S.A. *Journal of Ethnopharmacology* 5: 271–85.
Big D. Unlimited. 1976. *A Guidebook to the Psilocybin Mushrooms of Mexico.* Tucson: The Mother Duck Press.
Bigwood, Jeremy, and Beug, Michael W. 1982. Variation of Psilocybin and Psilocin Levels with Repeated Flushes (Harvests) of Mature Sporocarps of *Psilocybe cubensis* (Earle) Singer. *Journal of Ethnopharmacology* 5: 287–91.
Black, David. 2001. *Acid. A New Secret History of LSD*. London: Vision.
Blain, Jenny. 2002. *Nine Worlds of Seidr-Magic: Ecstasy and Neo-Shamanism in North European Paganism*. London: Routledge.
Blain, Jenny, Ezzy, Douglas, and Harvey, Graham, eds. 2004. *Researching Paganisms*. Walnut Creek, CA: Altamira Press.
Blanchette, Robert A., Compton, Brian D., and Turner, Nancy J. 1992. Nineteenth Century Grave Guardians are Carved *Fomitopsis officinalis* Sporophores. *Mycologia* 84: 119–24.
Blythe, R. H. 1973. Mushrooms in Japanese Verse. *Transactions of the Asiatic Society of Japan* 3rd Series XI: 93–106.
Borhegyi, Stephan F. de. 1961. Miniature Mushroom Stones from Guatemala. *American Antiquity* 26: 498–504.
Borowiak, Krzysztof S., Ciechanowski, Kazimierz, and Waloszczky, Piotr. 1998. Psilocybin Mushroom (*Psilocybe semilanceata*) Intoxication with Myocardial Infarction. *Clinical Toxicology* 36(1&2): 47–9.

Bowden, K., and Mogey, G. A. The Story of Muscarine. *Journal of Pharmacy and Pharmacology* 10: 145–56.

Bowman, M. 1995. The Noble Savage and the Global Village: Cultural Evolution in New Age and Neo-Pagan Thought. *Journal of Contemporary Religion* 10(2): 139–49.

Boyes, G. 1993. *The Imagined Village. Culture, Ideology and the English Folk Revival*. Manchester University Press, Manchester.

Bradley, Marion. 1984. *The Mists of Avalon*. London: Sphere Books Ltd.

Bradley, Richard. 1989. Deaths and Entrances: A Contextual Analysis of Megalithic Art. *Current Anthropology* 30: 69–75.

Brande, Everard. 1800. Letter to the Editors. *London Medical and Physical Journal* III: 41–4.

Bresser, Bonnie. 1969. *Troia: Mexican Memories*. New York: Croton Press.

Briggs, Raymond. 1993. *Fungus the Bogeyman*. London: Penguin.

Briggs, Robin. 2002. *Witches and Neighbours. The Social and Cultural Context of European Witchcraft*, 2nd edn. Oxford: Blackwell.

Britten, Masha Wasson. 1990. My Life with Gordon Wasson, in Riedlinger, Thomas, ed. *The Sacred Mushroom Seeker. Essays for R. Gordon Wasson*. Portland, OR: Dioscorides Press, pp. 31–42.

Brough, John. 1971. Soma and *Amanita muscaria*. *Bulletin of the School of Oriental and African Studies* 34: 331–62.

Brown, J. Christopher. 1990. R. Gordon Wasson: Brief Biography and Personal Appreciation, in Riedlinger, Thomas, ed. *The Sacred Mushroom Seeker. Essays for R. Gordon Wasson*. Portland, OR: Dioscorides Press, pp. 19–24.

Brown, R. E. 1968. *The Psychedelic Guide to the Preparation of the Eucharist*. Austin, Texas: Linga Sharira Incense Co.

Bullock, J. Ernest. 1879. Poisoning by 'Mushrooms'. *The Lancet* 11 October: 554.

Cailleux, Roger. 1958. Trois Essais D'Ingestion avec les Psilocybes Hallucinogènes, in Heim, Roger, and Wasson, R. Gordon, *Les Champignons Hallucinogènes de Mexique*. Paris: Éditions du Muséum National d'Histoire Naturelle, pp. 283–5.

Carmichael, Michael. 1997. Wonderland Revisited, in Melechi, Antonio, ed. *Psychedelia Britannica. Hallucinogenic Drugs in Britain*. London: Turnaround Books, pp. 5–20.

Carter, Angela. 1994 [1984]. *Nights at the Circus*. London: Vintage.

Carter, Michael. 1976. Will the Liberty Cap Cause Home Office Hallucinations? *New Scientist* 16 September: 599.

Castaneda, Carlos. 1987 [1968]. *The Teachings of Don Juan: A Yaqui Way of Knowledge*. London: Penguin.

Chilton, William S. 1975a. . . . And a Touch of Cyanide. *McIlvainea* 2(1): 11–12.

— 1975b. The Course of an Intentional Poisoning. *McIlvainea* 2(1): 17–18.

— 1994. The Chemistry and Mode of Action of Mushroom Toxins, in Spoerke, David G., and Rumack, Barry H., eds *Handbook of Mushroom Poisoning. Diagnosis and Treatment*. London: CRC Press, ch. 9, pp. 165–231.

Chilton, William S., Bigwood, Jeremy, and Jensen, Robert E. 1979. Psilocin, Bufotenine and Serotonin: Historical and Biosynthetic Observations. *Journal of Psychedelic Drugs* 11(1–2): 61–9.

Chipindale, Christopher. 2001. *Stonehenge Complete*, rev. edn. London: Thames and Hudson.

Christiansen, A. L., Rasmussen, K. E., and Høiland, K. 1981. The Content of Psilocybin in Norwegian *Psilocybe semilanceata. Planta Medica* 42: 229–35.

Circulus. 2005. *The Lick on the Tip of an Envelope Yet to be Sent*. Rise Above Records.

Clifton, Chas. 2001. If Witches No Longer Fly: Today's Pagans and the Solanaceous Plants. *The Pomegranate* 16: 17–23.

– 2004. A Pipe of Oyntment: *Some Reflections on the Witches' Flying Ointments*. Paper presented at the 'Exploring Consciousness' Conference, Bath, June 2004.

Cohen, Daniel. 2001. Letter. *The Pomegranate* 17: 1–2.

Cohen, Sidney. 1973. *Folk-Devils and Moral Panics. The Creation of the Mods and Rockers*. St Albans: Granada/Paladin.

Coleman, John. 1971. Music from Planet Gong. *Frendz* 14 June: 223.

Collin, Matthew. 1997. *Altered State. The Story of Ecstasy Culture and Acid House*. London: Serpent's Tale.

Cooke, Mordecai C. 1997[1860]. *The Seven Sisters of Sleep. The Celebrated Drug Classic [Popular History of the Seven Prevailing Narcotics of the World]*. Rochester, VT: Park Street Press.

– 1862 *A Plain and Easy Account of British Fungi*. London: Robert Hardwicke.

– 1894. *Edible and Poisonous Fungi. What to Eat and What to Avoid*. London: Society for Promoting Christian Knowledge.

– 1904. *A Plain and Easy Account of British Fungi with Especial Reference to the Esculent and Economic Species*, 3rd edn. Edinburgh: John Grant.

Cooles, Philip. 1980. Abuse of the Mushroom *Panaeolus foenisecii. British Medical Journal* 16 February: 446–7.

Cooper, Richard. 1994 [1977]. *A Guide to British Psilocybin Mushrooms*. Illustrations by Graeme Jackson, Alexandra King and Richard Cooper. Hassle Free Press, n.p.

Cox, Paul Allan. 1981. Use of a Hallucinogenic Mushroom, *Copelandia cyanescens*, in Samoa. *Journal of Ethnopharmacology* 4: 115–16.

Creighton, J. 1995. Visions of Power: Imagery and Symbols in Late Iron Age Britain. *Britannia* 26: 285–301.

– 2000. *Coins and Power in Late Iron Age Britain*. Cambridge: Cambridge University Press.

Curry, Steven C., and Rose, Mark C. 1985. Intravenous Mushroom Poisoning. *Annals of Emergency Medicine* 14 September: 125–7.

Curtis, William. 1777–98. *Flora Londinensis: Or Plates and Descriptions of Such Plants as Grow Wild in the Environs of London*. London.

Dallapiccola, Anna L. 2002. *Dictionary of Hindu Lore and Legend*. London: Thames and Hudson.

Darnton, Lynn. 1968a. Letter, in Appendix 1 of Keeler, Paul, ed. *Planted. A report of the events leading up to and surrounding the arrest and committal to be tried before a judge and jury of 3 members of the Exploding Galaxy charged with being in possession of dangerous drugs.* London: Privately Published, pp. 20–1.

Darnton, Lynn. 1968b. Icarus. *International Times* 33, 14–27 June: unpaged.

— 1970. Magick Mushroom. *Oz* 31: 10–13.

Dash, Santosh K., and Padhy, Sachidananda N. 1997. Mushrooms as Prohibited Food in *Manusmruti* vis-à-vis *Amanita muscaria* as Vedic Soma Plant. *Ethnobotany (New Delhi)* 9(1/2): 129.

Davenport-Hines, Richard. 2001. *The Pursuit of Oblivion. A Social History of Drugs.* London: Phoenix Press.

Davies, Wade. 2004. *The Lost Amazon. The Photographic Journey of Richard Evans Schultes.* London: Thames and Hudson.

Davis, Erik. 2004. *TechGnosis. Myth, Magic and Mysticism in the Age of Information*, updated edn. London: Serpent's Tail.

Dawkins, Richard. 1989. *The Selfish Gene*, 2nd edn. Oxford: Oxford University Press.

— 1998. *Unweaving the Rainbow. Science, Delusion and the Appetite for Wonder.* London: Penguin.

de Borhegyi, Stephan F. 1963. Pre-Columbian Pottery Mushrooms from Mesoamerica. *American Antiquity* 28: 328–38.

de Mille, Richard. 1976. *Castaneda's Journey: The Power and the Allegory.* Santa Barbara: Capra Press.

de Solier, René. 1965. *Curandera. Les Champignons Hallucinogènes.* Paris: Jean-Jacques Pauvert.

Death's Jester. 1968. Icarus. *International Times* 32, 31 May–13 June: 3.

Delay, Jean, Pichot, Pierre, Lempérière, Thérèse, Nicolas-Charles, Pierre J., and Quétin, Anne-Marie. 1958. Étude Psycho-Physiologique et Clinique de la Psilocybine, in Heim, Roger, and Wasson, R. Gordon. *Les Champignons Hallucinogènes de Mexique.* Paris: Éditions du Muséum National d'Histoire Naturelle, pp. 287–309.

Delay, Jean, Pichot, Pierre, Lempérière, Thérèse, Nicolas-Charles, Pierre J., and Quétin, Anne-Marie. 1959a. Les Effets Somatiques de la Psilocybine. *La Presse Médicale* 67(33): 368.

— 1959b. Les Effets Psychiques de la Psilocybine et les Perspectives Thérapeutiques. *Annales Médico-Psychologiques* 117: 899–907.

Delay, Jean, Pichot, Pierre, Lempérière, Thérèse, and Quétin, Anne-Marie. 1959c. Effet Thérapeutique de la Psilocybine sur une névrose compulsive. *Annales Médico-Psychologiques* 117: 509–15.

Delay, Jean, Pichot, Pierre, and Nicolas-Charles, Pierre J. 1959d. Premières Essais de la Psilocybine en Psychiatrie, in Bradley, P. B., Deniker, P., and Radouco-Thomas, C. eds *Neuro-Psychopharmacology. Proceedings of the 1st International Congress of Neuro-Psychopharmacology.* London: Elsevier, pp. 528–31.

Dent, Susie. 2004. *Larpers and Shroomers.* The Language Report. Oxford: Oxford University Press.

Devereux, Paul. 1997. *The Long Trip. A Prehistory of Psychedelia.* Harmondsworth: Penguin/Arkana.

Dighton, Ralph. 1960. A Second Mushroom, Then – Wham! *Asheville Citizen Times* Sunday 11 December 1960: 11.

Dobkin de Rios, Marlene. 1974. The Influence of Psychotropic Flora and Fauna on Maya Religion. *Current Anthropology* 15: 147–64.

Dobkin de Rios, Marlene, and Janiger, Oscar. 2003. *LSD: Spirituality and the Creative Process.* Rochester, VT: Park Street Press.

Doblin, Rick. 1991. Pahnke's 'Good Friday Experiment': A Long-Term Follow Up and Methodological Critique. *Journal of Transpersonal Psychology* 23(1). Accessed at www.psychedelic-library.org/doblin.htm

– 1999. Dr Leary's Concord Prison Experiment: A 34 Year Follow-Up Study. *Bulletin of the Multidisciplinary Association for Psychedelic Studies MAPS* 9(4): 10–18. Accessed at www.maps.org/news-letters/v09n4/09410con.html

Doniger O'Flaherty, Wendy. 1971. The Post-Vedic History of the Soma Plant, in Wasson, R. Gordon. *Soma. Divine Mushroom of Immortality.* New York: Harcourt Brace Jovanovich, pp. 528–31.

Doniger O'Flaherty, Wendy, trans. 1981. *The Rig Veda.* London: Penguin.

Douglas, Mary. 1994. *Purity and Danger. An Analysis of the Concepts of Pollution and Taboo.* London: Routledge.

Douglass, Beaman. 1917. Mushroom Poisoning. *Torreya* 17(10): 171–5; 207–21.

Dowson, Thomas A. 2003. Like People in Prehistory, in Harvey, Graham, ed. *Shamanism. A Reader.* London: Routledge, ch. 9, pp. 159–69.

Dronfield, Jeremy. 1995a. Subjective Vision and the Source of Irish Megalithic Art. *Antiquity* 69: 539–49.

– 1995b. Migraine, Light, and Hallucinogens: The Neurocognitive Basis of Irish Megalithic Art. *Oxford Journal of Archaeology* 14: 261–75.

– 1996. The Vision Thing: Diagnosis of Endogenous Derivation in Abstract Arts. *Current Anthropology* 37: 373–91.

Dunn, Ethel. 1973. Russian use of *Amanita muscaria*: a Footnote to Wasson's *Soma. Current Anthropology* 14(4): 488–92.

Edwards, D. O. 1836. On Poisoning with the Esculent Mushroom. *The Lancet* 2: 512.

Efron, Daniel H., ed. 1967. *Ethnopharmacologic Search for Psychoactive Drugs.* Washington: US Department of Health, Education and Welfare.

Egan, Timothy. 1973. Down on the Farm with the Slugs, Cows and Magic Mushrooms. *University of Washington Daily*, 29 October 1973. Accessed at www.mushroomjohn.org

Elferink, Jan G. R. 1988. Some Little-Known Hallucinogenic Plants of the Aztecs. *Journal of Psychoactive Drugs* 20: 427–35.

English, Mary P. 1987. *Mordecai Cubitt Cooke. Victorian Naturalist, Mycologist, Teacher and Eccentric.* Bristol: Biopress Ltd.

Enos, L. 1970. *A Key to the American Psilocybin Mushroom.* Lemon Grove, CA: Youniverse.

Estrada, Álvaro. 1981. *María Sabina. Her Life and Chants.* Santa Barbara: Ross-Erikson Inc.

— 2003a. The Life, in Rothenberg, Jerome, ed. *María Sabina. Selections.* Berkeley: University of California Press, pp. 1–79.
— 2003b. Introduction to the Life of María Sabina, in Rothenberg, Jerome, ed. *María Sabina. Selections.* Berkeley: University of California Press, pp. 127–32.
Farber, David. 2002. The Intoxicated State/Illegal Nation. Drugs in Sixties Counterculture, in Braunstein, Peter, and Doyle, Michael William. *Imagine Nation. The American Counterculture of the 1960s and '70s.* London: Routledge.
Farndon, Edward. A. 1879. Fungus Poisoning. *The Lancet* 4 October: 520.
Festi, Francesco, and Bianchi, Antonio. 1991. *Amanita muscaria*: Mycopharmacological Outline and Personal Experiences. *Psychedelic Monographs and Essays* 5: 209–52.
Findlay, W. P. K. 1982. *Fungi, Folklore, Fiction and Fact.* Richmond: The Richmond Publishing Co. Ltd.
Finkelstein, Nat. 1969. Honghi, Meester? *Psychedelic Review* 10: 52–63.
Finnegan, R. 1970. *Oral Literature in Africa.* Oxford: Oxford University Press.
Firsoff, George. 1980. *1968 Spring of Youth.* Bath: Privately Published.
Flaherty, Georgina. 1992. *Shamanism and the Eighteenth Century.* New Jersey: Princeton University Press.
Ford, William W. 1909. The Distribution of Poisons in Mushrooms. *Science* 30: 97–108.
Foucault, Michel. 1980. *Power/Knowledge: Selected Interviews and Other Writings 1972–1977,* ed. Colin Gordon. New York: Pantheon.
Fowler, Brenda. 2002. *Iceman. Uncovering the Life and Times of a Prehistoric Man Found in an Alpine Glacier.* London: Pan.
Francis, J., and Murray, V. S. G. 1983. Review of Enquiries Made to the NPIS Concerning Psilocybe Mushroom Ingestion 1978–1981. *Human Toxicology* 2: 349–52.
Frazer, James George. 1922. *The Golden Bough. A Study in Magic and Religion,* abridged edn. London: MacMillan.
Freeman, Jim. 1978. The 'Magic Mushroom' Craze That Could End in Tragedy. *Glasgow Herald,* 8 September: 7.
Furst, Peter T., ed. 1972. *Flesh of the Gods. The Ritual Use of Hallucinogens.* London: George Allen and Unwin Ltd.
Gable, Robert S. 1993. Skinner, Maslow, and Psilocybin. *Journal of Humanistic Psychology* 33(3): 42–51.
Garner, Alan. 2004. *Thursbitch.* London: Vintage.
Gartz, Jochen. 1996. *Magic Mushrooms Around the World. A Scientific Journey Across Cultures and Time,* trans. Claudia Taake. Los Angeles: Lis Publications.
Gelpke, Rudolf. 1981 [1962]. On Travels in the Universe of the Soul: Reports on Self-Experiments with Delysid (LSD) and Psilocybin (CY). *Journal of Psychoactive Drugs* 13(1): 81–9.
Gerth, H. H., and Wright Mills, C., eds. 1991. *From Max Weber: Essays in Sociology.* London: Routledge.
Ghouled, F. C. 1972. *Field Guide to the Psilocybin Mushroom-Species Common to North America.* New Orleans: Guidance Publications.

Ginzburg, Carlo. 1992. *Ecstasies. Deciphering the Witches' Sabbath.* Harmondsworth: Penguin.

Gladwell, Malcolm. 2000. *The Tipping Point. How Little Things Can Make a Big Difference.* London: Abacus.

Glen, G. 1816. A Case Proving the Deleterious Effects of the Agaricus Campanulatus, Which was Mistaken for the Agaricus Campestris, or Champignon. *London Medical and Physical Journal* 36: 451–3.

Goffman, Erwing. 1959. *The Presentation of Self in Everyday Life.* Garden City: Doubleday.

Goldschmidt, Walter. 1987. Foreword, in Castaneda, Carlos, *The Teachings of Don Juan: A Yaqui Way of Knowledge.* London: Penguin.

Goldsmith, Oliver. 1762. *The Citizen of the World. In a Series of Letters from a Chinese Philosopher at London to his Friends in the East. In Two Volumes.* London: Harrison & Co.

Goskar, Thomas A., Carty, Alistair, Cripps, Paul, Brayne, Chris, and Vickers, Dave. The Stonehenge Laser Scans. *British Archaeology* November 2003: 8–13.

Graves, Richard. 1996. *Robert Graves and the White Goddess 1940–1985.* London: Phoenix.

Graves, Robert. 1959. Mushrooms, Food of the Gods. *Atlantic Monthly*, August Edition: 73–7.

— 1960. *The Greek Myths: 1*, rev. edn. London: Penguin.

— 1961a. *Oxford Addresses on Poetry.* London: Cassell.

— 1961b. *The White Goddess.* London: Faber and Faber.

— 1969. *The Crane Bag, and Other Disputed Subjects.* London: Cassell.

— 1972. *Difficult Questions, Easy Answers.* London: Cassell.

Green, Jonathan. 1974. *The IT Book of Drugs.* London: Bloom Publications Ltd.

— 1999. *All Dressed Up. The Sixties and the Counterculture.* London: Pimlico.

Green, Miranda. 1997. *Exploring the World of the Druids.* London: Thames and Hudson.

— 2001. *Dying for the Gods. Human Sacrifice in Iron Age and Roman Europe.* Stroud: Tempus Publishing.

Greene, Mott T. 1992. *Natural Knowledge in Pre-Classical Antiquity.* Baltimore: Johns Hopkins University Press.

Guzmán, Gastón. 1983. *The Genus* Psilocybe. *A Systematic Revision of the Known Species Including the History, Distribution and Chemistry of the Hallucinogenic Species.* Vaduz: J. Cramer.

Guzmán, Gastón, and Ott, Jonathan. 1976. Description and Chemical Analysis of a New Species of Hallucinogenic Psilocybe from the Pacific Northwest. *Mycologia* 68(6): 1261–7.

Guzmán, Gastón, Allen, John W., and Gartz, Jochen. 1998. A Worldwide Geographical Distribution of the Neurotropic Fungi, An Analysis and Discussion. *Annali Museo Civico di Rovereto* 14: 189–280.

Hajicek-Dobberstein, Scott. 1995. Soma Siddhas and Alchemical Enlightenment: Psychedelic Mushrooms in Buddhist Tradition. *Journal of Ethnopharmacology* 48: 99–118.

Halifax, Joan. 1979. *Shamanic Voices. A Survey of Visionary Narratives.* Harmondsworth: Arkana.

— 1990. The Mushroom Conspiracy, in Riedlinger, Thomas, ed. *The Sacred Mushroom Seeker. Essays for R. Gordon Wasson.* Portland, OR: Dioscorides Press, pp. 111–14.

Hall, Alan H., and Hall, Priscilla K. 1994. Ibotenic Acid/Muscimol-Containing Mushrooms, in Spoerke, David G., and Rumack, Barry H., eds *Handbook of Mushroom Poisoning. Diagnosis and Treatment.* London: CRC Press, ch. 12, pp. 265–78.

Halstead, Bruce W. 1967. *Poisonous and Venomous Marine Animals of the World. Volume 2 – Vertebrates.* Washington DC: United States Government Printing Office.

Hammelev, Dorte. 1986. The Story of the Dyeing 'Witches', or a New Activity of the Danish Mycological Society. *Bulletin of the British Mycological Society* 20(1): 61–2.

Hancock, Stephen. 2005. *Cannibalise Legalism.* Oxford: Pig and Ink Books.

Harding, Jan, and Healy, Frances. 2003. Raunds. From Hunters to Farmers. *British Archaeology* 73: 16–21.

Harner, Michael. 1973. *Hallucinogens and Shamanism.* Oxford and New York: Oxford University Press.

— 1980. *The Way of the Shaman.* San Francisco: HarperSanFrancisco.

— 2003 [1980]. Discovering the Way, in Harvey, Graham, ed. *Shamanism. A Reader.* London: Routledge, pp. 41–56.

Harries, Anthony D., and Evans, Valmai. 1981. Sequelae of a 'Magic Mushroom Banquet'. *Postgraduate Medical Journal* 57: 571–2.

Harrison, Kathleen. 1999. Roads Where There Have Long Been Trails. *Terrain. A Journal of the Built and Natural Environments.* Accessed at www.terrain.org/essays/3/harrison.htm

Harvey, Graham. 1997. Listening People, Speaking Earth. Contemporary Paganism. London: Hurst & Co.

Harvey, Graham, ed. 2003. *Shamanism. A Reader.* London: Routledge, pp. 159–69.

Hasler Felix, Grimberg, Ulrike, Benz, Marco A., Huber, Theo, and Vollenweider, Franz X. 2004. Acute Psychological and Physiological Effects of Psilocybin in Healthy Humans: A Double-Blind, Placebo-Controlled Dose-Effect Study. *Psychopharmacology* 172: 145–56.

Hawk and Venus. 2003. *Soma Shamans.* Lakeport, CA: Red Angels Ltd.

Hawksworth, David L. 2001. Mushrooms: The Extent of the Unexplored Potential. *International Journal of Medicinal Mushrooms* 3: 333–7.

Hay, William D. 1887. *An Elementary Text-Book of British Fungi.* London: Swan Sonnenschein, Lowry & Co.

Hayter, Alethea. 1968. *Opium and the Romantic Imagination.* London: Faber and Faber.

Heberden, W. 1772. An Account of the Noxious Effects of Some Funguses. *Gentleman's Magazine* XLII: 380.

Heim, Roger. 1957. Analyse de Quelques Expériences Personelles Produites par

l'Ingestion des Agarics Hallucinogènes de Mexique. *Comptes Rendus Hebdomadaires des Séances de l'Académie des Sciences* 245: 597–603.
– 1963. *Les Champignons Toxiques et Hallucinogènes*. Paris: Éditions N. Boubée & Cie.
– 1971. À Propos des Propriétés Hallucinogènes du *Psilocybe semilanceata*. *Le Naturaliste Canadien* 98: 415–24.
Heim, Roger, and Wasson, R. Gordon. 1958. *Les Champignons Hallucinogènes de Mexique*. Paris: Éditions du Muséum National d'Histoire Naturelle.
– 1965. The 'Mushroom Madness' of the Kuma. Botanical Museum Leaflets Harvard University 21(1): 1–36.
Heim, Roger, Hofmann, Albert and Tscherter, Hans. 1966. Sur une Intoxication Collective à Syndrome Psilocybien Causé en France par un Copelandia. *Comptes Rendus Hebdomadaires des Séances de l'Académie des Sciences Série D*. 262: 519–23.
Heim, Roger, Cailleux, Roger, Wasson, R. Gordon, and Thévenard, Pierre. 1967. *Nouvelles Investigations sur les Champignons Hallucinogènes*. Paris: Éditions du Muséum National d'Histoire Naturelle.
Heim, Roger, and Thévenard, Pierre. 1967. Expériences Nouvelles d'Ingestion des Psilocybes Hallucinogènes, in Heim, Roger, Cailleux Roger, Wasson, R. Gordon, and Thévenard, Pierre, *Nouvelles Investigations sur les Champignons Hallucinogènes*. Paris: Éditions du Muséum National d'Histoire Naturelle, ch. 5, pp. 201–11.
Heinrich, Clark. 2002. *Magic Mushrooms in Religion and Alchemy*. Rochester, VT: Park Street Press.
Heitzer, Robert F. 1944. Mixtum Compositum. The Use of Narcotic Mushrooms by Primitive Peoples. *Ciba Symposia*, February Edition: 1713–16.
Helvenston, Patricia A., and Bahn, Paul G. 2003. Testing the 'Three Stages of Trance' Model. *Cambridge Archaeological Journal* 13: 213–24.
Hermans, H. G. M. [n.d.] Memories of a Maverick. Maassluis, Netherlands: Pi Publications. Accessed at www.uri-geller.com/books/maverick/maver.htm
Hetherington, Kevin. 2000. *New Age Travellers. Vanloads of Uproarious Humanity*. London: Cassell.
Hobsbawm, Eric, and Ranger, Terence, eds. 1983. *The Invention of Tradition*. Cambridge: Cambridge University Press.
Hofmann, Albert. 1958. Rapport sur une Auto-Expérience avec le *Psilocybe mexicana* Heim, in Heim, Roger, and Wasson, R. Gordon. *Les Champignons Hallucinogènes de Mexique*. Paris: Éditions du Muséum National d'Histoire Naturelle, pp. 278–80.
– 1983 [1979]. *LSD My Problem Child. Reflections on Sacred Drugs, Mysticism, and Science*, trans. Jonathan Ott. Los Angeles: J. P. Tarcher, Inc.
– 1990. Ride Through the Sierra Mazateca in Search of the Magic Plant 'Ska María Pastora', in Riedlinger, Thomas, ed. *The Sacred Mushroom Seeker. Essays for R. Gordon Wasson*. Portland, OR: Dioscorides Press, pp. 115–27.
Hofmann, Albert, Heim, Roger, Brack, A., and Kobel, H. 1958. Psilocybin, ein Psychotroper Wirkstoff aus dem Mexikanischen Rauschpilz *Psilocybe mexicana* Heim. *Experimentia* 14: 107–12.

Hofmann, Albert, Heim, Roger and Tscherter, Hans. 1963. Présence de la Psilocybine dans une Espèce Européenne d'Agaric, le *Psilocybe semilanceata* Fr. *Comptes Rendus Hebdomadaires des Séances de L'Académie des Sciences* 257: 10–12.

Holmstedt, Bo. 1967. Historical Survey, in Efron, Daniel H., Holmstedt, Bo, and Kline, Nathan S. *Ethnopharmacologic Search for Psychoactive Drugs*. Washington: US Department of Health, Education and Welfare, pp. 3–32.

Honigsbaum, Mark. 2003. High Times in Magic Mushroom Business – and It's Perfectly Legal. *The Guardian* Saturday 29 November: 3.

— 2004. Magic Mushroom Case Judge Tells Prosecutor: Chill Out. *The Guardian* Wednesday 15 December: 1.

— 2005. Headache Sufferers Flout New Drug Law. *The Guardian* Tuesday 2 August: 10.

Horne, C. H. W., and McCluskie, J. A. W. 1963. The Food of the Gods. *Scottish Medical Journal* 8: 489–91.

Horowitz, Michael, and Palmer, Cynthia, eds. 1999. *Moksha. Aldous Huxley's Classic Writings on Psychedelics and the Visionary Experience*. Rochester, VT: Park Street Press.

Hutton, Ronald. 1993. *The Pagan Religions of the Ancient British Isles. Their Nature and Legacy*. Oxford: Blackwell.

— 1996a. *The Stations of the Sun. A History of the Ritual Year in Britain*. Oxford: Oxford University Press.

— 1996b. Introduction – Who Possesses the Past?, in Carr-Gomm, Philip, ed. *The Druid Renaissance*. London: Thorsons, pp. 17–34.

— 1999. *The Triumph of the Moon. A History of Modern Pagan Witchcraft*. Oxford: Oxford University Press.

— 2001. *Shamans. Siberian Spirituality and the Western Imagination*. London: Hambledon and London.

— 2003. *Witches, Druids and King Arthur*. London: Hambledon and London.

Huxley, Aldous. 1994. *The Doors of Perception and Heaven and Hell*. London: Flamingo.

Hyde, C., Glancy, G., Omerod, P., Hall, D., and Taylor, G. S. 1978. Abuse of Psilocybin Mushrooms: A New Fashion and Some Psychiatric Complications. *British Journal of Psychiatry* 132: 602–4.

Imai, Sanshi. 1932. On *Stropharia caerulescens*, a New Species of Poisonous Toadstool. *Transactions of the Sapporo Natural History Society* 12(3): 148–51.

Imazeki, Rokuya. 1973. Japanese mushroom names. *Transactions of the Asiatic Society of Japan* 3rd Series XI: 26–80.

Inglis, Brian. 1975. *The Forbidden Game. A Social History of Drugs*. London: Hodder & Stoughton.

Jacobs, Keith W. 1975. Hallucinogenic Mushrooms in Mississippi. *Journal of the Mississippi State Medical Association* 16: 35–7.

Jay, Mike. 1999. *Blue Tide. The Search for Soma*. New York: Autonomedia.

— 2000. *Emperors of Dreams. Drugs in the Nineteenth Century*. Sawtry (UK): Dedalus.

Johnson, Dave. 2005. Underworld Connections. *Biologist* 52(3): 155–60.
Johnson, Jean Bassett. 1939. The Elements of Mazatec Witchcraft. *Ethnological Studies* (Gothenburg) 9: 128–50.
Johnson, P. C. 1995. Shamanism from Ecuador to Chicago: A Case of New Age Appropriation. *Religion* 25: 163–78.
Johnston, James. 1855. *The Chemistry of Common Life. In Two Volumes.* Edinburgh and London: William Blackwood & Sons.
Jordan, M. 1975. *A Guide to Mushrooms. The Edible and Poisonous Fungi of the Northern Hemisphere.* London: Millington.
Kaplan, Reid. 1975. The Sacred Mushroom in Scandinavia. *Man ns* 10: 72–9.
Karana. 1968. The Code of the Sacred Islands. *International Times* (12–25 July) 35: 5.
Keay, Susan M., and Brown, Averil E. 1990. Colonization by *Psilocybe semilanceata* of roots of grassland flora. *Mycological Research* 94(1): 49–56.
Keewaydinoquay. 1978. *Puhpohwee for the People. A Narrative Account of Some Uses of Fungi among the Ahnishinaubeg.* Cambridge, MA: Botanical Museum of Harvard University.
– 1979. The Legend of Miskwedo. *Journal of Psychedelic Drugs* 11(1/2): 29–31.
Keightley, Thomas. 1850. *The Fairy Mythology. Illustrative of the Romance and Superstition of Various Countries.* 3rd edn. London: H. G. Bohn.
Kendrick, Dave. 1976. The Magic is 'Mushrooming' These Days at Turnwater. *Sunday Olympian*, Sunday 14 November: 8–9. Accessed at www.mushroomjohn.org
Kibby, Geoffrey. 1977. *Colour Encyclopaedia of Mushrooms and Toadstools.* London: Cathay Books.
– 1979. *Mushrooms and Toadstools. A Field Guide.* Oxford: Oxford University Press.
King, John. 1970. *A Christian View of the Mushroom Myth.* London: Hodder & Stoughton.
Krippner, Stanley, and Winkelman, Michael. 1983. María Sabina: Wise Lady of the Mushrooms. *Journal of Psychoactive Drugs* 15: 225–8.
Kuhn, Thomas S. 1996. *The Structure of Scientific Revolutions.* 3rd edn. Chicago: University of Chicago Press.
La Barre, Weston. 1970. Book Review of *Soma: Divine Mushroom of Immortality* by R. Gordon Wasson. *American Anthropologist* 72: 368–73.
– 1970. Old and New World Narcotics: A Statistical Question and an Ethnological Reply. *Economic Botany* 24: 73–80.
– 1972. Hallucinogens and the Shamanic Origins of Religion, in Furst, Peter T., ed. *Flesh of the Gods. The Ritual Use of Hallucinogens.* London: George Allen and Unwin Ltd, ch. 10, pp. 261–78.
Lajoux, Jean-Dominique. 1963. *The Rock Paintings of Tassili.* London: Thames and Hudson.
Lasco, Peter. 1972. *Ars Sacra: 800–1200.* Harmondsworth: Penguin.
Lawson, A. A. H. 1981. 'Magic' Mushrooms. *Safety Education (ROSPA)* 152: unpaged.

Leary, Timothy. 1970. *The Politics of Ecstasy*. St Albans: Paladin.
— 1983. *Flashbacks. An Autobiography*. London: Heinemann.
— 1989. *Info-Psychology. A Manual on the Use of the Human Nervous System According to the Instructions of the Manufacturers*. Las Vegas: Falcon Press.
— 1995 [1968]. *High Priest*. Berkeley: Ronin Publications Inc.
Leary, Timothy, Litwin, George H., and Ralph Metzner. 1963. Reactions to Psilocybin Administered in a Supportive Environment. *Journal of Nervous and Mental Disease* 137: 561–73.
Lee, Martin A., and Shlain, Bruce. 1992. *Acid Dreams. The Complete Social History of LSD: The CIA, the Sixties, and Beyond*. New York: Grove Weidenfeld.
Letcher, Andy. 2001. *The Role of the Bard in Contemporary Pagan Movements*. Unpublished PhD thesis, King Alfred's College, Winchester.
— *'Mad Thoughts on Mushrooms': Discourse and Power in the Study of Psychedelic Consciousness*. Paper delivered at the 'Exploring Consciousness' conference, Bath, June 2004.
Levack, Brian. 1995. *The Witch-Hunt in Early Modern Europe*. 2nd edn. London: Longman.
Lévi-Strauss, Claude. 1994. Mushrooms in Culture: Apropos of a Book by R. G. Wasson, in *Structural Anthropology Volume* 2, trans. Monique Layton. Harmondsworth: Penguin, ch. 12, pp. 222–37.
Lewin, Louis. 1931. *Phantastica. Narcotic and Stimulating Drugs. Their Use and Abuse*. London: Kegan Paul.
Lewin, Roger. 1991. Stone Age Psychedelia. *New Scientist* 8 June: 30–4.
Lewis-Williams, J. D., and Dowson, T. A. 1988. The Signs of All Times. Entoptic Phenomena in Upper Palaeolithic Art. *Current Anthropology* 29: 201–45.
Li, Hui-Lin. 1978. Hallucinogenic Plants in Chinese Herbals. *Journal of Psychedelic Drugs* 10(1): 17–26.
Lightfoot, John. 1777. *Flora Scotica*. London: B. White.
Lincoff, Gary, and Mitchel, D. H. 1977. *Toxic and Hallucinogenic Mushroom Poisoning. A Handbook for Physicians and Mushroom Hunters*. New York and London: Van Nostrand Reinhold Company.
Lindley, John. 1853. *The Vegetable Kingdom or, The Structure, Classification, and Uses of Plants*. 3rd edn. London: Bradbury & Evans.
Lipton, Morris A. 1970. The Relevance of Chemically-Induced Psychoses to Schizophrenia, in Efron, Daniel H., ed. *Psychotomimetic Drugs*. New York: Raven Press, pp. 231–46.
Livings, Henry. 1965. *eh?* London: Methuen & Co.
Long, D. J., Tipping, R., Holden, T. G., Bunting, M. J., and Milburn, P. 2000. The Use of Henbane (*Hyoscyamus niger* L.) as a Hallucinogen at Neolithic 'Ritual' Sites: A Re-evaluation. *Antiquity* 74: 49–53.
Lowy, Bernard. 1971. New Records of Mushroom Stones from Guatemala. *Mycologia* 63: 983–93.
— 1972. Mushroom Symbolism in Maya Codices. *Mycologia* 64: 816–21.
— 1977. Hallucinogenic Mushrooms in Guatemala. *Journal of Psychedelic Drugs* 9(2): 123–5.

Malbon, Ben. 1999. *Clubbing. Dancing, Ecstasy and Vitality*. London: Routledge.

Mann, John. 1994. *Murder, Magic and Medicine*. Oxford: Oxford University Press.

Mantle, P. G., and Waight, E. S. 1969. Occurrence of Psilocybin in the Sporophores of *Psilocybe semilanceata*. *Transactions of the British Mycological Society* 53(2): 302–4.

Marcano, V., Morales Méndez, A., Castellano, F., Salazar, F. J., and Martinez, L. 1994. Occurrence of psilocybin and psilocin in *Psilocybe pseudobullacea* (Petch) Pegler from the Venezuelan Andes. *Journal of Ethnopharmacology* 43: 157–9.

Marks, John. 1979. *In Search of the Manchurian Candidate*. London: Allen Lane.

Marsh, R. W. 1973. Mycological Millinery. *Bulletin of the British Mycological Society* 7(1): 34–5.

Martin, Bradley. 1967. Bradley Martin. *International Times* 11 (April 21–28): 4.

Martineau, Jane, ed. 1997. *Victorian Fairy Painting*. London: Royal Academy of Arts.

Mayes, Richard ('Ric the Vic'). 1980. Hail Albion. *Home Grown* 1(6): 16.

McDonald, Angus. 1978. The Present Status of Soma: The Effects of California *Amanita muscaria* on Normal Human Volunteers, in Rumack, Barry H., and Saltzman, Emanuel, ed. *Mushroom Poisoning: Diagnosis and Treatment*. West Palm Beach, FL: CRC Press, ch. 13, pp. 215–23.

McIlvaine, Charles. 1900. *One Thousand American Fungi. How to Select and Cook the Edible; How to Distinguish and Avoid the Poisonous*. Indianapolis: The Bowen-Merrill Company.

McKay, George. 1996. *Senseless Acts of Beauty. Cultures of Resistance Since the Sixties*. London: Verso.

McKenna, Terence. 1991. *The Archaic Revival. Speculations on Psychedelic Mushrooms, the Amazon, Virtual Reality, UFOs, Evolution, Shamanism, the Rebirth of the Goddess, and the End of History*. San Francisco: HarperSanFrancisco.

— 1992a. *Food of the Gods. The Search for the Original Tree of Knowledge. A Radical History of Plants, Drugs and Human Evolution*. London: Rider.

— 1992b. *The Camden Centre Talk 15/6/92*. Accessed at http://www.deoxy.org/t_camden.htm

— 1994. *True Hallucinations. Being an Account of the Author's Extraordinary Adventures in the Devil's Paradise*. San Francisco: HarperSanFrancisco.

— n.d. *Under the Teaching Tree 1–4*. mp3 available at http://mckenna.otterly.com/ http://deoxy.org/mckenna.htm; http://mckenna.drugtext.org and http://mckenna.psychedelic-library.org

McKenna, Terence, and McKenna, Dennis. 1993 [1975]. *The Invisible Landscape. Mind, Hallucinogens and the I Ching*. San Francisco: HarperSanFrancisco.

McKenna, Terence, and Hunter, Robert. 1996. *Orfeo. An Ongoing Web Dialogue Between Robert Hunter and Terence McKenna*. Available at http://deoxy.org/mckenna.htm

Psilocybin: Magic Mushroom Grower's Guide O.T. Oss O.N. Oeric

McKenzie, Eric H. C. 1997. *Collect Fungi on Stamps*. London and Ringwood: Stanley Gibbons Ltd.

Melechi, Antonio, ed. 1997. *Psychedelia Britannica. Hallucinogenic Drugs in Britain*. London: Turnaround.

Merlin, M. D. 2003. Archaeological Evidence for the Tradition of Psychoactive Plant Use in the Old World. *Economic Botany* 57(3): 295–323.

Metcalfe, Stuart. 1997. Ecstasy Evangelists and Psychedelic Warriors, in Melechi, Antonio, ed. *Psychedelia Britannica. Hallucinogenic Drugs in Britain*. London: Turnaround, ch. 8, pp. 166–84.

Metzner, Ralph. 1971. Mushrooms and the Mind, in Aaronson, Bernard, and Osmond, Humphrey, eds. *Psychedelics. The Uses and Implications of Hallucinogenic Drugs*. London: The Hogarth Press, pp. 90–107.

Michelot, Didier, and Melendez-Howell, Leda Marina. 2003. *Amanita muscaria*: chemistry, biology, toxicology, and ethnomycology. *Mycological Research* 107: 131–46.

Mills, P. R., Lesinskas, D., and Watkinson, G. 1979. The Danger of Hallucinogenic Mushrooms. *Scottish Medical Journal* 24: 316–17.

Moffat, B. 1993. An Assessment of Residues on the Grooved Ware. In Barclay, G. J., and Russell-White, C. J. Excavations in the Ceremonial Complex of the Fourth to Second Millennium BC at Balfarg/Balbirnie, Glenrothes, Fife. *Proceedings of the Society of Antiquaries of Scotland* 123: 43–210.

Moorcraft, Colin. 1970. God and the Sacred Mushroom. *Friends* (29 May) 8: 27.

More, Sheila. 1968. The Restless Generation: 3. *The Times* 18 December: 13.

Moreno, F. A., Wiegand, C. B., Delgado, P. L., Taitano, K. 2003. Safety, Tolerability, and Efficacy of Psilocybin in Treatment-Refractory Obsessive-Compulsive Disorder: Preliminary Findings. *Biological Psychiatry* 53 Supplement: 46S.

Morgan, Adrian. 1986. Who put the toad in toadstool? *New Scientist* 25 December: 44–7.

Mount, P., Harris, G., Sinclair, R., Finlay, M., and Becker, G. J. 2002. Acute Renal Failure Following Ingestion of Wild Mushrooms. *Internal Medicine Journal* 32: 187–90.

Munn, Henry. 1973. The Mushrooms of Language, in Harner, Michael, ed. *Hallucinogens and Shamanism*. Oxford: Oxford University Press. Accessed at www.psychedelic-library.org/munn.htm

— 2003. The Uniqueness of María Sabina, in Rothenberg, Jerome, ed. *María Sabina. Selections*. Berkeley: University of California Press, pp. 140–63.

Murray, Margaret. 1921. *The Witch-Cult in Western Europe*. Oxford: Clarendon Press.

Murray, Virginia S. G. 1984. Would the consumption of the mushrooms *Psilocybe mexicana* and *P. semilanceata* in moderate (but frequent) amounts be likely to harm adults, children, or household pets? *British Medical Journal* 288: 46.

Nada. 1969. Ongos. *International Times* 9–22 May: 7.

Naranjo, Claudio. 1990. A Posthumous 'Encounter' with R. Gordon Wasson, in

Riedlinger, Thomas, ed. *The Sacred Mushroom Seeker. Essays for R. Gordon Wasson*. Portland, OR: Dioscorides Press, pp. 177–81.

Navet, Éric. 1988. Les Ojibway et L'Amanite Tue-Mouche (*Amanita muscaria*). Pour une Ethnomycologie des Indiens D'Amérique du Nord. *Journal de la Sociète des Américanistes* 74: 163–80.

Needham, Joseph. 1974. *Science and Civilisation in China. Volume 5, Part 2. Chemistry and Chemical Technology*. Cambridge: Cambridge University Press.

Noel, Daniel. 1987. Shamanic Ritual as Poetic Model: The Case of María Sabina and Anne Waldman. *Journal of Ritual Studies* 1: 57–71.

Nuttall, Jeff. 1970 [1968]. *Bomb Culture*. London: Paladin.

Ohenoja, E., Jokiranta, J., Makinen, T., Kaikkonen, A., and Airaksinen, M. 1987. The Occurrence of Psilocybin and Psilocin in Finnish Fungi. *Journal of Natural Products* 50(4): 741–44.

Olivelle, Patrick. 2005. *Manu's Code of Law. A Critical Edition and Translation of the Manava-Dharmasastra*. Oxford: Oxford University Press.

O'Prey, Paul, ed. 1984. *Between Moon and Moon. Selected Letters of Robert Graves 1946–1972* London: Hutchinson.

Osmond, Humphrey. 1957. A Review of the Clinical Effects of Psychotomimetic Agents. *Annals of the New York Academy of Sciences* 66: 418–34.

Oster, Gerald. 1970. Phosphenes. *Scientific American* 222: 83–7.

Ott, Jonathan. 1976a. *Hallucinogenic Plants of North America*. Berkeley: Wingbow Press.

– 1976b. Psycho-Mycological Studies of Amanita – From Ancient Sacrament to Modern Phobia. *Journal of Psychedelic Drugs* 8(1): 27–35.

– 1978. *Mr Jonathan Ott's Rejoinder to Dr Alexander H. Smith*. Cambridge, MA: Botanical Museum of Harvard University.

– 1990. A Twentieth Century Darwin, in Riedlinger, Thomas, ed. *The Sacred Mushroom Seeker. Essays for R. Gordon Wasson*. Portland, OR: Dioscorides Press, pp. 183–91.

– 1996. *Pharmacotheon. Entheogenic Drugs, Their Plant Sources and History*. 2nd edn. Kennewick, WA: Natural Products Co.

Ott, Jonathan, and Bigwood, Jeremy, eds. 1978. *Teonanácatl. Hallucinogenic Mushrooms of North America. Extracts from the Second International Conference on Hallucinogenic Mushrooms, held October 27–30 1977, near Port Townsend, Washington*. Seattle: Madrona Publications Inc.

Palmer, Cynthia, and Horowitz, Michael. 2000. *Sisters of the Extreme. Women Writing on the Drug Experience*. Rochester, VT: Park Street Press.

Passie, Torsten, Seifert, Juergen, Schneider, Udo, and Emrich, Hinderk M. 2002. The Pharmacology of Psilocybin. *Addiction Biology* 7: 357–64.

Pearson, J. L. 2002. *Shamanism and the Ancient Mind: A Cognitive Approach to Archaeology*. Walnut Creek: Altamira.

Peden, Norman R., Bissett, Ann F., Macauley, K. E. C., Crooks, James, and Pelosi, A. J. 1981. Clinical Toxicology of 'Magic Mushroom' Ingestion. *Postgraduate Medical Journal* 57: 543–5.

Peden, Norman R., Pringle, Stuart D., and Crooks, James. 1982. The Problem of Psilocybin Mushroom Abuse. *Human Toxicology* 1: 417–24.

Pfister, Donald R. 1988. R. Gordon Wasson – 1898–1986. *Mycologia* 80: 11–13.

Phillips, Roger. 1981. *Mushrooms and Other Fungi of Great Britain and Europe*. London: Pan Books.

Pickering, Roger. 1755. A Brief Dissertation upon Fungi in General, and Concerning the Poisonous Faculty of Some Species in Particular, Being a Supplement to the Papers on Poisonous English Plants by the Same Author. *Gentleman's Magazine* XV: 542–5, 585–7.

Piearce, G. D. 1981. Zambian Mushrooms – Customs and Folklore. *Bulletin of the British Mycological Society* 15(2): 139–42.

Piggott, S. 1968. *The Druids*. London: Thames and Hudson.

Pike, Eunice, and Cowan, Florence. 1959. Mushroom Ritual versus Christianity. *Practical Anthropology* 6(4): 145–50.

Pinchbeck, Daniel. 2002. *Breaking Open the Head. A Psychedelic Journey into the Heart of Contemporary Shamanism*. New York: Broadway Books.

Plant, Sadie. 1999. *Writings on Drugs*. London: Faber & Faber.

Pollock, Steven H. 1975a. The Psilocybin Mushroom Pandemic. *Journal of Psychedelic Drugs* 7(1): 73–84.

– 1975b. The Alaskan Amanita Quest. *Journal of Psychedelic Drugs* 7(4): 397–9.

– 1976. Psilocybian Mycetismus with Special Reference to Panaeolus. *Journal of Psychedelic Drugs* 8(1): 43–57.

– 1977. *Magic Mushroom Cultivation*. San Antonio: Herbal Medicine Research Foundation.

– 1977/8. Psychotropic Mushrooms and the Alteration of Consciousness, I: The Ascent of Psilocybian Mushroom Consciousness. *Journal of Altered States of Consciousness* 3(1): 15–34.

Powell, Simon G. 2003. Sacred Ground. Psilocybin Mushrooms and the Rebirth of Nature. Privately Published. Accessed at www.island.org/prescience

Prokhorov, Vadim. 2004. Play That Fungi Music. *The Guardian* Friday 30 July G2: 7.

Puharich, Andrija. 1959. *The Sacred Mushroom. Key to the Door of Eternity*. London: Victor Gollancz Ltd.

– 1962. *Beyond Telepathy*. London: Darton, Longman and Todd.

Q Magazine Special Edition. 2005. *Psychedelic!* London: Emap Metro Ltd.

Ramsbottom, J. 1923. *A Handbook of the Larger British Fungi. Based on the Guide to Sowerby's Models of British Fungi in the Department of Botany, British Museum (Natural History)*. London: British Museum.

– 1945. *Poisonous Fungi*. London: Penguin.

– 1953. *Mushrooms and Toadstools. A Study of the Activities of Fungi*. London: Collins.

Reay, Marie. 1960. 'Mushroom Madness' in the New Guinea Highlands. *Oceania* 31: 137–9.

Reichel-Dolmatoff, G. 1978. Drug-Induced Optical Sensations and Their Relationship to Applied Art Among Some Colombian Indians, in Greenhalgh, Michael, and Megaw, Vincent, eds *Art in Society*. London: Duckworth, pp. 289–304.

Release. 1979. *Hallucinogenic Mushrooms*. London: Release Guides.
Richardson, Allan B. 1990. Recollections of R. Gordon Wasson's 'Friend and Photographer', in Riedlinger, Thomas, ed. *The Sacred Mushroom Seeker. Essays for R. Gordon Wasson*. Portland, OR: Dioscorides Press, pp. 193–203.
Riedlinger, Thomas J., ed. 1990a. *The Sacred Mushroom Seeker. Essays for R. Gordon Wasson*. Portland, OR: Dioscorides Press.
Riedlinger, Thomas J. 1990b. A Latecomer's View of R. Gordon Wasson, in Riedlinger, Thomas J., ed. *The Sacred Mushroom Seeker. Essays for R. Gordon Wasson*. Portland, OR: Dioscorides Press, pp. 205–20.
Riedlinger, Thomas J. 1993. Wasson's Alternative Candidates for Soma. *Journal of Psychoactive Drugs* 25(2): 149–56.
Riper, Heleen, and de Kort, Marcel. 1999. Smart Policies for Smart Products and Ecodrugs? *Journal of Drug Issues* 29(3): 703–26.
Robbins, Tom. 1973. *Another Roadside Attraction*. London: W. H. Allen.
Rosenberger, Alex. 1990. Rave On! *Festival Eye* Summer: 13.
Rothenberg, Jerome, ed. 2003. *María Sabina. Selections*. Berkeley: University of California Press.
Ruck, Carl A. P. 1981. Mushrooms and Philosophers. *Journal of Ethnopharmacology* 4: 179–205.
— 1982. The Wild and the Cultivated: Wine in Euripides' *Bacchae*. *Journal of Ethnopharmacology* 5: 231–70.
— 1983. The Offerings from the Hyperboreans. *Journal of Ethnopharmacology* 8: 177–207.
— 1986. Poets, Philosophers, Priests: Entheogens in the Formation of the Classical Tradition, in Wasson, R. Gordon, Kramrisch, S., Ott, Jonathan, and Ruck, Carl A. P. *Persephone's Quest. Entheogens and the Origins of Religion*. New Haven and London: Yale University Press, pp. 151–256.
Rudgley, Richard. 1993. *The Alchemy of Culture. Intoxicants in Society*. London: British Museum Press.
Rushkoff, Douglas. 1994. *Cyberia. Life in the Trenches of Hyperspace*. San Francisco: HarperSanFrancisco.
Saar, Maret. 1991. Ethnomycological data from Siberia and North-East Asia on the effect of *Amanita muscaria*. *Journal of Ethnopharmacology* 31: 157–73.
Safford, William. 1915. An Aztec Narcotic. *Journal of Heredity* 6: 291–311.
Said, Edward. W. 1995. *Orientalism. Western Conceptions of the Orient*. Harmondsworth: Penguin.
Salisbury, W. 1815. Letter. *Gentleman's Magazine* LXXXV (ii): 103.
Salzman, Emanuel, Salzman, Jason, Salzman, Joanna, and Lincoff, Gary. 1996. In Search of *Mukhomor*, the Mushroom of Immortality. *Shaman's Drum* 41: 36–47.
Samorini, Giorgio. 1989. Etnomicologia nell'arte rupestre Sahariana (Periodo delle 'Teste Rotonde'). *Bollettino Camuno Notizie* 6: 18–22.
— 1997. A Peculiar Historical Document About Fly Agaric. *Eleusis* 4: 3–16.
— 1998. The Pharsalus Bas-Relief and the Eleusinian Mysteries. *Entheogen Review* 7: 60–3. Accessed at www.samorini.net/doc/sam/pharsal.htm
San Antonio, J. P. 1971. A Laboratory Method to Obtain Fruit from Cased

Grain Spawn of the Cultivated Mushroom *Agaricus bisporus. Mycologia* 50: 239–61.

Sandford, Jeremy. 1972. *In Search of the Magic Mushroom. A Journey Through Mexico*. London: Peter Owen.

Sarchin, Larry. 1973. Bumper Crop: Hallucinogenic Campus Mushrooms Harvested. *University of Washington Daily* 20 November, unpaged. Accessed at www.mushroomjohn.org

Saunders, Nicholas. 1970. *Alternative London*. London: Privately Published.

— 1997. *Ecstasy Reconsidered*. London: Privately Published.

Schechner, Richard. 1993. *The Future of Ritual. Writings on Culture and Performance*. London: Routledge.

Schieffelin, Edward. 1996. On Failure and Performance: Throwing the Medium out of the Séance, in Laderman, C., and Roseman, M., eds *The Performance of Healing*. London: Routledge.

Schultes, Richard Evans. 1940. Teonanacatl: The Narcotic Mushroom of the Aztecs. *American Anthropologist* n.s. 42: 429–43.

— 1990. Foreword, in Riedlinger, Thomas, ed. *The Sacred Mushroom Seeker. Essays for R. Gordon Wasson*. Portland, OR: Dioscorides Press, pp. 13–18.

Schultes, Richard Evans, and Hofmann, Albert. 1992. *Plants of the Gods. Their Sacred, Healing and Hallucinogenic Powers*. Rochester, VT: Healing Arts Press.

Schultes, Richard Evans, and von Ries, Siri, eds. 1995. *Ethnobotany: Evolution of a Discipline*. London: Chapman and Hall.

Self, Will. 1996. *Junk Mail*. London: Penguin.

Seymour-Smith, Martin. 1982. *Robert Graves. His Life and Work*. London: Hutchinson.

Shamen, The, with Terence McKenna. 1993. *Re-Evolution*. 118TP7CD. London: One Little Indian Records.

Sharpe, J. 1996. *Instruments of Darkness. Witchcraft in England 1550–1750*. London: Hamish Hamilton.

Sheldrake, Rupert, McKenna, Terence, and Abraham, Ralph. 2001 [1992]. *Chaos, Creativity and Cosmic Consciousness*. Rochester, VT: Park Street Press.

Sherratt, Andrew. 1991. Sacred and Profane Substances, the Use of Narcotics in Later Neolithic Europe, in Garwood, P., Jennings, D., Skeates, R., and Toms, J., eds *Sacred and Profane: Proceedings of a Conference on Archaeology, Ritual and Religion*. Oxford University Committee for Archaeology Monographs 32.

— 1995. Alcohol and its Alternatives: Symbol and Substance in Pre-Industrial Cultures, in Goodman, Jordan, Lovejoy, Paul E., and Sherratt, Andrew, eds *Consuming Habits. Drugs in History and Anthropology*. London: Routledge, ch. 1, pp. 11–46.

Shulgin, Alexander, and Shulgin, Ann. 1997. *Tihkal. The Continuation*. Berkeley: Transform Press.

— 2000 [1991]. *Pihkal. A Chemical Love Story*. Berkeley: Transform Press.

Siegel, R. K. 1989. *Intoxication. Life in Pursuit of Artificial Paradise*. New York: Dutton.

Simmons, I. G. 2001. *An Environmental History of Great Britain. From 10,000 Years Ago to the Present*. Edinburgh: Edinburgh University Press.

Singer, Rolf. 1958. Mycological Investigations on Teonanacatl, the Mexican Hallucinogenic Mushroom. Part 1. The History of Teonanacatl, Field Work and Culture Work. *Mycologia* 50: 239–61.

— 1982. *A Correction*. Cambridge, MA: Botanical Museum of Harvard University.

Sink, C. A., Beluhan, L. J., Schlemmer, R. F., Heinze, W. J., and Davis, J. M. 1983. Dose Dependent Behavioral Changes Induced by Psilocybin in Selected Members of a Primate Social Colony. *Federation Proceedings* 42: 1166.

Smith, Huston. 1972. Wasson's *Soma*. A Review Article. *Journal of the American Academy of Religion* 40(4): 480–99.

— 2003. *Cleansing the Doors of Perception. The Religious Significance of Entheogenic Plants and Chemicals*. Boulder, CO: Sentient Publications, LLC.

Smith, Myron L., Bruhn, Johann N., and Anderson, James B. 1992. The Fungus *Armillaria bulbosa* is Among the Largest and Oldest Living Organisms. *Nature* 356: 428–31.

Smith, Worthington G. 1891. *Outlines of British Fungology Supplement*. London: L. Reeve & Co.

— 1910. *Guide to Mr Worthington Smith's Drawings of Field and Cultivated Mushrooms and Poisonous or Worthless Fungi Often Mistaken for Mushrooms*. London: British Museum.

Snow, Christopher P. 1999. *The Last of the Hippies*. London: Faber and Faber.

Southcott, Ronald V. 1974. Notes on Some Poisonings and Other Clinic Effects Following Ingestion of Australian Fungi. *South Australian Clinics* 6(5): 441–78.

Southworth, John. 1998. *Fools and Jesters at the English Court*. Stroud: Sutton Publishing.

Sowerby, James. 1803. *Coloured Figures of English Fungi or Mushrooms*. London: J. Davies.

Sowerby, J., Junior. 1832. *The Mushroom and Champignon Illustrated, Compared with, and Distinguished from, the Poisonous Fungi That Resemble Them*. London: J. Sowerby Junior.

Spitzer, Manfred, Thimm, Markus, Hermle, Leo, Holzmann, Petra, Kovar, Karl-Artur, Heinmann, Hans, Gouzoulis-Mayfrank, Euphrosyne, Kischka, Udo, and Schneider, Frank. 1996. Increased Activation of Indirect Semantic Associations Under Psilocybin. *Biological Psychiatry* 39: 1055–7.

Spooner, Brian. 1996. *Collins Wild Guide. Mushrooms and Toadstools*. London: HarperCollins.

Stabell-Kulø, Arnt. 1980. *Amanita muscaria* (the fly-agaric): A Positivistic Approach. *Temenos* 16: 122–31.

Stallybrass, P., and White, A. 1986. *The Politics and Poetics of Transgression*. Ithaca, NY: Cornell University Press.

Stamets, Paul. 1996. *Psilocybin Mushrooms of the World. An Identification Guide*. Berkeley: Ten Speed Press.

Stamets, Paul, and Chilton, Jeff C. 1983. *The Mushroom Cultivator. A Practical*

Guide to Growing Mushrooms at Home. Olympia, WA: Agarikon Press.
Stein, Sam. 1958. An Unusual Effect from a Species of Mexican Mushrooms *Psilocybe cubensis*. *Mycopathologia et Mycologia Applicata* 9(4): 263–7.
Stein, Sam, Closs, Gerhard L., and Gabel, Norman W. 1959. Observations on Psychoneurophysiologically Significant Mushrooms. *Mycopathologia et Mycologia Applicata* 11(3): 205–16.
Stevens, Jay. 1989. *Storming Heaven. LSD and the American Dream*. London: Paladin.
Sticht, G., and Käferstein, H. 2000. Detection of Psilocin in Body Fluids. *Forensic Science International* 113: 403–7.
Stresser-Péan, Guy. 1990. Travels with R. Gordon Wasson in Mexico, 1956–1962, in Riedlinger, Thomas, ed. *The Sacred Mushroom Seeker. Essays for R. Gordon Wasson*. Portland, OR: Dioscorides Press, pp. 231–7.
Stijve T., and Kuyper W. 1985. Occurrence of Psilocybin in Various Higher Fungi from Several European Countries. *Planta Medica* 51(5): 385–7.
Swain, Frederick. 1963. 'One', pp. 219–29, in Anon. Four Psilocybin Experiences. *Psychedelic Review* 1: 219–43.
Syal, Rajeev. 1996. Raves in the Caves: Stone Age Britons Took Drugs. *Sunday Times* 28 January: 5.
Taylor, Gordon Rattray. 1961. Eye on Research. The First Programme of a New Series. *Radio Times* 15–21 April: 39.
Taylor-Hawksworth, P. A. 2001. Exploiting fungi in images, artefacts and culture, in Pointing, S. P., and Hyde, K. D., eds *Bio-exploitation of Filamentous Fungi*. Hong Kong: Fungal Diversity Press, pp. 437–51.
Thomas, Benjamin. 2002. 'Mushroom Madness' in the Papua New Guinea Highlands: A Case of Nicotine Poisoning? *Journal of Psychoactive Drugs* 34(3): 321–3.
Thomas, Dylan. 2000 [1954]. *Under Milk Wood. A Play for Voices*. London: Penguin.
Travis, Alan. 2005. Revealed: How Drugs War Failed. *The Guardian* 5 July: 1–2.
Trubshaw, Bob. 2002. *Explore Folklore*. Wymeswold, Loughborough: Heart of Albion Press.
Tsunoda Koujun, Inoue Noriko, Aoyagi Yasuo, and Sugahara Tatsuyuki. 1993a. Changes in Concentrations of Ibotenic acid and Muscimol in the Fruit Body of *Amanita muscaria* during the Reproduction Stage. *Journal of the Food Hygienic Society of Japan* 34(1): 18–24.
— 1993b. Change in Ibotenic Acid and Muscimol Contents in *Amanita muscaria* During Drying, Storing or Cooking. *Journal of the Food Hygienic Society of Japan* 34(2): 153–60.
Tyldesley, J. A., and Bahn, P. G. 1983. Use of Plants in the European Palaeolithic: A Review of the Evidence. *Quaternary Science Reviews* 2: 53–81.
Tylor, Edward. 1871. *Primitive Culture: Researches into the Development of Mythology, Philosophy, Religion, Art and Custom (in two volumes)*. London: John Murray.

Vastokas, J. M. 1988. Comment, pp. 229–30, in Lewis-Williams, J. D., and Dowson, T. A. The Signs of All Times. Entoptic Phenomena in Upper Palaeolithic Art. *Current Anthropology* 29: 201–45.
Verkaik, Robert. 2004. Customs to Rake in £1m from VAT on Magic Mushrooms. *Independent* 10 August.
Verrill, A. E. 1914. A Recent Case of Mushroom Intoxication. *Science* 40: 408–10.
Vollenweider, F. X., Leenders, K. L., Scharfetter, C., Maguire, P., Stadelmann, O., and Angst, J. 1997. Positron Emission Tomography and Fluorodeoxyglucose Studies of Metabolic Hyperfrontality and Psychopathology in the Psilocybin Model of Psychosis. *Neuropsychopharmacology* 16(5): 357–72.
Wallis, Robert J. 2003. *Shamans/neo-Shamans: Ecstasy, Alternative Archaeologies and Contemporary Pagans*. London: Routledge.
Waser, Peter G. 1967. The Pharmacology of *Amanita muscaria*, in Efron, Daniel H., Holmstedt Bo, and Cline, Nathan S. *Ethnopharmacologic Search for Psychoactive Drugs. Proceedings of a Symposium held in San Francisco, California, January 28–30 1967*. Washington DC: US Department of Health, Education, and Welfare, pp. 419–39.
Wasson, R. Gordon. 1957. Seeking the Magic Mushroom. *Life* 13 May 49: 100. Accessed at www.psychedelic-library.org/lifep2.htm
– 1958. The Divine Mushroom: Primitive Religion and Hallucinatory Agents. *Proceedings of the American Philosophical Society*. 102: 221–3.
– 1963. Notes on the Present Status of Ololiuhqui and the Other Hallucinogens of Mexico. *Botanical Museum Leaflets of Harvard University* 20: 161–212.
– 1969. Review of *The Teachings of Don Juan* [Castaneda. 1968]. *Economic Botany* 23(2): 197.
– 1970. Drugs: The Sacred Mushroom. *New York Times* Saturday September 26: 29.
– 1971. *Soma: Divine Mushroom of Immortality*. New York: Harcourt Brace Jovanovich.
– 1972a. *Soma and the fly-agaric. Mr Wasson's rejoinder to Professor Brough*. Cambridge, MA: Botanical Museum of Harvard University.
– 1972b. Review of Castaneda. 1971. *A Separate Reality. Economic Botany* 26(1): 98–9.
– 1973a. Review of *Journey to Ixtlan* [Castaneda. 1972]. *Economic Botany* 27(2): 151–2.
– 1973b. Mushrooms and Japanese Culture. *Transactions of the Asiatic Society of Japan 3rd Series*. XI: 5–25.
– 1974. Review of *Tales of Power* [Castaneda. 1974]. *Economic Botany* 28(3): 245–6.
– 1979. Traditional Use in North America of *Amanita muscaria* for Divinatory Purposes. *Journal of Psychedelic Drugs* 11(1/2): 25–8.
– 1980. *The Wondrous Mushroom. Mycolatry in Mesoamerica*. New York: McGraw-Hill Book Company.
– 1981. A Retrospective Essay, in Estrada, Álvaro. *María Sabina: Her Life and Chants*. Santa Barbara: Ross-Erikson Inc., pp. 13–20.

– 1982a. The Last Meal of the Buddha. *Journal of the American Oriental Society* 102: 591–603.
– 1982b. *R. Gordon Wasson's Rejoinder to Dr Rolf Singer.* Cambridge, MA: Botanical Museum of Harvard University.
– 1986. 'Persephone's Quest', in Wasson, R. Gordon, Kramrisch, S., Ott, Jonathan, and Ruck, Carl, A. P. *Persephone's Quest. Entheogens and the Origins of Religion.* New Haven and London: Yale University Press.
Wasson, R. Gordon, Cowan, George, Cowan, Florence, and Rhodes, Willard. 1974. *María Sabina and her Mazatec Mushroom Velada.* New York: Harcourt Brace Jovanovich.
Wasson, R. Gordon, Hofmann, Albert, and Ruck, Carl A. P. 1978. *The Road to Eleusis. Unveiling the Secret of the Mysteries.* New York: Harcourt Brace Jovanovich.
Wasson, Valentina Pavlovna. 2000. I Ate the Sacred Mushroom, in Palmer, Cynthia, and Horowitz, Michael, eds *Sisters of the Extreme. Women Writing on the Drug Experience.* Rochester, VT: Park Street Press, pp. 157–60.
Wasson, Valentina Pavlovna, and Wasson, R. Gordon. 1957. *Mushrooms, Russia and History.* New York: Pantheon Books.
Watkins, Matthew. n.d. *Autopsy for a Mathematical Hallucination?* Accessed at www.fourmilab.ch/rpkp/autopsy.html
Watling, Roy. 1975. Prehistoric Puffballs. *Bulletin of the British Mycological Society.* 9(2): 112–14.
Watson, William. 1744. A Letter from Mr Wm. Watson F. R. S. to the Royal Society; Containing Further Remarks Concerning Mushrooms: Occasioned by the Reverend Mr. Pickering's F. R. S. Paper in the Preceding Transact p. 96 with Observations Upon the Poisonous Faculty of Some Sorts of Fungi. *Philosophical Transactions of the Royal Society of London* 43: 51–7.
Weil, Andrew T. 1972. *The Natural Mind. An Investigation of Drugs and the Higher Consciousness.* Boston, MA: Houghton-Mifflin.
– 1975. Mushroom Hunting in Oregon. *Journal of Psychedelic Drugs* 7(1): 89–102.
– 1977. The Use of Psychoactive Mushrooms in the Pacific Northwest: An Ethnopharmacologic Report. *Botanical Museum Leaflets, Harvard University* 25(5): 131–49.
– 1980. *The Marriage of the Sun and Moon. A Quest for Unity in Consciousness.* Boston, MA: Houghton-Mifflin.
– 1988. Review of *Persephone's Quest. Journal of Psychoactive Drugs* 20: 489–90.
Weitlaner-Johnson, Irmgard. 1990. Remembrances of Things Past, in Riedlinger, Thomas, ed. *The Sacred Mushroom Seeker. Essays for R. Gordon Wasson.* Portland, OR: Dioscorides Press, pp. 135–40.
White, Jean, and White, Andrew. 1995. Mushrooming Habit. *Practical Nursing (Supplement on Substance Abuse)* 6(20): 28–30.
White, Timothy. 2004. Enchanted Realms of a Zapotec *Curandero. Shaman's Drum* 66: 38–47.
Whitley, David S. 1992. Shamanism and Rock Art in Far Western North America. *Cambridge Archaeological Journal* 2: 89–113.

Wilson, Robert Anton. 1990. *Prometheus Rising*. Las Vegas: Falcon Press.
Wright, Mary Anna. 1998. The Great British Ecstasy Revolution, in McKay, George, ed. *DiY Culture. Party and Protest in Nineties Britain*. London: Verso, ch. 10, pp. 228–42.
Yépez, Heriberto. n.d. Clock Woman in the Land of Mixed Feelings: The Place of María Sabina in Mexican Culture. Accessed at www.ubu.com/ethno/discourses/yepez_clock.html
Young, Richard E., Milroy, Robert, Hutchinson, Stephen, and Kesson, Colin M. 1982. The Rising Price of Mushrooms. *The Lancet* 23 January: 213–15.
Young, David E., and Goulet, Jean-Guy. eds. 1994. *Being Changed by Cross-Cultural Encounters. The Anthropology of Extraordinary Experience.* Ontario: Broadview Press.

Index

Illustrations are entered in **bold** type